MYWORKBOOK

BEVERLY FUSFIELD

INTERMEDIATE ALGEBRA
ELEVENTH EDITION

Margaret L. Lial
American River College

John Hornsby
University of New Orleans

Terry McGinnis

Addison-Wesley
is an imprint of

ISBN-13: 978-0-321-71586-9
ISBN-10: 0-321-71586-1

1 2 3 4 5 6 BRR 15 14 13 12 11

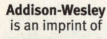

Addison-Wesley
is an imprint of

www.pearsonhighered.com

CONTENTS

Chapter 1 REVIEW OF THE REAL NUMBER SYSTEM

1.1   Basic Concepts

Learning Objectives
1   Write sets using set notation.
2   Use number lines.
3   Know the common sets of numbers.
4   Find additive inverses.
5   Use absolute value.
6   Use inequality symbols.
7   Graph sets of real numbers.

Key Terms

Use the vocabulary terms listed below to complete each statement in exercises 1–18.

set	elements	finite set	infinite set
empty set	variable	set-builder notation	number line
coordinate	graph	additive inverses	signed numbers
absolute value	equation	inequality	interval
interval notation		three-part inequality	

1.   A(n) _____ is a symbol, usually a letter, used to represent an unknown number.

2.   The _____ of a number is the point on a number line that corresponds to the number.

3.   _____ are the objects that belong to a set.

4.   Two numbers that are the same distance from 0 on a number line but on opposite sides of 0 are called _____.

5.   If all the elements of a set cannot be listed, the set is called a(n) _____.

6.   A(n) _____ is a statement that two algebraic expressions are equal.

7.   A(n) _____ is a portion of a number line.

8.   A(n) _____ is a statement that two expressions are not equal.

9.   A(n) _____ is a line with a scale that is used to show how numbers relate to each other.

10. If the number of elements in a set can be counted, then the set is a(n) _____.

11. Each number on a number line is called the _____ of the point it labels.

12. _____ is used to describe a set of numbers without actually having to list all of the elements.

13. _____ are numbers that can be written with a positive or negative sign.

14. The _____ of a number is the distance between 0 and the number on a number line.

15. The statement $0 < x \leq 4$ is an example of a(n) _____.

16. The _____, denoted by { } or $\emptyset$, is the set containing no elements.

17. A(n) _____ is a collection of objects.

18. _____ is a simplified notation that uses parentheses ( ) and/or brackets [ ] to describe an interval on a number line.

Guided Examples

Review these examples for Objective 1:

1. List the elements in the set
$\{w|w$ is a natural number greater than 6 but less than 11$\}$

$\{7, 8, 9, 10\}$.

2. Use set-builder notation to describe the set $\{-3, -2, -1, 0, 1\}$.

One answer is
$\{x|x$ is an integer greater than -4 but less than 2$\}$.

Another answer is
$\{x|x$ is an integer greater than or equal to -3 but less than or equal to 1$\}$.

Now Try:

1. $\{z|z$ is an odd natural number less than 9$\}$

_____

2. Use set-builder notation to describe the set $\{0, 2, 4, 6, 8\}$.

_____

Name: Date:

Instructor: Section:

Review these examples for Objective 3:

3. List the numbers in the set

$\left\{-8, -\frac{2}{3}, 0, \sqrt{2}, 3.5, \pi\right\}$ that are elements of each set.

 a. Whole numbers

 b. Integers

 c. Rational numbers

 d. Irrational numbers

 a. $\{0\}$

 b. $\{-8, 0\}$

 c. $\left\{-8, -\frac{2}{3}, 0, 3.5\right\}$

 d. $\left\{\sqrt{2}, \pi\right\}$

4. Decide whether each statement is *true* or *false*. If it is false, tell why.

 a. All rational numbers are real numbers.

 b. All whole numbers are natural numbers.

 a. True

 b. False. All natural numbers are whole numbers. 0 is a whole number, but not a natural number.

Now Try:

3. List the numbers in the following set that are elements of each set.

$\left\{-9, -\frac{4}{3}, 0, 0.\overline{3}, \sqrt{3}, \pi\right\}$

 a. Natural numbers _____

 b. Integers _____

 c. Irrational numbers _____

 d. Real numbers_____

4. Decide whether each statement is *true* or *false*. If it is true, tell why.

 a. Some rational numbers are whole numbers. _____

 b. Some irrational numbers are rational numbers._____

Review these examples for Objective 5:

5. Simplify by finding the absolute value.

 a. $|-7|$ **b.** $-|-7|$

 c. $|7|-|-7|$

 a. $|-7| = 7$ **b.** $-|-7| = -7$

 c. $|7|-|-7| = 7 - 7 = 0$

Now Try:

5. Simplify by finding the absolute value.

 a. $|-2|$ _____

 b. $-|-2|$ _____

 c. $-|-2|-|2|$_____

6. The table below shows the changes in population for five cities.

City	1980–1990	1990–2000	2000–2009
New York	250,925	685,714	383,603
Los Angeles	518,548	209,422	137,048
Chicago	−221,346	112,290	−44,748
Houston	35,415	323,078	304,295
Philadelphia	−102,633	−68,027	29,747

Source: factfinder.census.gov

Which city had the greatest change in population in which year?

New York from 1990–2000.

6. Use the table shown at the left to determine which city has the least change in population in which year.

_____

Review these examples for Objective 6:

7. Use a number line to determine whether each statement is *true* or *false*.

    a. $-1 > -5$ **b.** $0 < -1$

    a. The smaller of two numbers is to the left of the other number on a number line, so the statement is true.

    b. 0 is to the right of − 1 on the number line, so the statement is false.

8. Simplify each side of the statement to tell whether the resulting statement is *true* or *false*.

    a. $4(7) - 3 \geq 3(5)$ **b.** $-|7-4| \leq -4$

    a. Simplify the left side:
    $4(7) - 3 = 28 - 3 = 25$
    Simplify the right side: $3(5) = 15$
    $4(7) - 3 \geq 3(5)$ is true since $25 \geq 15$ is true.

    b. Simplify the left side: $-|7-4| = -3$
    $-|7-4| \leq -4$ is false since $-3 \leq -4$ is false.

Now Try:

7. Use a number line to determine whether each statement is *true* or *false*.

    a. $0 > -6$ _____

    b. $-8 \leq -9$ _____

8. Simplify each side of the statement to tell whether the resulting statement is *true* or *false*.

    a. $3 + 5 \leq |-7|$ _____

    b. $|-5| + |-14| - |-2| < 0$

    _____

Name: Date:
Instructor: Section:

Review these examples for Objective 7:

9. Write $x < -2$ in interval notation and graph.

In interval notation, $x < -2$ is written $(-\infty, -2)$.

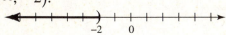

10. Write $x > -3$ in interval notation and graph.

In interval notation, $x > -3$ is written $(-3, \infty)$.

11. Write $\{x \mid -3 < x \le 2\}$ in interval notation and graph.

In interval notation, $\{x \mid -3 < x \le 2\}$ is written $(-3, 2]$.

Now Try:

9. Write $x < -10$ in interval notation and graph.

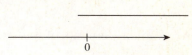

10. Write $x \ge 1$ in interval notation and graph.

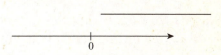

11. Write $\{x \mid -3 \le x \le 0\}$ in interval form, and graph the interval.

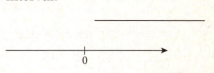

Objective 1 Write sets using set notation.

For extra help, see Examples 1 and 2 on page 3 of your text, the Section Lecture video for Section 1.1, and Exercise Solution Clips 1 and 13.

Decide whether each statement is true *or* false.

1. *True or false*:
   $\{x \mid x$ is a month beginning with the letter J$\}$ defines the set {January, June, July}.

1. _____

2. List the elements of the set
   $\{y \mid y$ is a natural number divisible by 5$\}$

2. _____

Objective 2 Use number lines.

For extra help, see pages 3–4 of your text, the Section Lecture video for Section 1.1.

3. Graph the elements of the set $\left\{-4, -\dfrac{3}{2}, \dfrac{1}{2}, 5\right\}$ on a number line.

3. _____

Objective 3 Know common sets of numbers.

For extra help, see Examples 3 and 4 on pages 5–6 of your text, the Section Lecture video for Section 1.1, and Exercise Solution Clips 21 and 25.

4. Which elements of the set $\left\{-4, -\sqrt{2}, -\frac{1}{3}, 0, \frac{4}{5}, \sqrt{7}, 5\right\}$

    4. a. _____

    b. _____

are elements of the following sets?

    a. Irrational numbers

    b. Rational numbers

5. *True* or *False*.

    a. All integers are whole numbers.

    b. Some irrational numbers are whole numbers.

    5. a. _____

    b. _____

Objective 4 Find additive inverses.

For extra help, see page 6 of your text and the Section Lecture video for Section 1.1.

6. Find each additive inverse.

    a. 0

    b. −2

    c. $\sqrt{2}$

    d. −0.75

    6. a. _____

    b. _____

    c. _____

    d. _____

Objective 5 Use absolute value.

For extra help, see Examples 5 and 6 on page 7 of your text, the Section Lecture video for Section 1.1, and Exercise Solution Clip 43.

Find the value of each expression.

7. $\left| 6 \right| - \left| -1 \right|$

    7. _____

8. $\left| -2 \right| + \left| 17 \right| - \left| -4 \right|$

    8. _____

Objective 6 Use inequality symbols.

For extra help, see Examples 7 and 8 on pages 8–9 of your text and the Section Lecture video for Section 1.1.

9. Use a number line to identify each inequality as *true* or *false*.

    a. −3 < −5   **b.** −2 < 7

    9. a. _____

    b. _____

10. Use inequality symbols to write each statement.

    a. −8 is less than 2.

    b. 0 is greater than or equal to x.

10. a. _____

    b. _____

Simplify each side of the statement to tell whether the resulting statement is true *or* false.

11. $-11 \geq 8 + 2$

11. _____

12. $-|7 - 4| \leq -4$

12. _____

Objective 7 Graph sets of real numbers.

For extra help, see Examples 9–11 on pages 10–11 of your text, the Section Lecture video for Section 1.1, and Exercise Solution Clips 101, 103, and 109.

Write in interval notation and graph.

13. $\{x \mid x \geq 3\}$

13. _____

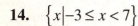

14. $\{x \mid -3 \leq x < 7\}$

14. _____

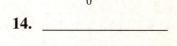

Use an inequality symbol to write the statement.

15.

    a. x is between −2 and 4, inclusive.

    b. x is between −3 and 5, excluding 5 and including −3.

15. a. _____

    b. _____

Chapter 1 REVIEW OF THE REAL NUMBER SYSTEM

1.2   Operations on Real Numbers

Learning Objectives
1   Add real numbers.
2   Subtract real numbers.
3   Find the distance between two points on a number line.
4   Multiply real numbers.
5   Find reciprocals and divide real numbers.

Key Terms

Use the vocabulary terms listed below to complete each statement in exercises 1–5.

sum   difference   product   quotient   reciprocals

1.   The answer to a multiplication problem is called the _____.

2.   The answer to an addition problem is called the _____.

3.   The answer to a division problem is called the _____.

4.   Pairs of numbers whose product is 1 are called _____ of each other.

5.   The answer to a subtraction problem is called the _____.

Guided Examples

Review these examples for Objective 1:
1.   Find each sum.

   a. $-3 + (-4)$ **b.** $-6.25 + (-3.51)$

   c. $-\dfrac{3}{5} + \left(-\dfrac{3}{10}\right)$

   a. $-3 + (-4) = -(3 + 4)$ Add the absolute values.

            $= -7$

   b. $-6.25 + (-3.51) = -(6.25 + 3.51)$

                Add the absolute values.

              $= -9.76$

Now Try:
1.   Find each sum.

   a. $-2 + (-7)$ _____

   b. $-12.5 + (-8.3)$ _____

   c. $-\dfrac{4}{11} + \left(-\dfrac{7}{22}\right)$ _____

c. $-\dfrac{3}{5}+\left(-\dfrac{3}{10}\right)=-\left(\dfrac{3}{5}+\dfrac{3}{10}\right)$

$\qquad\qquad\qquad = -\left(\dfrac{6}{10}+\dfrac{3}{10}\right)$

$\qquad\qquad\qquad = -\dfrac{9}{10}$

2. Find each sum.

   a. $-23+4$ **b.** $-16.32+2.27$

   c. $\dfrac{5}{6}+\left(-\dfrac{5}{8}\right)$

   a. $-23+4$
      First find the absolute values: $|-23|=23$
      and $|4|=4$. Because -23 and 4 have
      different signs, subtract their absolute
      values.
      $23-4=19$
      The number -23 has a greater absolute
      value than 4, so the answer is negative.
      $-23+4=-19$.

   b. $-16.32+2.27$
      Subtract the absolute values:
      $16.32-2.27=14.05$
      Because -16.32 has the greater absolute
      value, the sum is negative.
      $-16.32+2.27=-14.05$.

   c. $\dfrac{5}{6}+\left(-\dfrac{5}{8}\right)=\dfrac{5}{6}-\dfrac{5}{8}=\dfrac{20}{24}-\dfrac{15}{24}=\dfrac{5}{24}$

2. Find each sum.

   a. $2+(-7)$ _____

   b. $-12.5+9.6$ _____

   c. $-\dfrac{4}{5}+\dfrac{2}{3}$ _____

Review these examples for Objective 2:

3. Find each difference.

   a. $14-20$ **b.** $-7-(-12)$

   c. $\dfrac{3}{8}-\left(-\dfrac{1}{2}\right)$

      For all real numbers a and b,
      $a-b=a+(-b)$.

                  — Change − to +.

No change ————   ————— Additive inverse of 20

   a. $14-20=14+(-20)=-6$

Now Try:

3. Find each difference.

   a. $22-(-24)$ _____

   b. $-7.2-8.9$_____

   c. $\dfrac{1}{10}-\dfrac{1}{2}$ _____

Change − to +.

No change — Additive inverse
of −12

b. $-7-(-12)=-7+12=5$

c. $\dfrac{3}{8}-\left(-\dfrac{1}{2}\right)=\dfrac{3}{8}+\dfrac{1}{2}=\dfrac{3}{8}+\dfrac{4}{8}=\dfrac{7}{8}$

4. Perform the indicated operations.

a. $3-\left[-4+(11-19)\right]$
$=3-\left[-4+(-8)\right]$
Start within the innermost parenthesis.

$=3-\left[-12\right]$ Add.
$=3+12$ Definition of subtraction.
$=15$ Add.

b. $-2+4\left|(-12+10)-(-6+2)\right|$
Start within the innermost parenthesis.

$=-2+4\left|-2-(-4)\right|$ Add.
$=-2+4\left|-2+4\right|$ Definition of subtraction.
$=-2+4\left|2\right|$ Add.
$=-2+4\cdot2$ Evaluate absolute value.
$=-2+8$ Multiply.
$=6$ Add.

4. Perform each indicated operation.

a. $-9+\left[5+(21-30)\right]$

_____

b. $\left|-6+9\right|-2\left|-4+(11-19)\right|$

_____

Review this example for Objective 3:

5. Find the distance between the points −7 and 3.

Find the absolute value of the difference of the numbers, taken in either order.
$\left|-7-3\right|=10$ or $\left|3-(-7)\right|=10$

Now Try:

5. Find the distance between the points −8 and −3.

_____

Review this example for Objective 4:

6. Find each product.
a. $-12(-8)$ **b.** $-1.2(0.7)$

c. $\left(\dfrac{1}{5}\right)\left(-\dfrac{2}{3}\right)$

a. The product of two negative numbers is positive, so $-12\,(-8)=12(8)=96.$

Now Try:

6. Find each product.
a. $(-7)(-16)$ _____

b. $(-3.2)(4.1)$ _____

c. $\left(-\dfrac{3}{8}\right)\left(\dfrac{14}{9}\right)$ _____

b. The product of a negative number and a positive number is negative, so $-1.2(0.7) = -0.84$

c. $\left(\dfrac{1}{5}\right)\left(-\dfrac{2}{3}\right) = -\dfrac{2}{15}$

Review this example for Objective 5:	**Now Try:**

7. Find each quotient.

a. $\dfrac{48}{-8}$ **b.** $-\dfrac{1}{5} \div \left(-\dfrac{2}{3}\right)$

c. $\dfrac{-\frac{7}{12}}{\frac{14}{3}}$

a. $\dfrac{48}{-8} = 48\left(-\dfrac{1}{8}\right) = -6$ $\dfrac{x}{y} = x \cdot \dfrac{1}{y}$

b. $-\dfrac{1}{5} \div \left(-\dfrac{2}{3}\right) = -\dfrac{1}{5} \cdot \left(-\dfrac{3}{2}\right) = \dfrac{3}{10}$

c. $\dfrac{-\frac{7}{12}}{\frac{14}{3}} = -\dfrac{7}{12} \cdot \dfrac{3}{14} = -\dfrac{1}{8}$

7. Find each quotient.

a. $\dfrac{-120}{-20}$ _____

b. $-\dfrac{3}{16} \div \dfrac{9}{8}$ _____

c. $\dfrac{\frac{9}{2}}{-\frac{3}{2}}$ _____

Objective 1 Add real numbers.

For extra help, see Examples 1 and 2 on pages 14–15 of your text, the Section Lecture video for Section 1.2, and Exercise Solution Clips 11 and 13.

Find each sum.

1. $-5.1 + (-7.3)$ **1.** _____

2. $-\dfrac{1}{2} + \dfrac{3}{8}$ **2.** _____

3. $\left(-\dfrac{7}{12}\right) + \left(-\dfrac{3}{4}\right)$ **3.** _____

Objective 2 Subtract real numbers.
For extra help, see Examples 3 and 4 on pages 16–17 of your text, the Section Lecture video for Section 1.2, and Exercise Solution Clips 19 and 39.

Find each difference.

4. $3 - 7$ 4. _____

5. $-6.25 - (-2.47)$ 5. _____

6. $\dfrac{3}{5} - \left(-\dfrac{1}{3}\right)$ 6. _____

Objective 3 Find the distance between two points on a number line.
For extra help, see Example 5 on page 17 of your text, the Section Lecture video for Section 1.2, and Exercise Solution Clip 51.

Find the distance between the pair of points on the number line.

7. -5 and -12 7. _____

8. 12 and -7 8. _____

9. -4 and 8 9. _____

Objective 4 Multiply real numbers.
For extra help, see Example 6 on page 18 of your text, the Section Lecture video for Section 1.2, and Exercise Solution Clip 59.

Find each product.

10. $(-5)(-3)$ 10. _____

11. $(-7)(3)(-5)$ 11. _____

12. $\left(-\dfrac{3}{8}\right)\left(-\dfrac{4}{5}\right)$

12. _____ .

Objective 5 Find reciprocals and divide real numbers.

For extra help, see Example 7 on pages 19–20 of your text, the Section Lecture video for Section 1.2, and Exercise Solution Clip 73.

Divide where possible.

13. $\dfrac{-3}{12}$

13. _____

14. $\dfrac{-7}{0}$

14. _____

15. $-\dfrac{2}{3} \div \dfrac{2}{9}$

15. _____

Chapter 1 REVIEW OF THE REAL NUMBER SYSTEM

1.3   Exponents, Roots, and Order of Operations

Learning Objectives
1   Use exponents.
2   Find square roots.
3   Use the order of operations.
4   Evaluate algebraic expressions for given values of variables.

Key Terms

Use the vocabulary terms listed below to complete each statement in exercises 1–8.

factors	**exponent**	**base**	**exponential expression**
square root	**principal square root**		**negative square root**
algebraic expression			

1.   We write the _____ of a number with the symbol $\sqrt{}$, called a radical sign.

2.   We use exponents as a way of writing products of repeated _____.

3.   Any sequence of numbers, variables, operation symbols, and/or grouping symbols formed in accordance with the rules of algebra is called a(n) _____.

4.   The _____ is the number that is a repeated factor when written with an exponent.

5.   A(n)_____ is a number or a variable written with an exponent.

6.   A(n) _____ is a number that indicates how many times a factor is repeated.

7.   A number b is a _____ of a if $b^2 = a$.

8.   The symbol $-\sqrt{}$ is used for the _____.

Guided Examples

Review these examples for Objective 1:

1. Write using exponents.
   a. $(-5)(-5)(-5)(-5)$
   b. $a \cdot a \cdot a \cdot a \cdot a \cdot a \cdot a \cdot a \cdot a$

   a. $(-5)(-5)(-5)(-5) = (-5)^4$
   b. $a \cdot a \cdot a \cdot a \cdot a \cdot a \cdot a \cdot a \cdot a = a^9$

2. Evaluate.
   a. $-(-4)^2$ b. $-(-4)^3$
   c. $-(4)^2$

   a. $-(-4)^2 = -(-4)(-4) = -16$
   b. $-(-4)^3 = -(-4)(-4)(-4) = -(-64) = 64$
   c. $-(4)^2 = -(4)(4) = -16$

Now Try:

1. Write using exponents.
   a. $\left(\dfrac{2}{3}\right)\left(\dfrac{2}{3}\right)\left(\dfrac{2}{3}\right)\left(\dfrac{2}{3}\right)\left(\dfrac{2}{3}\right)$

   _____

   b. $x \cdot x \cdot x \cdot x \cdot x \cdot x$ _____

2. Evaluate.
   a. $\left(\dfrac{4}{3}\right)^3$ _____

   b. $\left(-\dfrac{4}{3}\right)^4$ _____

   c. $-\left(\dfrac{4}{3}\right)^4$ _____

Review this example for Objective 2:

3. Find each square root that is a real number.
   a. $-\sqrt{16}$ b. $\sqrt{0.81}$
   c. $\sqrt{-\dfrac{121}{25}}$

   a. $-\sqrt{16} = -4$
   b. $\sqrt{0.81} = 0.9$
   c. $\sqrt{-\dfrac{121}{25}}$ is not a real number.

Now Try:

3. Find each square root that is a real number.
   a. $\sqrt{10,000}$ _____

   b. $-\sqrt{\dfrac{4}{49}}$ _____

   c. $-\sqrt{-0.25}$ _____

Review these examples for Objective 3:

4. Simplify $12 \div 4 \cdot 3 + 2$.

   Multiplications and division are done in the order in which they appear from left to right. The multiplications and divisions are performed before addition.
   $$12 \div 4 \cdot 3 + 2 = 3 \cdot 3 + 2$$
   $$= 9 + 2$$
   $$= 11$$

Now Try:

4. Simplify $-5 \cdot 2 - (-7)(-3)$.

   _____

5. Simplify $8\left[14+3^2(9-4)\right]$.

$8\left[14+3^2(9-4)\right]$

$\quad=8\left[14+9(9-4)\right]$

                Apply exponent.

$\quad=8\left[14+9(5)\right]$

                Subtract inside parentheses.

$\quad=8\left[14+45\right]$ Multiply.

$\quad=8\left[59\right]$ Add inside brackets.

$\quad=472$ Multiply

6. Simplify $\dfrac{6\cdot10+2\sqrt{81}}{3(4-2)}$.

$\dfrac{6\cdot10+2\sqrt{81}}{3(4-2)}=\dfrac{60+2\cdot9}{3(2)}$

             Simplify numerator and
             denominator separately.

$\quad\quad\quad\quad=\dfrac{60+18}{6}=\dfrac{78}{6}$

$\quad\quad\quad\quad=13$

5. Simplify $(-3-3)(-7-5)-5^2$.

_____

6. Simplify $\dfrac{\left(\frac{3}{4}\cdot12\right)+7}{9-\frac{3}{4}\cdot8}$

_____

Review this example for Objective 4:

7. Evaluate each expression for $x=-2$, $y=4$, and $z=-1$.

  a. $\dfrac{4x+3\sqrt{y}}{z-1}$ **b.** $3x^4-2y^2$

  a. $\dfrac{4x+3\sqrt{y}}{z-1}=\dfrac{4(-2)+3\sqrt{4}}{-1-1}$

                 $x=-2,\ y=4,\ z=-1$

            $=\dfrac{-8+3\cdot2}{-1-1}=\dfrac{-8+6}{-2}$

             Work separately above and
             below the fraction bar.

            $=\dfrac{-2}{-2}=1$

Now Try:

7. Find the value of each algebraic expression for $x=-4$ and $y=2$.

  a. $\dfrac{3x-2y}{2x-3y+2}$ _____

  b. $\dfrac{\sqrt[3]{xy}}{\frac{1}{2}x^4-\frac{1}{4}y^2}$ _____

---

b. $3x^4 - 2y^2 = 3(-2)^4 - 2(4)^2$

$ x = -2,\ y = 4$

$ = 3(16) - 2(16)$

$ $ Apply the exponents.

$ = 48 - 32 \quad$ Multiply.

$ = 16 \qquad$ Subtract.

Objective 1 Use exponents.

For extra help, see Examples 1 and 2 on pages 24–25 of your text, the Section Lecture video for Section 1.3, and Exercise Solution Clips 13, 21 and 29.

Write each expression without exponents.

1. $(-2)^3$ 1. _____

2. $\left(\dfrac{3}{5}\right)^2$ 2. _____

3. $(-0.5)^2$ 3. _____

Objective 2 Find square roots.

For extra help, see Example 3 on page 26 of your text, the Section Lecture video for Section 1.3, and Exercise Solution Clip 37.

Find each square root that is a real number.

4. $\sqrt{\dfrac{81}{4}}$ 4. _____

5. $-\sqrt{-121}$ 5. _____

6. $-\sqrt{0.01}$ 6. _____

Objective 3 Use the order of operations.
For extra help, see Examples 4–6 on pages 27–28 of your text, the Section Lecture video for Section 1.3, and Exercise Solution Clips 53, 65, and 75.

Simplify.

7. $-3 \cdot \dfrac{7}{6} - (-3)$ 7. _____

8. $\dfrac{4(-3) + (-4)(-3)}{-5 + 6 + 1}$ 8. _____

9. $4^3 \div 2^5 + 3\sqrt{36}$ 9. _____

10. $-\dfrac{1}{2}[4(-2) + 5(-7) + (-4)(-2)]$ 10. _____

11. $\dfrac{-[2 - (-4 + 1) + 1]}{\sqrt{9}(-1 + 3)}$ 11. _____

Objective 4 Evaluate algebraic expressions for given values of variables.
For extra help, see Example 7 on page 28 of your text, the Section Lecture video for Section 1.3, and Exercise Solution Clip 79.

Evaluate if a = –1, b = 3, and c = –5.

12. $-3(a - 2c)$ 12. _____

13. $-a^2 + b^2$

13. _____

14. $\dfrac{a + 3b}{2c}$

14. _____

15. $\dfrac{2a + 4b^2}{3c - 2a}$

15. _____

Chapter 1 REVIEW OF THE REAL NUMBER SYSTEM

1.4   Properties of Real Numbers

Learning Objectives
1   Use the distributive property.
2   Use the identity properties.
3   Use the inverse properties.
4   Use the commutative and associative properties.
5   Use the multiplication property of 0.

Key Terms

Use the vocabulary terms listed below to complete each statement in exercises 1–7.

identity element for addition   identity element for multiplication

term   numerical coefficient   like terms

unlike terms   combine like terms

1.   In order to simplify an expression, we must _____.

2.   The terms $7x^2$ and $-4x$ are _____.

3.   Since multiplying a number by 1 does not change the number, 1 is called the
      _____.

4.   Terms with exactly the same variables, including the same exponents, are called
      _____.

5.   A _____ is the numerical factor of a term.

6.   Since adding 0 to a number does not change the number, 0 is called the
      _____.

7.   A _____ is a number, a variable, or the product or quotient of a
      number and one or more variables raised to powers.

Guided Examples

Review this example for Objective 1:

1. Use the distributive property to rewrite each expression.

   a. $2(7y - 3z)$

   b. $4y - 9y$

   c. $4x + 3y$

   a. $2(7y - 3z) = 2(7y) + 2(-3z)$

   Distributive property

   $= 14y - 6z$ Multiply.

   b. $4y - 9y = 4y + (-9y)$

   Definition of subtraction

   $= [4 + (-9)]y$

   Distributive property

   $= -5y$

   c. $4x + 3y$ cannot be rewritten

Now Try:

1. Use the distributive property to rewrite each expression.

   a. $-9(-5x + 8y)$

   _____

   b. $6y - 15y$ _____

   c. $7z - 3a$ _____

Review this example for Objective 2:

2. Simplify each expression.

   a. $8x - x$ **b.** $-(-10 + n)$

   a. $8x - x = 8x - 1x$ Identity property

   $= (8 - 1)x$ Distributive property

   $= 7x$ Subtract.

   b. $-(-10 + n)$

   $= -1 \cdot (-10 + n)$ Identity property

   $= -1 \cdot (-10) + (-1) \cdot n$ Distributive property

   $= 10 - n$ Multiply.

Now Try:

2. Simplify each expression.

   a. $-x - 9x$ _____

   b. $-(-2a - 3b + c)$

   _____

Review these examples for Objective 4:

3. Simplify $9a - 12 - 7a + 4 + 2a$.

   $9a - 12 - 7a + 4 + 2a$

   $= 9a - 7a + 2a - 12 + 4$

   Commutative property

   $= a(9 - 7 + 2) - 12 + 4$

   Distributive property

   $= a(4) - 8$

   Combine like terms.

   $= 4a - 8$

   Commutative property

Now Try:

3. Simplify $-25 + 6x - 5 - 11x + x$.

   _____

4. Simplify each expression.
- **a.** $4x + 7 - 4x$
- **b.** $(4m + 8) - (2m + 8)$

a. $4x + 7 - 4x$

$= (4x + 7) - 4x$ Order of operations

$= (7 + 4x) - 4x$ Commutative property

$= 7 + (4x - 4x)$ Associative property

$= 7 + [4x + (-4x)]$

                   Definition of subtraction

$= 7 + 0$ Inverse property

$= 7$ Identity property

b. $(4m + 8) - (2m + 8)$

$= 4m + 8 - 2m - 8$

                Distributive property

$= 4m - 2m + 8 - 8$

                Commutative property

$= (4m - 2m) + (8 - 8)$

                Associative property

$= m(4 - 2) + 0$

                Distributive property

$= m(2)$ Subtract.

$= 2m$ Commutative property

4. Simplify each expression
- **a.** $-\dfrac{3}{5}x - 12 + \dfrac{3}{5}x$

     _____

- **b.** $5(3 - 2p) + 3(5p + 1)$

     _____

Objective 1 Use the distributive property.

For extra help, see Example 1 on pages 32–33 of your text, the Section Lecture video for Section 1.4, and Exercise Solution Clip 11.

Simplify each expression.

1. $-3(z - 7)$

1. _____

2. $4k + 9k$

2. _____

3. $7a + 3b$

3. _____

Objective 2 Use the identity properties.

For extra help, see Example 2 on pages 33–34 of your text, the Section Lecture video for Section 1.4, and Exercise Solution Clip 21.

Complete each statement.

4. $\frac{4}{3}a + \underline{\hspace{1cm}} = \frac{4}{3}a$

4. _____

5. $-12y \cdot 1 = \underline{\hspace{1cm}}$

5. _____

6. $-\frac{4}{3}x \cdot \underline{\hspace{1cm}} = -\frac{4}{3}x$

6. _____

Objective 3 Use the inverse properties.

For extra help, see page 34 in the text and the Section Lecture video for Section 1.4.

Complete each statement.

7. $7 + \underline{\hspace{1cm}} = 0$

7. _____

8. $-1.3 + 1.3 = \underline{\hspace{1cm}}$

8. _____

9. $-\frac{3}{4}\left(-\frac{4}{3}\right) = \underline{\hspace{1cm}}$

9. _____

Objective 4 Use the commutative and associative properties.

For extra help, see Examples 3 and 4 on pages 35–36 of your text, the Section Lecture video for Section 1.4, and Exercise Solution Clips 29 and 31.

Simplify each expression.

10. $\frac{1}{2}(2x - 4) - \frac{3}{4}(8x + 12)$

10. _____

11. $-(2w - 7) + 11 + 3(4w - 6) - 5w$ **11.** _____

12. $2(x - 3) - 7 - 6(2x - 5) + 7x$ **12.** _____

13. $8 - 2(4d - 1) + 3(d - 6) + 2d$ **13.** _____

Objective 5 Use the multiplication property of 0.
For extra help, see page 36 in the text and the Section Lecture video for Section 1.4.

Complete each statement.

14. $0(x + y) =$ _____   **14.** _____

15. $x \cdot$ _____ $= 0$ **15.** _____

Chapter 2   LINEAR EQUATIONS, INEQUALITIES, AND APPLICATIONS

2.1   Linear Equations in One Variable

Learning Objectives
1   Distinguish between expressions and equations.
2   Identify linear equations and decide whether a number is a solution of a linear equation.
3   Solve linear equations by using the addition and multiplication properties of equality.
4   Solve linear equations by using the distributive property.
5   Solve linear equations with fractions or decimals.
6   Identify conditional equations, contradictions, and identities.

Key Terms

Use the vocabulary terms listed below to complete each statement in exercises 1–7.

linear equation in one variable   **solution**   **solution set**

equivalent equations   **conditional equation**   **contradiction**

identity

1.   A(n) _____ can be written in the form $Ax + B = C$, where A, B, and C are real numbers with $A \neq 0$.

2.   A(n) _____ is true for some replacements of the variable and false for others.

3.   A(n) _____ is an equation that is never true. It has no solution.

4.   The _____ of an equation is the set of all solutions of the equation.

5.   _____ are equations that have the same solution set.

6.   A(n) _____ is an equation that is true for all replacements of the variable. It has an infinite number of solutions.

7.   A(n) _____ of an equation is any replacement for the variable that makes the equation true.

Guided Examples

Review this example for Objective 1:

1. Decide whether each of the following is an *expression* or an *equation*.

   a. $5x + 6$ **b.** $5x + 6 = 10$

   a. $5x + 6$ is an *expression* because there is no equals symbol.

   b. $5x + 6 = 10$ is an *equation* because there is an equals symbol.

Now Try:

1. Decide whether each of the following is an *expression* or an *equation*.

   a. $9b - 20 = 98$ _____

   b. $9b - 20$ _____

Review this example for Objective 3:

2. Solve $9r - 4r + 8r - 6 = 10r - 12 + 5r$.

$$9r - 4r + 8r - 6 = 10r - 12 + 5r$$
$$13r - 6 = 15r - 12 \quad \text{Combine like terms.}$$
$$13r - 6 + 6 = 15r - 12 + 6$$
$$\text{Add 6 to each side.}$$
$$13r = 15r - 6 \quad \text{Combine like terms.}$$
$$13r - 15r = 15r - 6 - 15r$$
$$\text{Subtract } 15r$$
$$\text{from each side.}$$
$$-2r = -6 \quad \text{Combine like terms.}$$
$$\frac{-2r}{-2} = \frac{-6}{-2} \quad \begin{array}{l}\text{Divide each side}\\ \text{by } -2.\end{array}$$
$$r = 3$$

Now Try:

2. Solve $7t - 11t - 6 + 15 = -t + 17$.

   _____

Review this example for Objective 4:

3. Solve $4m - 3(5 - 2m) = 6(m - 3) + 2m + 1$.

$$4m - 3(5 - 2m) = 6(m - 3) + 2m + 1$$
$$4m - 3(5) - 3(-2m) = 6m - 6(3) + 2m + 1$$
$$\text{Distributive property}$$
$$4m - 15 + 6m = 6m - 18 + 2m + 1$$
$$\text{Multiply.}$$
$$10m - 15 = 8m - 17$$
$$\text{Combine like terms.}$$
$$10m - 15 + 15 = 8m - 17 + 15$$
$$\text{Add 15 to each side.}$$
$$10m = 8m - 2$$
$$\text{Combine like terms.}$$

Now Try:

3. Solve
$$5z - 2(3 - 3z) = 4(z - 2) - 7.$$

   _____

$$10m - 8m = 8m - 2 - 8m$$

Subtract $8m$ from each side.

$$2m = -2 \quad \text{Combine like terms.}$$

$$\frac{2m}{2} = \frac{-2}{2} \quad \text{Divide each side by 2.}$$

$$m = -1$$

Review these examples for Objective 5:

4. Solve $\dfrac{p-2}{3} + \dfrac{p}{4} = \dfrac{1}{2}$.

Eliminate the fractions by multiplying each side by the LCD, 12.

$$12\left(\frac{p-2}{3} + \frac{p}{4}\right) = 12\left(\frac{1}{2}\right)$$

$$12\left(\frac{p-2}{3}\right) + 12\left(\frac{p}{4}\right) = 12\left(\frac{1}{2}\right)$$

Distributive property

$$4p - 8 + 3p = 6 \quad \text{Multiply.}$$

$$7p - 8 = 6 \quad \text{Combine like terms.}$$

$$7p - 8 + 8 = 6 + 8$$

Add 8 to each side.

$$7p = 14 \quad \text{Combine like terms.}$$

$$\frac{7p}{7} = \frac{14}{7} \quad \text{Divide each side by 7.}$$

$$p = 2$$

5. Solve $0.04(x+6) - 0.02x = 3.16$.

Clear the decimals by multiplying each term by 100.

$$0.04(x+6) - 0.02x = 3.16$$

$$4(x+6) - 2x = 316$$

Multiply each term by 100.

$$4x + 4(6) - 2x = 316 \quad \text{Distributive property}$$

$$4x + 24 - 2x = 316 \quad \text{Multiply.}$$

$$2x + 24 = 316 \quad \text{Combine like terms.}$$

$$2x + 24 - 24 = 316 - 24$$

Subtract 24 from each side.

$$2x = 292 \quad \text{Combine like terms.}$$

$$\frac{2x}{2} = \frac{292}{2} \quad \text{Divide each side by 2.}$$

$$x = 146$$

Now Try:

4. Solve $\dfrac{x-5}{2} - \dfrac{x+6}{3} = -4$.

_____

5. Solve
$$0.06x + 0.14(x+500) = 130.$$

_____

Review this example for Objective 6:

6. Solve each equation. Decide whether it is a *conditional equation*, an *identity*, or a *contradiction*.

   a. $6(3-4x)+10=-15x+3(2-3x)$

   b. $28-4k=36+2(2k-4)$

   c. $2(2y-5)-3(4-y)=7y-22$

   a. $6(3-4x)+10=-15x+3(2-3x)$
   $18-24x+10=-15x+6-9x$

   Distributive property
   $$28-24x=-24x+6$$

   Combine like terms.
   $$28-24x-6=-24x+6-6$$

   Subtract 6 from each side.
   $$22-24x+24x=-24x+24x$$

   Combine like terms.
   Add $24x$ to each side.
   $$22=0 \quad \text{False}$$

   The equation has no solution. The solution set is $\varnothing$. This is a contradiction.

   b. $\quad 28-4k=36+2(2k-4)$
   $$28-4k=36+4k-8$$

   Distributive property
   $$28-4k=28+4k$$

   Combine like terms.
   $$28-4k+4k=28+4k+4k$$

   Add $4k$ to each side.
   $$28=28+8k$$

   Combine like terms.
   $$28-28=28+8k-28$$

   Subtract 28 from each side.
   $$0=8k \quad \text{Combine like terms.}$$
   $$\frac{0}{8}=\frac{8k}{8} \quad \text{Divide each side by 8.}$$
   $$0=k$$

   The solution set is {0}. This is a conditional equation.

Now Try:

6. Solve each equation. Decide whether it is a *conditional equation*, an *identity*, or a *contradiction*.

   a.
   $7(2-5b)-32=10b-3(6+15b)$

   _____

   b. $4(x-12)-8(x+1)=56-4x$

   _____

   c. $5x-7x+14=4-11-9x$

   _____

c. $2(2y - 5) - 3(4 - y) = 7y - 22$
$\qquad 4y - 10 - 12 + 3y = 7y - 22$
$\qquad\qquad\qquad$ Distributive property
$\qquad\qquad 7y - 22 = 7y - 22$
$\qquad\qquad\qquad$ Combine like terms.
Since both sides of the equation are the same, the equation is an identity, and the solution set is {all real numbers}.

Objective 1 Distinguish between expressions and equations.

For extra help, see Example 1 on page 48 of your text and the Section Lecture video for Section 2.1.

Determine whether each of the following is an expression *or an* equation.

1. $2(x + 5) = 7$

$2x + 10 = 7$
$\quad -10 \quad -10$

1. $\underline{x = -3/2\ \text{equation}}$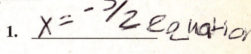

2. $2(3t - 8) - 7t$

2. $\underline{\text{expression}}$

Objective 2 Identify linear equations and decide whether a number is a solution of a linear equation.

For extra help, see page 48 of your text and the Section Lecture video for Section 2.1.

Decide whether the given value for the variable is a solution for the equation.

3. $3d - 5 = 5(d - 3)$; $d = -5$

$3(-5) - 5 = 5(-5 - 3)$
$-20 \overset{?}{=} -25 - 15$

3. $\underline{\text{no}}$

4. $2(4c - 3) = 7(c - 2)$; $c = -8$

$8c - 6 = 7c - 14$

4. $\underline{\text{yes}}$

Objective 3 Solve linear equations by using the addition and multiplication properties of equality.

For extra help, see Example 2 on page 49 of your text, the Section Lecture video for Section 2.1, and Exercise Solution Clips 15 and 17.

Solve the equation.

5. $11x - 14x - 7 + 8 = 4x + 5 - 2$

5. _____

6. $9q + 114 - 2q = 15 - 6q + 8$ **6.** _____

Objective 4   Solve linear equations by using the distributive property.

For extra help, see Example 3 on pages 50–51 of your text, the Section Lecture video for Section 2.1, and Exercise Solution Clip 19.

Solve the equation.

7. $-9y - (5 + y) = -(3y - 1) - 6$ **7.** _____

8. $k + 2(-k + 4) - 3(k + 5) = -4$ **8.** _____

Objective 5   Solve linear equations with fractions or decimals.

For extra help, see Examples 4 and 5 on pages 51–52 of your text, the Section Lecture video for Section 2.1, and Exercise Solution Clips 59 and 63.

Solve the equation.

9. $\dfrac{x-5}{2} - \dfrac{x+6}{3} = -4$ **9.** _____

10. $\dfrac{2m-1}{3} - \dfrac{3m}{4} = \dfrac{5}{6}$ **10.** _____

11. $0.02(x+4) = 0.03x - 0.02$

11. _____

12. $0.6x - 2(0.5x + 0.2) = 0.4 - 0.3x$

12. _____

Objective 6 **Identify conditional equations, contradictions, and identities.**
For extra help, see Example 6 on page 53 of your text, the Section Lecture video for
Section 2.1, and Exercise Solution Clips 23 and 33.

Solve each equation. Decide whether it is a conditional equation, *an* identity, *or a*
contradiction.

13. $6(3 - 4x) + 10 = -15x + 3(2 - 3x)$

13. _____

14. $28 - 4k = 36 + 2(2k - 4)$

14. _____

15. $6z - 3(8z - 4) = 2(4 - 7z) - 4(z - 1)$

15. _____

Chapter 2 LINEAR EQUATIONS, INEQUALITIES, AND APPLICATIONS

2.2   Formulas and Percent

Learning Objectives
1   Solve a formula for a specified variable.
2   Solve applied problems by using formulas.
3   Solve percent problems.
4   Solve problems involving percent increase or decrease.

Key Terms

Use the vocabulary terms listed below to complete each statement in exercises 1–3.

mathematical model   **formula**   **percent**

1.   An equation in which variables are used to describe a relationship is a

_____.

2.   _____ means "one per hundred."

3.   A _____ is an equation or inequality that describes a real situation.

Guided Examples

Review these examples for Objective 1:
1.   Solve the formula $V = lwh$ for h.

Solve the formula for h by treating V, l, and w as constants and treating h as the only variable.

$$V = lwh$$
$$V = (lw)h \quad \text{Associative property}$$
$$\frac{V}{lw} = \frac{(lw)h}{lw} \quad \text{Divide by } lw.$$
$$\frac{V}{lw} = h$$

Now Try:
1.   Solve the formula $V = lwh$ for w.

_____

2. Solve $F = \dfrac{9}{5}C + 32$ for C.

$$F = \dfrac{9}{5}C + 32$$

$$5F = 9C + 160 \qquad \text{Multiply by 5 to clear the fraction.}$$

$$5F - 160 = 9C \qquad \text{Subtract 160 from each side.}$$

$$\dfrac{5F - 160}{9} = C \qquad \text{Divide each side by 9.}$$

3. Solve $S = (n - 2)180$ for n.

$$S = (n - 2)180$$

$$\dfrac{S}{180} = \dfrac{(n-2)180}{180} \qquad \text{Divide each side by 180.}$$

$$\dfrac{S}{180} = n - 2$$

$$\dfrac{S}{180} + 2 = n - 2 + 2 \qquad \text{Add 2 to each side.}$$

$$\dfrac{S}{180} + \dfrac{360}{180} = n \qquad 2 = \dfrac{360}{180}$$

$$\dfrac{S + 360}{180} = n \qquad \text{Combine fractions.}$$

4. Solve $4x - 3y = 20$ for y.

$$4x - 3y = 20$$

$$4x - 3y - 4x = 20 - 4x \qquad \text{Subtract } 4x \text{ from each side.}$$

$$-3y = 20 - 4x \qquad \text{Combine like terms.}$$

$$\dfrac{-3y}{-3} = \dfrac{20 - 4x}{-3} \qquad \text{Divide each side by } -3.$$

$$y = \dfrac{20 - 4x}{-3} \quad \text{or} \quad y = \dfrac{4x - 20}{3}$$

2. Solve $A = \dfrac{1}{2}h(B + b)$ for h.

_____

3. Solve $A = \dfrac{1}{2}h(B + b)$ for B.

_____

4. Solve $3x + 5y = 4$ for y.

_____

Review this example for Objective 2:	**Now Try:**

Review this example for Objective 2:

5. Maria travels 45 miles at a rate of 60 miles per hour. How long did she travel?

Apply the formula $d = rt$ for and solve for t. Substitute 45 miles for d and 60 miles per hour for r.

$45 = 60t$

$\dfrac{45}{60} = \dfrac{3}{4} = t$

Maria traveled for $\dfrac{3}{4}$ hr.

Now Try:

5. Tyler travels 260 miles in 5 hours. What is his rate?

_____

Review these examples for Objective 3:

6. Solve each problem.

a. At a particular college 120 students are enrolled in organic chemistry. Of this group 18 are freshman, 52 are sophomores, 48 are juniors, and 2 are seniors. What percent of the organic chemistry students is made up of juniors?

b. A salesperson earns 15% commission each month on all that she sells. In a particular month she sold $12,000 in goods. What was her commission for that month?

a. There are 48 juniors out of 120 students in the class. Let x represent the percent of juniors in the class.

$x = \dfrac{48}{120}$ ← partial amount
       ← whole amount

$= 0.40$ or 40%

40% of the class are junior.

b. Let x represent the commission earned. Since 15% = 0.15, the equation is

$\dfrac{x}{12,000} = 0.15$ $\dfrac{\text{partial amount}}{\text{base } b} = \text{percent}$

$x = 0.15(12,000)$
                 Multiply by 12,000.

$x = 1800$

The salesperson earned $1800 in commission.

Now Try:

6. Solve each problem.

a. A 75 liter beaker of solution contains 15 liters of acid. What is the percent of acid in the solution?

_____

b. If a bank account earns 2.3% interest on a balance of $11,500, how much interest is earned? _____

7. The graph shows spending by department at a college. The college's total budget is $280 million. How much was budgeted for the business department?

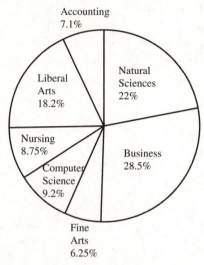

Accounting
7.1%

Natural Sciences
22%

Liberal Arts
18.2%

Nursing
8.75%

Computer Science
9.2%

Business
28.5%

Fine Arts
6.25%

Since 28.5% was budgeted for the business department, let x = this amount in millions of dollars. 28.5% = 0.285.

$$\frac{x}{280} = 0.285 \qquad \frac{\text{partial amount}}{\text{base } b} = \text{percent}$$

$$x = 0.285(280)$$

Multiply by 280.

$$x = 79.8$$

$79.8 million was budgeted for the business department.

7. Refer to the graph at the left. If the college's total budget is $330 million, how much is budgeted for fine arts?

_____

Review this example for Objective 4:

8. Solve each problem. Round your answers to the nearest tenth.

a. Students at Withrow College were charged $1560 for tuition this semester. If the tuition was $1480 last semester, find the percent of increase.

b. At the end of 2008, there were 124 wolves in Yellowstone National Park, compared to 171 wolves in 2007. What was the percent of decrease in the wolf population? (*Source*: Yellowstone Wolf Project Annual Report, 2008.)

Now Try:

8. Solve each problem. Round your answers to the nearest tenth.

a. Over the last three years, Calvin's salary has increased from $2700 per month to $3200. What is the percent increase? _____

b. Tracy works as a massage therapist. During July, she cut her price on massages from $54 to $45.50. By what percent did she decrease the price? _____

a. The amount of increase is
$1560 − $1480 = $80. Let x represent the percent increase.

$$\text{percent increase} = \frac{\text{amount of increase}}{\text{base}}$$

$$x = \frac{80}{1480} \approx 0.054 \text{ or } 5.4\%$$

The tuition increase by 5.4%

b. The amount of decrease is
$171 − 124 = −47$. Let x represent the percent decrease.

$$\text{percent decrease} = \frac{\text{amount of decrease}}{\text{base}}$$

$$x = \frac{-47}{171}$$
$$\approx -0.275 \text{ or } -27.5\%$$

The wolf population decreased by 27.5%.

Objective 1 Solve a formula for a specified variable.
For extra help, see Examples 1–4 on pages 57–59 of your text, the Section Lecture video for Section 2.2, and Exercise Solution Clips 3 and 9.

Solve the formula for the specified variable.

1. $V = \frac{1}{3}Bh$ for B **1.** _____

2. $C = \frac{5}{9}(F - 32)$ for F **2.** _____

3. $\frac{x+y}{3} = \frac{y}{5}$ for y **3.** _____

Objective 2 Solve applied problems by using formulas.
For extra help, see Example 5 on page 59 of your text and the Section Lecture video for Section 2.2.

Solve using the appropriate formula.

4. A cord of wood contains 128 cubic feet of wood. A stack of wood is 4 feet high and 8 feet long. How wide must it be to contain a cord?

4. _____

5. The area of a trapezoid is 60 square feet. If the bases are 6 feet and 14 feet, find the altitude of the trapezoid.

5. _____

6. Find the width of a rectangle if the perimeter is 20 inches and the length is 6 inches.

6. _____

7. If $700 earns $112 simple interest in 2 years, find the rate of interest.

7. _____

Objective 3 Solve percent problems.
For extra help, see Examples 6 and 7 on pages 60–61 of your text, the Section Lecture video for Section 2.2, and Exercise Solution Clip 33.

Solve each problem.

8. Rachael earned 7% interest on a $2,500 savings account for one year. How much money did she earn in interest?

8. _____

9. A salesperson earned $33,250 on annual sales of $950,000. What is her rate of commission?

9. _____

10. An alcohol and water mixture measures 36 liters of water and 9 liters of alcohol. What percent of the mixture is alcohol?

10.  _____

11. The graph shows spending by department at a college. The college's total budget is $280 million. How much was budgeted for the accounting department?

11.  _____

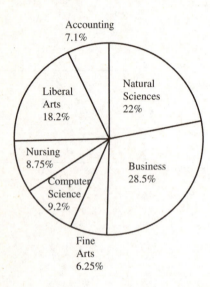

Objective 4 Solve problems involving percent increase or decrease.
For extra help, see Example 8 on pages 61–62 of your text and the Section Lecture video for Section 2.2.

Solve each problem. Round money answers to the nearest cent and percent answers to the nearest tenth, if necessary.

12. The sale price of a leather coat is $471.25. The regular price is $650. What is the percent of decrease?

12. _____

13. Mark Johnson invested $2500 in stock one year ago. During the year the stock increased in value by $162.50. What was the percent of increase?

13. _____

14. Lewis has increased his exercise schedule from 5 hours per week to $10\frac{1}{2}$ hours per week. What is the percent increase?

14.

15. The number of days employees of Acme Manufacturing Company were absent from their jobs decreased from 96 days last month to 72 days this month. Find the percent of decrease.

15.

Chapter 2 LINEAR EQUATIONS, INEQUALITIES, AND APPLICATIONS

2.3   Applications of Linear Equations

Learning Objectives
1   Translate from words to mathematical expressions.
2   Write equations from given information.
3   Distinguish between simplifying expressions and solving equations.
4   Use the six steps in solving an applied problem.
5   Solve percent problems.
6   Solve investment problems.
7   Solve mixture problems.

Key Terms

Use the vocabulary terms listed below to complete each statement in exercises 1–4.

sum	of	product	increased by
quotient	less than	per	double
decreased by	more than	difference	ratio

1.   _____, _____, and _____
      are words that mean addition.

2.   _____, _____, and _____
      are words that mean multiplication.

3.   _____, _____, and _____
      are words that mean division.

4.   _____, _____, and _____
      are words that mean subtraction.

Guided Examples

Review this example for Objective 2:
1.   Translate each verbal sentence into an equation, using x as the variable.
      a.   The quotient of twice a number and 3 is 8.
      b.   The product of −2 and the difference between 4 and a number is 24.

Now Try:
1.   Translate each verbal sentence into an equation, using x as the variable.
      a.   The product of a number and the number decreased by 6 is 18. _____

c. If 2 is subtracted from four times a number, the result is three more than six times the number.

a. The quotient of twice a number and 3 is 8.

$$\frac{2x}{3} = 8$$

b. The product of −2 and the difference between 4 and a number is 24.
$$-2(4-x) = 24$$

c. If 2 is subtracted from four times a number, the result is three more than six times the number.
$$4x - 2 = 3 + 6x$$

b. The quotient of 4 more than a number and 9 is the same as the number increased by 5.

_____

c. If twice a number is added to 50, the result is the number decreased by 6.

_____

Review this example for Objective 3:

2. Decide whether each is an expression or an equation. Simplify any expressions and solve any equations.

  a. $3(z-9) = 8(2z-4)$

  b. $\dfrac{k}{3} - \dfrac{k+5}{4} - 12$

  a. $3(z-9) = 8(2z-4)$
    This is an equation because there is an equals symbol.

$$3(z-9) = 8(2z-4)$$
$$3z - 27 = 16z - 32 \quad \text{Distributive property}$$
$$3z + 5 = 16z \quad\quad \text{Add 32.}$$
$$5 = 13z \quad\quad\quad \text{Subtract } 3z.$$
$$\frac{5}{13} = z \quad\quad\quad \text{Divide by 13.}$$

The solution set is $\left\{\dfrac{5}{13}\right\}$.

  b. $\dfrac{k}{3} - \dfrac{k+5}{4} - 12$
    This is an expression because there is no equals symbol.

Now Try:

2. Decide whether each is an expression or an equation. Simplify any expressions and solve any equations.

  a. $3(4x-5) - 7x + 2$

_____

  b. $7x - 8 + 9(x-2) = 1$

_____

$$\frac{k}{3} - \frac{k+5}{4} - 12$$

$$= \frac{4k}{12} - \frac{3(k+5)}{12} - \frac{12(12)}{12} \qquad \text{The LCD is 12.}$$

$$= \frac{4k}{12} - \frac{3k+15}{12} - \frac{144}{12} \qquad \begin{array}{l}\text{Distributive}\\\text{property.}\end{array}$$

$$= \frac{4k - 3k - 15 - 144}{12} \qquad \begin{array}{l}\text{Combine}\\\text{fractions.}\end{array}$$

$$= \frac{k - 159}{12} \qquad \begin{array}{l}\text{Combine like}\\\text{terms.}\end{array}$$

Review these examples for Objective 4:

3. In a triangle with two sides of equal length, the third side is 6 feet less than the sum of the lengths of the two equal sides. The perimeter of the triangle is 26 feet. Find the lengths of the three sides of the triangle.

   ***Step 1* Read the problem.** We must find the lengths of the three sides of the triangle. Two of the sides have equal length and the length of the third side is 6 feet less than the sum of the lengths of the two equal sides. The perimeter is 26 feet.

   ***Step 2* Assign a variable.** Let x = the length of each of the equal sides. Then $2x - 6$ = the length of the third side.

   ***Step 3* Write an equation.** The perimeter is equal to the sum of the lengths of the three sides.
   $$26 = x + x + (2x - 6)$$

   ***Step 4* Solve the equation.**

   $26 = x + x + (2x - 6)$

   $26 = 4x - 6$ Combine like terms.

   $32 = 4x$ Add 6 to each side.

   $8 = x$ Divide each side by 4.

   ***Step 5* State the answer.**
   The length of the two equal sides is 8 feet and the length of the third side is $2(8) - 6 = 10$ feet.

   ***Step 6* Check.**
   The length of the third side, 10 feet, is 6 feet less than twice the length of each of the equal sides, 8 feet. The perimeter is $8 + 8 + 10 = 26$ feet, as required.

Now Try:

3. The width of a rectangle is $\frac{1}{4}$ of the difference between the length and 1 meter. If the perimeter of the rectangle is 62 meters, find the length and width.

   _____

4. In a local government election, two candidates received 3215 votes. The losing candidate received 637 fewer votes than the winning candidate. How many votes did each candidate receive?

    ***Step 1* Read the problem.** We must how many votes each candidate received.

    ***Step 2* Assign a variable.** Let x = the number of votes the winning candidate received. Then $x - 637$ = the number of votes the losing candidate received.

    ***Step 3* Write an equation.** There were 3215 votes in all.

    $x + (x - 637) = 3215$

    ***Step 4* Solve the equation.**

    $x + (x - 637) = 3215$
    $2x - 637 = 3215$ Combine like terms.
    $2x = 3852$ Add 637 to each side.
    $x = 1926$ Divide each side by 2.

    ***Step 5* State the answer.**
    The winning candidate received 1926 votes and the losing candidate received
    $1926 - 637 = 1289$ votes.

    ***Step 6* Check.**
    The losing candidate received 637 fewer votes than did the winning candidate. The total number of votes is $1926 + 1289 = 3215$ votes, as required.

4. A lake manager wants to stock a small lake with 11,550 fish, some bass and some walleyes. He wants to use 1830 more bass. How many of each kind of fish does he need?

    _____

Review this example for Objective 5:

5. A brokerage account earns interest at an annual rate of 4%. At the end of one year Sheri would like to have a total of $5200 in the account. How much should she deposit now?

    ***Step 1* Read the problem.** We must find out how much Sherri needs to deposit in order to have $5200 in her account at the end of one year. The account earns 4% each year.

    ***Step 2* Assign a variable.** Let x = the amount to be deposited. Then $0.04x$ = the amount of interest earned in one year.

Now Try:

5. A merchant in a small store took in $508.80 in one day. This included merchandise sales and the 6% sales tax. Find the amount of merchandise sold.

    _____

***Step 3* Write an equation.** The total at the end of the year equals the original amount deposited plus the interest earned.

$x + 0.04x = 5200$

***Step 4* Solve the equation.**

$x + 0.04x = 5200$

$\quad 1.04x = 5200$ Combine like terms.

$\dfrac{1.04x}{1.04} = \dfrac{5200}{1.04}$ Divide each side by 1.04.

$\qquad x = 5000$

***Step 5* State the answer.**

Sherri must deposit $5000.

***Step 6* Check.**

At 4% per year, the interest on $5000 is $0.04(5000) = \$200$. Thus, at the end of one year, the account will have $5000 + \$200 = \$5200$, as required.

Review this example for Objective 6:

6. Dan has $45,000 invested in bonds paying 10%. How much additional money should he invest at 5% to have an average return on the two investments of 8%?

   ***Step 1* Read the problem.** We must find out how much Dan needs to invest at 5% in order to have an average return of 8% on his two investments.

   ***Step 2* Assign a variable.** Let x = the amount to be invested at 5%. Then $45,000 + x$ = the total amount invested. The formula for interest is $I = prt$. Here the time t is one year. Use a table to organize the given information.

Principle	Rate	Interest
45,000	0.10	0.10(45,000)
x	0.05	0.05x
45,000 + x	0.08	0.08(45,000 + x)

   ***Step 3* Write an equation.**

   Interest at 10% + interest at 5% = interest at 8%.

   $0.10(45,000) + 0.05x = 0.08(45,000 + x)$

Now Try:

6. Fred received an inheritance of $13,500. He wishes to divide the amount between investments at 15% and 12 % to receive an average return on both investments of 13%. How much should he invest at each rate?

   _____

Step 4 Solve the equation.

$$0.10(45,000) + 0.05x = 0.08(45,000 + x)$$
$$4500 + 0.05x = 3600 + 0.08x$$

Multiply; distributive property

$$900 + 0.05x = 0.08x$$

Subtract 3600.

$$900 = 0.03x$$

Subtract $0.05x$.

$$\frac{900}{0.03} = \frac{0.03x}{0.03}$$

Divide by 0.03.

$$30,000 = x$$

Step 5 State the answer.
Dan should invest $30,000 at 5%.

Step 6 Check.
The $45,000 invested at 10% earns $4500 in interest, and the $30,000 invested at 5% earns $1500 in interest, for a total of $6000. If $75,000 is invested at 8%, the interest for one year is $0.08(75,000) = \$6000$, as required.

Review these examples for Objective 7:

7. How many liters of 20% alcohol solution must be mixed with 10 liters of a 50% solution to get a 30% solution?

Step 1 Read the problem. The problem asks for the number of liters of the 20% solution.

Step 2 Assign a variable. Let $x =$ the amount of 20% solution.

Number of liters	Percent	Liters of Pure Acid
x	0.20	$0.20x$
10	0.50	$0.50(10) = 5$
$10 + x$	0.30	$0.30(10 + x)$

Step 3 Write an equation.
$$0.20x + 5 = 0.30(10 + x)$$

Now Try:

7. How many pounds of peanuts that sell for $2 a pound must be mixed with 6 pounds of cashew nuts that sell for $4 a pound to get a mixture that can sell for $2.50 a pound?

Step 4 **Solve the equation.**

$$0.20x + 5 = 0.30(10 + x)$$

$0.20x + 5 = 3 + 0.30x$ Distributive property

$0.20x + 2 = 0.30x$ Subtract 3.

$2 = 0.10x$ Subtract 0.20x.

$$\frac{2}{0.10} = \frac{0.10x}{0.10}$$ Divide by 0.10.

$20 = x$

Step 5 **State the answer.**

20 liters of 20% solution should be used.

Step 6 **Check.**

20 liters of 20% solution contains $20(0.20) = 4$ liters of alcohol, and 10 liters of 50% solution contains $10(0.50) = 5$ liters of alcohol, for a total of $4 + 5 = 9$ liters of alcohol. There are $10 + 20 = 30$ liters of 30% solution, containing $30(0.30) = 9$ liters of alcohol, as required.

8. How many liters of pure alcohol must be mixed with 20 liters of an 18% alcohol solution to obtain a 20% alcohol solution?

Step 1 **Read the problem.** The problem asks for the number of liters of pure alcohol.

Step 2 **Assign a variable.** Let $x =$ the amount of pure alcohol.

Number of liters	Percent	Liters of Pure Alchohol
x	1	x
20	0.18	$0.18(20) = 3.6$
$x + 20$	0.20	$0.20(x + 20)$

Step 3 **Write an equation.**

$x + 3.6 = 0.20(x + 20)$

Step 4 **Solve the equation.**

$x + 3.6 = 0.20(x + 20)$

$x + 3.6 = 0.20x + 4$ Distributive property

$0.80x + 3.6 = 4$ Subtract 0.20x.

$0.80x = 0.4$ Subtract 4.

$$\frac{0.80x}{0.80} = \frac{0.4}{0.80}$$ Divide by 0.80.

$x = 0.5$

8. How many milliliters of water must be mixed with 25 milliliters of a 32% sulfuric acid solution to obtain a 20% sulfuric acid solution?

_____

> ***Step 5* State the answer.**
> 0.5 liters of pure alcohol should be added.
> ***Step 6* Check.**
> Show that $0.5 + 0.18(20) = 0.20(0.5 + 20)$ is true.

Objective 1 Translate from words to mathematical expressions.

Objective 2 Write equations from given information.

For extra help, see Example 1 on page 68 of your text, the Section Lecture video for Section 2.3, and Exercise Solution Clip 7.

Translate the verbal sentence into an equation. Do not solve. *Use x to represent the unknown number.*

1. The product of a number and 6 is 7 plus the product of 5 times the number.

    1._____

2. Twice a number is 3 times the sum of the number and 7.

    2._____

3. The ratio of a number and the difference between the number and 3 is 17.

    3._____

Objective 3 Distinguish between simplifying expressions and solving equations.

For extra help, see Example 2 on pages 68–69 of your text and the Section Lecture video for Section 2.3.

Decide whether each is an expression or an equation. Simplify any expressions and solve any equations.

4. $7x - 8 + 9(x - 2) + 1$

    4. _____

5. $3(4x - 5) = 7x + 2$

    5. _____

Objective 4 Use the six steps in solving an applied problem.
For extra help, see Examples 3 and 4 on pages 69–70 of your text, the Section Lecture video for Section 2.3, and Exercise Solution Clip 29.

Use the six-step process discussed in the textbook to solve the following problems.

6. The length of a rectangle is 3 inches less than twice the width. The perimeter of the rectangle is 90 inches. Find the dimensions of the rectangle.

6. _____

7. In the controversial 2000 presidential election, George W. Bush and Al Gore received a total of 537 electoral votes. Bush received five more votes than Gore. How many votes did each candidate receive?

7. _____

8. The sum of two numbers is 41. The larger number is 1 more than 4 times the smaller number. What are these numbers?

8. _____

Objective 5 Solve percent problems.
For extra help, see Example 5 on page 71 of your text and the Section Lecture video for Section 2.3.

Solve the problem.

9. At the end of a day, a storekeeper had $642 in the cash registers, counting both the sale of goods and the sales tax of 7%. Find the amount of the tax.

9. _____

10. A total of 168 votes were cast in an election. The loser received 75% as many votes as the winner. How many votes did the winner and loser each get?

10. _____

Objective 6 Solve investment problems.
For extra help, see Example 6 on page 72 of your text, the Section Lecture video for Section 2.3, and Exercise Solution Clip 49.

Solve each problem.

11. Mae invested some money at 8% and $300 more than twice this amount at 12%. Her total annual income from the two investments is $548. How much is invested at each rate?

11. _____

12. Ann invested some money at 8% and $500 less than this amount at 13%. Her total annual income from interest was $460. How much was invested at each rate?

12. _____

Objective 7 Solve mixture problems.
For extra help, see Examples 7 and 8 on pages 73–74 of your text, the Section Lecture video for Section 2.3, and Exercise Solution Clips 55 and 59.

Solve each problem.

13. How many liters of a 20% alcohol solution must be mixed with 50 liters of a 50% solution to get a 25% solution?

13. _____

14. Candy worth $2.35 per pound is mixed with candy worth $1.85 per pound. How many pounds of each are mixed to have 50 pounds of candy worth $2 per pound?

14. _____

15. How many liters of water should be added to a 25% antifreeze solution to obtain 30 liters of a 20% solution?

15. _____

Chapter 2   LINEAR EQUATIONS, INEQUALITIES, AND APPLICATIONS

2.4   Further Applications of Linear Equations

Learning Objectives
1 Solve problems about different denominations of money.
2 Solve problems about uniform motion.
3 Solve problems about angles.

Key Terms

Use the vocabulary terms listed below to complete each statement in exercises 1–2.

denomination   **uniform motion**

1. _____ problems use the distance formula, $d = rt$.

2. In problems involving money, use the basic fact that the number of monetary units of the same kind times the _____ equals the total monetary value.

Guided Examples

Review this example for Objective 1:

1. A collection of dimes and nickels has a total value of $2.70. The number of nickels is 2 more than twice the number of dimes. How many of each type of coin are in the collection?

    Step 1 **Read the problem.** The problem asks for the number dimes and nickels in the collection of coins.

    Step 2 **Assign a variable.** Let $x =$ the number of dimes. Then $2x + 2 =$ the number of nickels.

	Number of coins	Denomination	Value
dimes	x	0.10	$0.10x$
nickels	$2x + 2$	0.05	$0.05(2x + 2)$
Total value			2.70

    Step 3 **Write an equation.**
    $0.10x + 0.05(2x + 2) = 2.70$

Now Try:

1. Eric's piggy bank has quarters and nickels in it. The total number of coins in the piggy bank is 50 and the total value is $8.90. How many of each type of coin are in the piggy bank?

    quarters   _____

    nickels   _____

Step 4 Solve the equation.

$$0.10x + 0.05(2x + 2) = 2.70$$

$0.10x + 0.10x + 0.10 = 2.70$ Distributive property

$0.20x + 0.10 = 2.70$ Combine like terms.

$0.20x = 2.60$ Subtract 0.10.

$\dfrac{0.20x}{0.20} = \dfrac{2.60}{0.20}$ Divide by 0.20.

$x = 13$

Step 5 State the answer.

There are 13 dimes and $2(13) + 2 = 28$ nickels.

Step 6 Check.

The value of the coins is

$13(0.10) + 28(0.05) = \$2.70$, as required.

Review these examples for Objective 2:

2. Two boats leave a port at the same time, one traveling north and the other south. The one traveling north steams at 13 miles per hour, and the one traveling south steams at 19 miles per hour. In how many hours will they be 232 miles apart?

   Step 1 Read the problem. We are looking for the time it takes for the boats to be 232 miles apart.

   Step 2 Assign a variable. The boats are going in opposite directions. Let t = the time that each boat travels.

	rate	time	distance
Northbound boat	13	t	$13t$
Southbound boat	19	t	$19t$
Total distance			232

   Step 3 Write an equation.
   $$13t + 19t = 232$$

Now Try:

2. Two cars leave a city at the same time, but in opposite directions. One travels at 56 miles per hour, and the other travels at 60 miles per hour. In how many hours will they be 522 miles apart?

   _____

***Step 4* Solve the equation.**

$$13t + 19t = 232$$

$$32t = 232 \qquad \text{Combine like terms.}$$

$$\frac{32t}{32} = \frac{232}{32} \qquad \text{Divide by 32.}$$

$$x = 7.25$$

***Step 5* State the answer.**

Each boat traveled for 7.25 hours.

***Step 6* Check.**

The northbound boat traveled
$13(7.25) = 94.25$ miles. The southbound boat traveled $19(7.25) = 137.75$ miles. They are $94.25 + 137.73 = 232$ miles apart, as required.

3. Kathy can get to school in 1/4 hour if she rides her bike. It takes her 3/4 hour if she walks. Her speed when walking is 6 miles per hour slower than her speed when riding. What is her speed when riding?

***Step 1* Read the problem.** We must find Kathy's speed when she rides her bike to school.

***Step 2* Assign a variable.** Let r = Kathy's speed when she rides her bike. Then $r - 6$ = Kathy's speed when she walks.

	rate	time	distance
Bike	r	$\frac{1}{4}$	$\frac{1}{4}r$
Walk	$r - 6$	$\frac{3}{4}$	$\frac{3}{4}(r - 6)$

***Step 3* Write an equation.**

The two distances are equal.

$$\frac{1}{4}r = \frac{3}{4}(r - 6)$$

***Step 4* Solve the equation.**

$$\frac{1}{4}r = \frac{3}{4}(r - 6)$$

$$\frac{1}{4}r = \frac{3}{4}r - \frac{18}{4} \qquad \text{Distributive property}$$

$$r = 3r - 18 \qquad \text{Multiply by the LCD, 4.}$$

$$-2r = -18 \qquad \text{Subtract } 2r.$$

$$r = 9 \qquad \text{Divide by } -2.$$

3. Libby's drive to work takes 1 hour. If she increases her average speed by 10 miles per hour, the trip takes 3/4 hour. Find the distance she drives to work.

_____

Step 5 State the answer.
Kathy's speed when riding her bike is 9 miles per hour.

Step 6 Check.
The distance Kathy travels when riding her bike is $\frac{1}{4}(9) = 2\frac{1}{4}$ miles. Her walking speed is $9 - 6 = 3$ miles per hour, so her walking distance is $\frac{3}{4}(3) = 2\frac{1}{4}$ miles, as required.

Review this example for Objective 3:

4. Find the measures of each angle in the triangle.

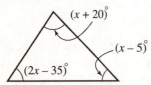

The sum of the measures of the angles of a triangle is 180°.

$$(x - 5) + (x + 20) + (2x - 35) = 180$$
$$4x - 20 = 180$$
Combine like terms.

$$4x = 200$$
Add 20.

$$x = 50$$
Divide by 4.

The measures of the angles are $50 - 5 = 45°$, $50 + 20 = 70°$, and $2(50) - 35 = 65°$.
Check by confirming that
$45° + 70° + 65° = 180°$.

Now Try:

4. One angle of a triangle measures 30° larger than a second angle. The third angle measures 4 times the second angle. Find the measure of each angle.

_____

Objective 1 Solve problems about different denominations of money.
For extra help, see Example 1 on page 81 of your text, the Section Lecture video for Section 2.4, and Exercise Solution Clip 11.

Solve the problem.

1. Stan has 14 bills in his wallet worth $95 altogether. If the wallet contains only $5 and $10 bills, how many bills of each denomination does he have?

1.
$5-bills ___9___

$10-bills ___5___

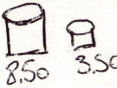

2. Shopwell Grocery sells two differently sized jars of peanut butter. The large size sells for $8.50 and the small size sells for $3.50. There are 80 jars worth $405. How many of each size jar are there?

2.

large_____

small _____

3. At the end of each day, Helen throws all the change from her purse into a box. The box contains only pennies, nickels, and dimes. At the end of one week she found that the total value of the coins was $4.80. If the number of dimes was 1 more than the number of nickels and the number of pennies was 6 more than the number of nickels, how many of each type of coin was in the box?

3.

dimes_____

nickels _____

pennies_____

4. The total receipts for a basketball game were $4690.50. There were 723 tickets sold, some for children and some for adults. If the adult tickets cost $9.50 and the children's tickets cost $4, how many of each type were there?

4.

adults _____

children _____

5. Christopher purchased some stamps for $102. He purchased three kinds of stamps in denominations of $0.44, $0.61, and $4.95. The number of $0.61 stamps was the same as the number of $0.44 stamps, which was 5 times the number of $4.95 stamps. How many of each kind of stamp did he purchase?

5.

44¢-stamps _____

61¢-stamps _____

$4.95-stamps _____

Objective 2 Solve problems about uniform motion.

For extra help, see Examples 2 and 3 on pages 82–83 of your text, the Section Lecture video for Section 2.4, and Exercise Solution Clips 21 and 25.

Solve the problem.

6. On a 690-mile automobile journey, Sanjay averaged 55 miles per hour on the first part of the trip and 60 miles per hour on the second part of the trip. How long did the entire journey take if the two parts each took the same number of hours?

6. _____

7. A 280-mile automobile trip took Max a total of 5 hours. His average speed for the last 3 hours was 10 miles per hour faster than his average speed on the first 2 hours. Find his average speed for the last 3 hours of the trip.

7. _____

8. Mary and Greta live 21 miles apart. At 1:00 P.M. they start riding their bicycles toward each other and meet at 1:45 P.M. If Greta's average speed is 4 miles per hour faster than Mary's average speed, find Greta's average speed.

8. _____

9. Two planes left Philadelphia traveling in opposite directions. Plane A left 15 minutes before plane B. After plane B had been flying for 1 hour, the planes were 860 miles apart. What were the speeds of the two planes if plane A was flying 40 miles per hour faster than plane B?

9.

plane A _____

plane B _____

10. John left Louisville at noon on the same day that Mike left Louisville at 1 P.M. Both were traveling in the same direction. At 5 P.M., Mike was 62 miles behind John. If John was traveling 2 miles per hour faster than Mike, what were their speeds?

10.
John _____

Mike _____

Objective 3 Solve problems about angles.

For extra help, see Example 4 on page 84 of your text, the Section Lecture video for Section 2.4, and Exercise Solution Clip 31.

11. Find the measure of each angle in the triangle.

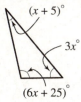

11. _____

Solve each problem.

12. The smallest angle in a triangle has a measure which is one-half the measure of the largest angle. The measure of the third angle is 20° less than the measure of the largest angle. Find the measures of the three angles in the triangle.

12. _____

13. One angle of a triangle measures twice the second angle. The measure of the third angle equals the sum of the measures of the first two angles. Find the measure of each angle.

13. _____

14. The two acute angles in a right triangle have measures such that one is 6° less than five times the measure of the other. What are the measures of the angles? (Hint: One angle in a right triangle measures 90°.)

14. _____

15. The measure of one angle of a triangle is 2 more than twice the smallest angle of the triangle. The measure of the third angle is 18 more the measure of the smallest angle. Find the measures of the angles.

15. _____

Chapter 2 LINEAR EQUATIONS, INEQUALITIES, AND APPLICATIONS

2.5   Linear Inequalities in One Variable

Learning Objectives
1   Solve linear inequalities by using the addition property.
2   Solve linear inequalities by using the multiplication property.
3   Solve linear inequalities with three parts.
4   Solve applied problems by using linear inequalities.

Key Terms

Use the vocabulary terms listed below to complete each statement in exercises 1–4.

inequality   **linear inequality in one variable**

equivalent inequalities   **three-part inequality**

1.   A(n) _____ can be written in the form $Ax + B < C$ or $Ax + B > C$ (or with $\leq$ or $\geq$), where A, B, and C are real numbers with $A \neq 0$.

2.   _____ are inequalities that have the same solution set.

3.   An inequality that states that one number is between two other numbers is called a(n) _____.

4.   An algebraic expression related by $>$, $\geq$, $<$, or $\leq$ is called a(n) _____.

Guided Examples

Review these examples for Objective 1:
1.   Solve $x + 7 \leq 5$ and graph the solution set.

$$x + 7 \leq 5$$
$$x + 7 - 7 \leq 5 - 7 \quad \text{Subtract 7.}$$
$$x \leq -2$$

Check: Substitute -2 for x in the equation $x + 7 = 5$.
$$x + 7 = 5$$
$$\overset{?}{-2 + 7 = 5}$$
$$5 = 5 \checkmark$$

Now Try:
1.   Solve $p - 3 > 6$ and graph the solution set.

_____

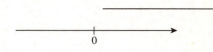

This shows that −2 is the boundary point. Now test a number on each side of −2 to verify that numbers less than or equal to −2 make the inequality true.

$-3+7 \leq 5$ Let $x = -3$.

$\quad\quad 4 \leq 5$ True

$\overline{0+7 \leq 5}$ Let $x = 0$.

$\quad\quad 7 \leq 5$ False

The check confirms that $(-\infty, -2]$ is the solution set.

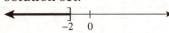

2. Solve $6x + 3 > 7x$ and graph the solution set.

$\quad\quad 6x + 3 > 7x$

$6x + 3 - 6x > 7x - 6x$ Subtract $6x$.

$\quad\quad\quad 3 > x$ Combine like terms.

$\quad\quad\quad x < 3$ Reverse the inequality.

Verify that 3 is the boundary point by substituting 3 for x in the equation $6x + 3 = 7x$. Then choose a test value on either side of 3 to confirm that $(-\infty, 3)$ is the solution set.

$6(0) + 3 > 7(0)$ Let $x = 0$.

$\quad\quad 3 > 0$ True

$\overline{6(5) + 3 > 7(5)}$ Let $x = 5$.

$\quad\quad 33 > 35$ False

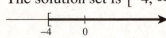

2. Solve $8x + 2 < 7x - 3$ and graph the solution set.

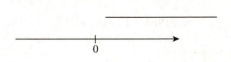

Review these examples for Objective 2:

3. Solve each inequality and graph the solution set.

    a. $4k \geq -16$ **b.** $-\dfrac{3}{2}x < -3$

    a. $4k \geq -16$

    $\quad k \geq -4$ Divide each side by 4.

    The solution set is $[-4, \infty)$.

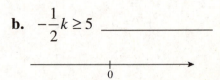

Now Try:

3. Solve each inequality and graph the solution set.

    a. $3p > 9$ _____

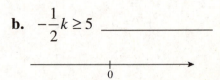

    b. $-\dfrac{1}{2}k \geq 5$ _____

b. $\quad -\dfrac{3}{2}x < -3$

$$-\dfrac{3}{2}x\left(-\dfrac{2}{3}\right) > -3\left(-\dfrac{2}{3}\right)$$

Multiply each side by $-\dfrac{2}{3}$.
Reverse the direction of the inequality symbol.

$$x > 2$$

The solution set is $(4, \infty)$.

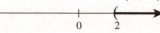

4. Solve and graph the solution set.

$$4(y-3)+2 < 3(y-2)$$

$$4(y-3)+2 < 3(y-2)$$

$\quad 4y-12+2 < 3y-6$ Distributive property

$\quad 4y-10 < 3y-6$ Combine like terms.

$\quad\quad 4y < 3y+4$ Add 10 to each side.

$\quad\quad\quad y < 4$ Subtract $3y$ from each side.

The solution set is $(-\infty, 2)$.

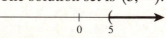

5. Solve and graph the solution set.

$$-\dfrac{2}{3}(x+1)+2 < \dfrac{1}{2}(1-x)$$

Clear the fractions by multiplying both sides by the LCD, 6.

$$-\dfrac{2}{3}(x+1)+2 < \dfrac{1}{2}(1-x)$$

$$6\left[-\dfrac{2}{3}(x+1)+2\right] < 6\left[\dfrac{1}{2}(1-x)\right]$$

$$-4(x+1)+6(2) < 3(1-x)$$

Distributive property

$\quad -4x-4+12 < 3-3x$ Multiply.

$\quad\quad -4x+8 < 3-3x$ Combine like terms.

$\quad\quad\quad -x+8 < 3$ Add $3x$ to each side.

$\quad\quad\quad\quad -x < -5$ Subtract 8.

$\quad\quad\quad\quad\quad x > 5$ Multiply each side by -1; reverse the order of the inequality.

The solution set is $(5, \infty)$.

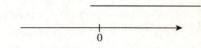

4. Solve and graph the solution set.

$$-3(m+4)+1 \geq -4(m-2)$$

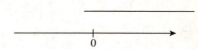

5. Solve and graph the solution set.

$$5x-\dfrac{3}{2}(x+2) > -\dfrac{5}{4}(x+1)+3$$

Review these examples for Objective 3:

6. Solve $-2 < y - 3 < 6$ and graph the solution set.

$$-2 < y - 3 < 6$$

$$-2 + 3 < y - 3 + 3 < 6 + 3 \quad \text{Add 3 to each of the parts of the inequality.}$$

$$1 < y < 9$$

The solution set is $(1, 9)$.

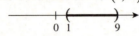

7. Solve $2 < 8 - 3x \leq 5$ and graph the solution set.

$$2 < 8 - 3x \leq 5$$

$$2 - 8 < 8 - 3x - 8 \leq 5 - 8 \quad \text{Subtract 8 from each of the parts of the inequality.}$$

$$-6 < -3x \leq -3 \quad \text{Combine terms.}$$

$$2 > x \geq 1 \quad \text{Divide each side by } -3; \text{ reverse the order of the inequalities.}$$

$$1 \leq x < 2 \quad \text{Rewrite in the order on the number line.}$$

The solution set is $[1, 2)$.

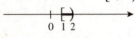

Now Try:

6. Solve $10 < z + 5 < 14$ and graph the solution set.

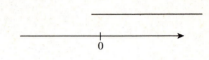

7. Solve $1 \leq -2z + 3 \leq 19$ and graph the solution set.

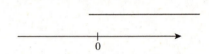

Review these examples for Objective 4:

8. A car rental company offers two option. Option A is $25 per day plus 15¢ per mile. Option B is $10 per day plus 40¢ per mile. If a car is rented for one day, for how many miles will Option A be cheaper?

Step 1 Read the problem. We must determine the number of mile that will make Option A cheaper.

Step 2 Assign a variable. Let x = the number of miles driven.

Step 3 Write an inequality. Option A costs $25 + 0.15x$, while Option B costs $10 + 0.40x$.

Thus, the inequality is
$$25 + 0.15x < 10 + 0.40x.$$

Now Try:

8. Ruth tutors mathematics in the evenings in an office for which she pays $600 per month rent. If rent is her only expense and she charges each student $40 per month, how many students must she teach to make a profit of at least $1600 per month?

Step 4 **Solve the inequality.**

$25 + 0.15x < 10 + 0.40x$

$15 + 0.15x < 0.40x$ Subtract 10 from each side.

$15 < 0.25x$ Subtract $0.15x$ from each side.

$60 < x$ Divide each side by 0.25.

$x > 60$ Rewrite the inequality.

Step 5 **State the answer.**

Option A will be cheaper if more than 60 miles are driven.

Step 6 **Check.**

If 60 miles are driven, Option A costs $\$25 + 0.15(60) = \$34$ and Option B costs $10 + 0.40(60) = \$34$. If 61 miles are driven, Option A costs $\$25 + 0.15(61) = \$34.15$, while Option B costs $10 + 0.40(61) = \$34.40$. Thus, Option A is cheaper.

9. Sara scored 78, 72, 87, and 90 on the first four quizzes this semester. What grade must she earn on the fifth quiz so that her average is at least 82?

Let x = the score on the fifth quiz. To find the average of five numbers, add them and then divide by 5.

$$\frac{78 + 72 + 87 + 90 + x}{5} \geq 82$$

$$\frac{327 + x}{5} \geq 82$$

$327 + x \geq 410$ Multiply by 5.

$x \geq 83$ Subtract 327.

Sara must score 83 or more on the fifth quiz so that her average will be at least 82.

9. Margaret gets scores of 88 and 78 on her first two tests. What score must she make on her third test to keep an average of 80 or greater?

_____

Objective 1 Solve linear inequalities by using the addition property.

For extra help, see Examples 1 and 2 on pages 92–93 of your text, the Section Lecture video for Section 2.5, and Exercise Solution Clip 9.

Solve the inequality, giving its solution set in interval form.

1. $x - 3 \geq -5$

1. _____

2. $y - 7 \leq 8$

2. _____

3. $8s < 7s - 3$

3. _____

Objective 2 Solve linear inequalities by using the multiplication property.
For extra help, see Examples 3–5 on pages 94–96 of your text, the Section Lecture video for Section 2.5, and Exercise Solution Clip 13, 25, 29, and 35.

Solve the inequality, giving its solution set in both interval and graph forms.

4. $-3a < 9$

4. _____

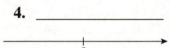

5. $2z < -8$

5. _____

6. $-\dfrac{3}{4}r \geq 27$

6. _____

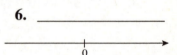

Objective 3 Solve linear inequalities with three parts.
For extra help, see Examples 6 and 7 on pages 96–97 of your text, the Section Lecture video for Section 2.5, and Exercise Solution Clip 57.

Solve the inequality, giving its solution set in both interval and graph forms.

7. $-6 < k + 2 < 8$

7. _____

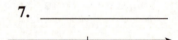

8. $-4 \leq a + 5 < -2$

8. _____

9. $-1 \leq 4r + 2 \leq 5$

9. _____

10. $-3 \le \dfrac{6q-1}{4} \le 0$

10. _____

$$\xrightarrow{\hspace{3cm}\underset{0}{|}\hspace{2cm}}$$

Objective 4 Solve applied problems by using linear inequalities.
For extra help, see Examples 8 and 9 on page 98 of your text, the Section Lecture video for Section 2.5, and Exercise Solution Clips 65 and 67.

Solve. Give your answer as a verbal statement.

11. When 3 times a number is subtracted from 8, the result is greater than or equal to 5. Find all such numbers.

11. _____

12. If twice the sum of a number and 7 is subtracted from three times the number, the result is more than −9. Find all such numbers.

12. _____

13. Two sides of a triangle are equal in length, with the third side 8 feet longer than one of the equal sides. The perimeter of the triangle cannot be more than 38 feet. Find the largest possible value for the length of the equal sides.

13. _____

14. Sally has 3 times as many nickels as dimes and she has at least 20 coins. At least how many dimes does she have?

14. _____

15. Jim must have an average of 80% of the points on 4
exams to receive a B in the class. He has earned 78%,
83%, and 75% on the first three exams. What is the
lowest score he can earn on a 100-point test to guarantee
a B in the class?

15. _____

Chapter 2 LINEAR EQUATIONS, INEQUALITIES, AND APPLICATIONS

2.6   Set Operations and Compound Inequalities

Learning Objectives

1   Find the intersection of two sets.
2   Solve compound inequalities with the word *and*.
3   Find the union of two sets.
4   Solve compound inequalities with the word *or*.

Key Terms

Use the vocabulary terms listed below to complete each statement in exercises 1–3.

intersection   **compound inequality**   **union**

1.   The _____ A and B, written $A \cup B$, is the set of elements that belong to either A or B (or both).

2.   The _____ A and B, written $A \cap B$, is the set of elements that belong to both A and B.

3.   A _____ is formed by joining two inequalities with a connective word such as *and* or *or*.

Guided Examples

Review this example for Objective 1:	**Now Try:**
1. Let $A = \{0,\ 1,\ 2,\ 3\}$ and $B = \{2,\ 3,\ 4,\ 5\}$. Find $A \cap B$. $A \cap B = \{2,\ 3\}$	1. Let $A = \{-6, -5, -4\}$ and $B = \{-3, -2, -1\}$. Find $A \cap B$. _____

Review these examples for Objective 2:	**Now Try:**
2. Solve the compound inequality $r + 2 < 5$ and $r + 3 > -3$, and graph the solution set. Solve each inequality individually. $r + 2 < 5$ and $r + 3 > -3$ $r < 3$ $r > 0$ The solution set is $(0, 3)$. 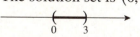	2. Solve the compound inequality $r - 5 \le 3$ and $r + 5 \ge 3$, and graph the solution set. _____ 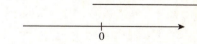

3. Solve and graph the solution set.

$-4x - 2 \le 10$ and $1 - x > 0$

Solve each inequality individually.

$-4x - 2 \le 10$ and $1 - x > 0$
$\quad -4x \le 12 \qquad\qquad -x > -1$
$\quad\quad x \ge -3 \qquad\qquad\quad x < 1$

The solution set is $[-3, 1)$.

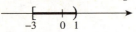

3. Solve and graph the solution
set.

$2 - n > 2$ and $-4n + 1 \le 21$

_____

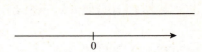

4. Solve.

$9 + x \ge 16$ and $x + 7 < 11$

Solve each inequality individually.

$9 + x \ge 16$ and $x + 7 < 11$
$\quad\quad x \ge 7 \qquad\qquad\quad x < 4$

The graphs of $x \ge 7$ and $x < 4$ are shown below.

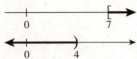

There is no number that is both greater than or equal to 7 and less than 4, so the compound inequality has no solution.
The solution set is $\varnothing$.

4. Solve.

$6x > -36$ and $3x \le -24$

_____

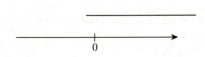

Review this example for Objective 3:

5. Let $A = \{0,\ 1,\ 2,\ 3\}$ and $B = \{2,\ 3,\ 4,\ 5\}$. Find $A \cup B$.

$A \cup B = \{0, 1, 2, 3, 4, 5\}$

Now Try:

5. Let $A = \{-6, -5, -4\}$ and
$B = \{-3, -2, -1\}$. Find $A \cup B$.

_____

Review these examples for Objective 4:

6. Solve the compound inequality
$m + 1 > 5$ or $-2m > 2$ and graph the solution set.

Solve each inequality individually.
$m + 1 > 5$ or $-2m > 2$
$\quad m > 4 \qquad\qquad\quad m < -1$

The solution set is $(-\infty, -1) \cup (4, \infty)$.

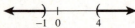

Now Try:

6. Solve the compound inequality
$x + 1 \ge 3$ or $6 + x < 4$ and
graph the solution set.

_____

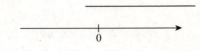

7. Solve the compound inequality $2x + 1 \le 7$ or $3x - 4 \le 2$ and graph the solution set.

Solve each inequality individually.

$$2x + 1 \le 7 \quad \text{or} \quad 3x - 4 \le 2$$
$$2x \le 6 \qquad\qquad 3x \le 6$$
$$x \le 3 \qquad\qquad\;\; x \le 2$$

The graphs of $x \le 3$ and $x \le 2$ are shown below.

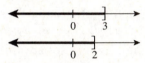

By taking the union, we find that the solution set is $(-\infty, 3]$.

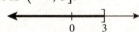

8. Solve the compound inequality $2 - n < 8$ or $-4n + 1 \ge 21$ and graph the solution set.

Solve each inequality individually.

$$2 - n < 8 \quad \text{or} \quad -4n + 1 \ge 21$$
$$-n < 6 \qquad\qquad -4n \ge 20$$
$$n > -6 \qquad\qquad n \le -5$$

The graphs of $n > -6$ and $n \le -5$ are shown below.

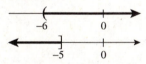

By taking the union, we obtain every real number as a solution, since every real number satisfies at least one of the two inequalities. The solution set is $(-\infty, \infty)$.

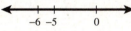

7. Solve the compound inequality $x - 4 \ge 3$ or $2x > 18$ and graph the solution set.

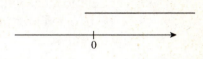

8. Solve the compound inequality $4n - 6 \le 6n$ or $-n > 4n - 10$ and graph the solution set.

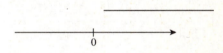

9. The table below lists the land area and 2009 population estimate for ten states. List the elements that satisfy each set.

    a. The set of states with population less than 5,000,000 and area greater than 5000 square miles.

    b. The set of states with population greater than 10,000,000 or area less than 10,000 square miles.

9. Use the table at the left to list the elements that satisfy each set.

    a. The set of states with population greater than 8,000,000 and area less than 45,000 square miles.

        _____

    b. The set of states with population greater than 5,000,000 or area greater than 35,000 square miles.

        _____

State	Area (square miles)	2009 Population Estimate
New York	47,214	19,541,453
Pennsylvania	44,817	12,604,767
New Jersey	7417	8,707,739
Delaware	1954	885,122
Maryland	9774	5,699,478
Virginia	39,594	7,882,590
West Virginia	24,078	1,819,777
Ohio	40,948	11,542,645
Indiana	35,867	6,423,113
Kentucky	39,728	4,314,113

Source: U.S. Census Bureau

    a. {West Virginia, Kentucky}

    b. {New York, Pennsylvania, Ohio, New Jersey, Delaware, Maryland}

Objective 1 Find the intersection of two sets.
For extra help, see Example 1 on page 103 of your text, the Section Lecture video for Section 2.6, and Exercise Solution Clip 7.

Let $A = \{1, 2, 3, 4, 5, 6\}$, $B = \{0, 2, 4, 6, 8, 10\}$, $C = \{1, 3, 5, 7, 9\}$, $D = \{1, 2, 3\}$, and $E = \{0\}$. Specify each set.

1. $A \cap C$

1. _____

2. $A \cap D$

2. _____

3. $B \cap C$

3. _____

Objective 2 Solve compound inequalities with the word *and*.

For extra help, see Examples 2–4 on pages 104–105 of your text, the Section Lecture video for Section 2.6, and Exercise Solution Clips 23, 25, and 29.

For the compound inequality, give the solution set in both interval and graph forms.

4. $x - 3 \leq 6$ and $x + 2 \geq 7$

4. _____

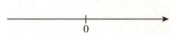

5. $m - 7 \leq -3$ and $m + 2 < -3$

5. _____

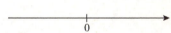

6. $2q < -2$ and $q + 3 > 1$

6. _____

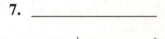

7. $5t > 0$ and $5t + 4 \leq -1$

7. _____

Objective 3 Find the union of two sets.

For extra help, see Example 5 on page 105 of your text, the Section Lecture video for Section 2.6, and Exercise Solution Clip 13.

Let $A = \{1, 2, 3, 4, 5, 6\}$, $B = \{0, 2, 4, 6, 8, 10\}$, $C = \{1, 3, 5, 7, 9\}$, $D = \{1, 2, 3\}$, *and* $E = \{0\}$. *Specify each set.*

8. $A \cup D$

8. _____

9. $B \cup C$

9. _____

10. $B \cup D$ 10. _____

Objective 4 Solve compound inequalities with the word *or*.
For extra help, see Examples 6–9 on pages 106–108 of your text, the Section Lecture video for Section 2.6, and Exercise Solution Clips 35, 41, and 45.

For the compound inequality, give the solution set in both interval and graph forms.

11. $q + 3 > 7$ or $q + 1 \leq -3$ 11. _____

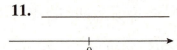

12. $2r + 3 \geq 1$ or $3r \geq 12$ 12. _____

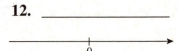

13. $s - 5 > 0$ or $s + 7 < 6$ 13. _____

14. $3 > 4m + 2$ or $7m - 3 \geq -2$ 14. _____

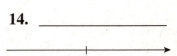

The table below lists the land area and 2009 population estimate for ten states. List the elements that satisfy each set.

State	Area (square miles)	2009 Population Estimate	State	Area (square miles)	2009 Population Estimate
New York	47,214	19,541,453	Virginia	39,594	7,882,590
Pennsylvania	44,817	12,604,767	West Virginia	24,078	1,819,777
New Jersey	7417	8,707,739	Ohio	40,948	11,542,645
Delaware	1954	885,122	Indiana	35,867	6,423,113
Maryland	9774	5,699,478	Kentucky	39,728	4,314,113

Source: U.S. Census Bureau

15.a. The set of states with area greater than 25,000 square miles and population less than 10,000,000.

15.a. _____

   b. The set of states with area greater than 25,000 square miles or population less than 10,000,000.

b. _____

Chapter 2   LINEAR EQUATIONS, INEQUALITIES, AND APPLICATIONS

2.7   Absolute Value Equations and Inequalities

Learning Objectives

1   Use the distance definition of absolute value.

2   Solve equations of the form $|ax + b| = k$, for $k > 0$.

3   Solve inequalities of the form $|ax + b| < k$ and of the form $|ax + b| > k$, for $k > 0$.

4   Solve absolute value equations that involve rewriting.

5   Solve equations of the form $|ax + b| = |cx + d|$.

6   Solve special cases of absolute value equations and inequalities.

Key Terms

Use the vocabulary terms listed below to complete each statement in exercises 1–2.

absolute value equation   **absolute value inequality**

1.   A(n) _____ is an equation that involves the absolute value of a variable expression.

2.   A(n) _____ is an inequality that involves the absolute value of a variable expression.

Guided Examples

Review this example for Objective 2:

1.   Solve $|t - 2| = 5$.

    For $|t - 2|$ to equal 5, $t - 2$ must be 5 units from 0 on the number line. This can happen only when $t - 2 = 5$ or $t - 2 = -5$. Solve the compound inequality.

$$t - 2 = 5 \quad \text{or} \quad t - 2 = -5$$
$$t = 7 \qquad\qquad t = -3$$

    Check by substituting 7 and then -3 into the original absolute value equation to verify that the solution set is $\{-3, 7\}$.

Now Try:

1.   Solve $|3 - q| = 7$.

         _____

Name: _____ Date: _____

Instructor: _____ Section: _____

Review these examples for Objective 3:

2. Solve $|x - 8| > 8$ and graph the solution set.

This absolute value inequality is rewritten as $x - 8 > 8$ or $x - 8 < -8$. Solve the compound inequality.

$x - 8 > 8$ or $x - 8 < -8$

    $x > 16$ $x < 0$

The solution set is $(-\infty, 0) \cup (16, \infty)$.

3. Solve $|2x + 3| < 7$ and graph the solution set.

This absolute value inequality is rewritten as $-7 < 2x + 3 < 7$. Solve the compound inequality.

  $-7 < 2x + 3 < 7$

$-10 < 2x \quad\quad < 4$ Subtract 3.

  $-5 < x \quad\quad\quad < 2$ Divide by 2.

The solution set is $(-5, 2)$.

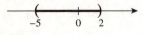

Review these examples for Objective 4:

4. Solve $|2x - 1| + 7 = 12$.

$|2x - 1| + 7 = 12$

   $|2x - 1| = 5$ Subtract 7.

$2x - 1 = -5$ or $2x - 1 = 5$

    $2x = -4$ $2x = 6$ Add 1.

     $x = -2$ $x = 3$ Divide by 2.

Check each solution by substituting each in the original equation.

The solution set is $\{-2, 3\}$.

Now Try:

2. Solve $|-6 + x| > 2$ and graph the solution set.

_____

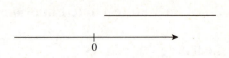

3. Solve $|3x + 15| \leq 6$ and graph the solution set.

_____

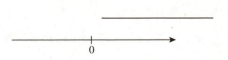

Now Try:

4. Solve $|4x + 3| + 8 = 10$.

_____

5. Solve each inequality.

   a. $|5 + x| - 7 \geq 5$ **b.** $|5 + x| - 7 \leq 5$

   a. $|5 + x| - 7 \geq 5$

$$|5 + x| \geq 12$$

$$5 + x \geq 12 \quad \text{or} \quad 5 + x \leq -12$$

$$x \geq 7 \qquad\qquad x \leq -17$$

The solution set is $(-\infty, -17] \cup [7, \infty)$.

   b. $|5 + x| - 7 \leq 5$

$$|5 + x| \leq 12$$

$$-12 \leq 5 + x \leq 12$$

$$-17 \leq \quad\; x \leq 7$$

The solution set is $[-17, 7]$.

Review this example for Objective 5:

6. Solve $|2x - 8| = |6x + 7|$.

This equation is satisfied either if $2x - 8$ and $6x + 7$ are equal to each other or if $2x - 8$ and $6x + 7$ are negatives of each other.

$$2x - 8 = 6x + 7 \quad \text{or} \quad 2x - 8 = -(6x + 7)$$

$$2x - 15 = 6x \qquad\qquad 2x - 8 = -6x - 7$$

$$-15 = 4x \qquad\qquad 2x = -6x + 1$$

$$-\frac{15}{4} = x \qquad\qquad 8x = 1$$

$$x = \frac{1}{8}$$

Check that the solution set is $\left\{ -\dfrac{15}{4}, \dfrac{1}{8} \right\}$.

Review these examples for Objective 6:

7. Solve each equation.

   a. $|7 + 3x| = -6$ **b.** $|x + 2| = 0$

   a. The absolute value of an expression can never be negative, so there are no solutions for this equation. The solution set is $\varnothing$.

   b. The expression $|x + 2|$ will equal 0 only if $x + 2 = 0$.

$$x + 2 = 0$$

$$x = -2$$

The solution set is $\{-2\}$.

---

5. Solve each inequality.

   a. $|-8x + 2| - 5 \geq 9$

     _____

   b. $|-8x + 2| - 5 \leq 9$

     _____

Now Try:

6. Solve $|y + 5| = |3y + 1|$.

     _____

Now Try:

7. Solve each equation.

   a. $|2x + 4| = -9$ _____

   b. $|3x + 9| = 0$ _____

8. Solve each inequality

a. $|x| > -20$ **b.** $|3 - 2x| + 5 < 1$

c. $1 + |x + 1| \leq 1$

a. $|x| > -20$

The absolute value of a number is always greater than or equal to 0. Therefore, $|x| > -20$ is true for all real numbers. The solution set is $(-\infty, \ \infty)$.

b. $|3 - 2x| + 5 < 1$

     $|3 - 2x| < -4$ Subtract 5.

There is no number whose absolute value is less than −4, so this inequality has no solution. The solution set is ∅.

c. $1 + |x + 1| \leq 1$

     $|x + 1| \leq 0$

We know that $|x + 1|$ will never be less than 0. However, $|x + 1|$ will equal 0 when $x = -1$. Therefore, the solution set is $\{-1\}$.

8. Solve each inequality.

a. $|x| > -5$ _____

b. $|3y - 1| + 5 \leq 3$ _____

c. $|x - 3| + 2 \leq 2$ _____

Objective 1 Use the distance definition of absolute value.
For extra help, see pages 112–113 of your text and the Section Lecture video for Section 2.7.

Graph the solution set of the equation or inequality.

1. $|p| \geq -2$

1.

2. $|x| \leq 10$

2.

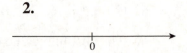

Objective 2 Solve equations of the form $|ax + b| = k$, for $k > 0$.
For extra help, see Example 1 on page 114 of your text, the Section Lecture video for Section 2.7, and Exercise Solution Clip 11.

Solve.

3. $|2x + 3| = 10$

3. _____

4. $|5r - 15| = 0$

4. _____

Objective 3 Solve inequalities of the form $|ax + b| < k$ and of the form $|ax + b| > k$, for $k > 0$.

For extra help, see Examples 2 and 3 on pages 114–115 of your text, the Section Lecture video for Section 2.7, and Exercise Solution Clip 27.

Solve the absolute value inequality. Graph the solution set.

5. $|n + 5| < 8$

5. _____

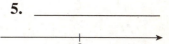

6. $|3q - 5| \geq 4$

6. _____

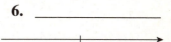

7. $|4y - 1| \leq 2$

7. _____

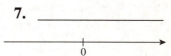

Objective 4 Solve absolute value equations that involve rewriting.

For extra help, see Examples 4 and 5 on pages 115–116 of your text, the Section Lecture video for Section 2.7, and Exercise Solution Clip 69.

Solve.

8. $|5 + y| + 3 = 7$

8. _____

9. $|5 - 2w| + 7 = 5$

9. _____

0. $\left|2 - \dfrac{1}{2}x\right| - 5 = 18$

10. _____ _

Objective 5 Solve equations of the form $|ax + b| = |cx + d|$.

For extra help, see Example 6 on page 116 of your text, the Section Lecture video for Section 2.7, and Exercise Solution Clip 79.

Solve.

11. $|2w + 4| = |3w - 2|$ 11. _____

12. $|y + 3| = |2y - 5|$ 12. _____

13. $|3 - a| = |a + 5|$ 13. _____

Objective 6 Solve special cases of absolute value equations and inequalities.

For extra help, see Examples 7 and 8 on page 117 of your text, the Section Lecture video for Section 2.7, and Exercise Solution Clips 87 and 89.

Solve.

14. $|4 + t| < 0$ 14. _____

15. $|m - 2| \geq -1$ 15. _____

Chapter 3 GRAPHS, LINEAR EQUATIONS, AND FUNCTIONS

3.1   The Rectangular Coordinate System

Learning Objectives
1 Interpret a line graph.
2 Plot ordered pairs.
3 Find ordered pairs that satisfy a given equation.
4 Graph lines.
5 Find x- and y-intercepts.
6 Recognize equations of horizontal and vertical lines and lines passing through the origin.
7 Use the midpoint formula.

Key Terms

Use the vocabulary terms listed below to complete each statement in exercises 1–15.

ordered pair   **components**   **origin**   **x-axis**   **y-axis**

rectangular coordinate system   **plot**   **coordinates**

quadrant   **graph of an equation**   **first-degree equation**

linear equation in two variables   **standard form**   **x-intercept**   **y-intercept**

1.   The x- and y-axis placed at a right angle at their zero points form a

_____.

2.   A(n) _____ is one of the four regions in the plane determined by a
rectangular coordinate system.

3.   The point at which the x-axis and y-axis of a coordinate system intersect is called the

_____.

4.   A point where a graph intersects the x-axis is called a(n) _____.

5.   The numbers in an ordered pair are called the _____ of the
corresponding point in the plane.

6.   The horizontal number line in a rectangular coordinate system is called the

_____.

7.   A(n) _____ is a pair of numbers written within parentheses in which
the order of the numbers is important.

8.   A point where a graph intersects the y-axis is called a(n) _____.

9. A(n) _____ is an equation that can be written in the form
    $Ax + By = C$, where A, B, and C are real numbers and A and B are both not zero. This
    form is called _____.

10. The vertical number line in a rectangular coordinate system is called the
    _____.

11. The _____ is the set of all points that correspond to all of the
    ordered pairs that satisfy the equation.

12. To _____ an ordered pair is to locate it on a rectangular coordinate
    system.

13. A(n) _____ has no term with the variable to a power other
    than 1.

14. The pair of numbers that make up an ordered pair are called its _____.

15. A linear equation in two variables that is written in the form $Ax + By = C$ is written in
    _____.

Guided Examples

Review this example for Objective 3:

1. Complete each ordered pair for $2x + y = 6$.
    Then write the results as a table of ordered
    pairs.

    a. (0, ____)   **b.** (____, 0)
    c. (4, ____)   **d.** (____, 5)

    a. $2x + y = 6$
             $2(0) + y = 6$ Let $x = 0$.
                  $y = 6$
         The ordered pair is (0, 6).

    b. $2x + y = 6$
             $2x + 0 = 6$ Let $y = 0$.
                  $x = 3$
         The ordered pair is (3, 0).

    c. $2x + y = 6$
             $2(4) + y = 6$ Let $x = 4$
               $8 + y = 6$ Multiply.
                   $y = -2$ Subtract 8.
         The ordered pair is (4, −2).

Now Try:

1. Complete the table of ordered
    pairs for $x - 2y = 1$.

x	y
0	
	0
3	
	−1

d.
$$2x + y = 6$$
$$2x + 5 = 6 \qquad \text{Let } y = 5$$
$$2x = 1 \qquad \text{Subtract 5.}$$
$$x = \frac{1}{2} \qquad \text{Divide by 2.}$$

The ordered pair is $\left(\frac{1}{2}, 5\right)$.

The table of ordered pairs is

x	y
0	6
3	0
4	−2
$\frac{1}{2}$	5

Review this example for Objective 5:

2. Find the x- and y-intercepts and graph the equation $4x + y = -6$.

To find the x-intercept, let $y = 0$.
$$4x + y = -6$$
$$4x + 0 = -6 \qquad \text{Let } y = 0.$$
$$4x = -6$$
$$x = -\frac{6}{4} = -\frac{3}{2} \qquad \text{Divide by 4.}$$

The x-intercept is $\left(-\frac{3}{2}, 0\right)$.

To find the y-intercept, let $x = 0$.
$$4x + y = -6$$
$$4(0) + y = -6 \qquad \text{Let } x = 0.$$
$$y = -6$$

The y-intercept is $(0, -6)$.
Verify by substitution that $(-2, 2)$ also satisfies the equation. Use these ordered pairs to draw the graph.

Now Try:

2. Find the x- and y-intercepts and graph the equation $3x - 5y = 15$.

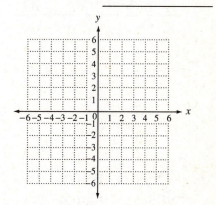

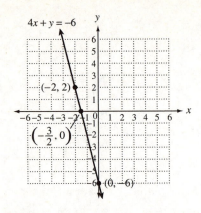

$4x + y = -6$

Review these examples for Objective 6:

3. Graph $y = 4$.

Writing $y = 4$ as $0x + y = 4$ shows that any value of x, including $x = 0$, gives $y = 4$. Thus, the y-intercept is $(0, 4)$. Since y is always 4, there is no value of x corresponding to $y = 0$, so the graph has no x-intercept. Two additional points that lie on the line are $(1, 4)$ and $(-2, 4)$.

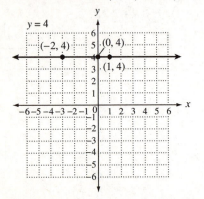

$y = 4$

4. Graph $x - 1 = 0$.

Writing $x - 1 = 0$ as $x + 0y = 1$ shows that any value of y gives $x = 1$. Thus, the x-intercept is $(1, 0)$. Since x is always 1, there is no value of y corresponding to $x = 0$, so the graph has no y-intercept. Two additional points that lie on the line are $(1, 3)$ and $(1, -4)$.

Now Try:

3. Graph $y = -4$.

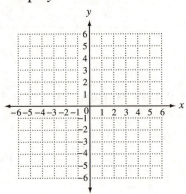

4. Graph $x - 2 = 0$.

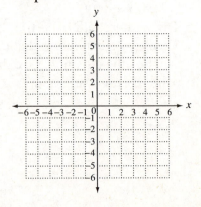

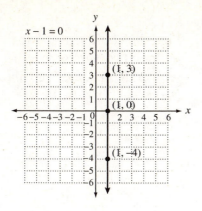

$x - 1 = 0$

5. Graph $4x - y = 0$.

Find the x-intercept.
$4x - y = 0$
$4x - 0 = 0$ Let $y = 0$.
$4x = 0$
$x = 0$

The x-intercept is $(0, 0)$.

Find the y-intercept.
$4x - y = 0$
$4(0) - y = 0$ Let $x = 0$.
$0 - y = 0$
$y = 0$

The y-intercept is $(0, 0)$.

Since both intercepts are $(0, 0)$, the graph passes through the origin. We must find two other points that lie on the line. If $x = 1$, then $y = 4$, so the point $(1, 4)$ lies on the line. If $y = -4$, then $x = -1$, so the point $(-1, -4)$ lies on the line.

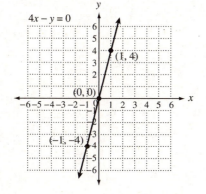

$4x - y = 0$

5. Graph $2x + y = 0$.

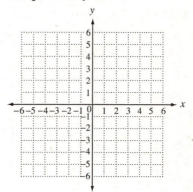

Review this example for Objective 7:

6. Find the coordinates of the midpoint of line segment PQ with endpoints $P(-4, -7)$ and $Q(-2, 3)$.

   Use the midpoint formula with $x_1 = -4$, $x_2 = -2$, $y_1 = -7$, and $y_2 = 3$.

   $$\left(\frac{-4+(-2)}{2}, \frac{-7+3}{2}\right) = \left(\frac{-6}{2}, \frac{-4}{2}\right) = (-3, -2)$$

   The midpoint of segment PQ is $(-3, -2)$.

Now Try:

6. Find the coordinates of the midpoint of line segment PQ with endpoints $P(-5, 3)$ and $(-3, -3)$.

   _____

Objective 1 Interpret a line graph.

For extra help, see page 136 of your text and the Section Lecture video for Section 3.1.

Refer to the graph below to answer Exercises 1–3.

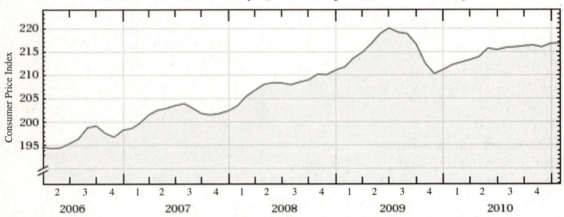

Consumer Price Index by Quarters, April 2005 to February 2010

Source: Wolfram Alpha LLC. 2009. Wolfram|Alpha.
http://www.wolframalpha.com/input/?i=consumer+price+index+u.s. (access April 1, 2010).

1.a. What is the highest consumer price index shown on the graph? When did it occur?

1. a. _____

b. What is the lowest consumer price index shown on the graph? When did it occur?

b. _____

c. During which quarter did the greatest decrease in the consumer price index occur?

c. _____

Objective 2 Plot ordered pairs.

For extra help, see pages 136–137 of your text and the Section Lecture video for Section 3.1.

Plot each point on the rectangular coordinate system below. Label each point with it coordinates.

2. (2, 1), (−3, 5),(−5, −4), (−3, 0)

2.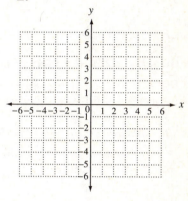

Objective 3 Find ordered pairs that satisfy a given equation.

For extra help, see Example 1 on page 138 of your text, the Section Lecture video for Section 3.1, and Exercise Solution Clip 25.

Complete the given table for each equation.

3. Equation: $2x − 5y = 10$

x	y
0	
	0
−5	
	−3

3. _____

4. Equation: $2x + 3y = 6$

x	y
0	
	0
−3	
	−4

4. _____

5. Equation: $-5x + 2y = 10$

x	y
0	
	0
3	
	-5

5. _____

Objectives 4 and 5 Graph lines. Find *x*- and *y*-intercepts.
For extra help, see Example 2 on pages 139–140 of your text, the Section Lecture video for Section 3.1, and Exercise Solution Clip 35.

Find the x-intercept and y-intercepts. Then graph each equation.

6. $5x - 4y = 20$

6. _____

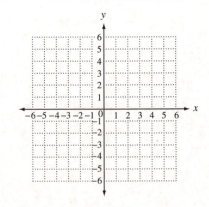

7. $3x + 7y = -8$

7. _____

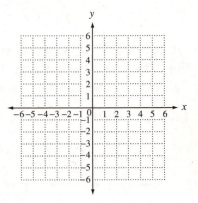

8. $4x - 3y = -12$

8. _____

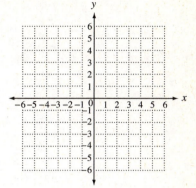

9. $-x - y = 4$

9. _____

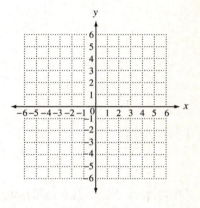

Objective 6 Recognize equations of horizontal and vertical lines and lines passing through the origin.

For extra help, see Examples 3–5 on pages 140–142 of your text, the Section Lecture video for Section 3.1, and Exercise Solution Clips 43, 47, and 51.

Graph each equation.

10. $y = -3$

10.

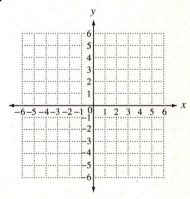

11. $x = -5$

11.

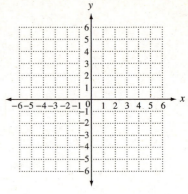

12. $4x - 3y = 0$

12.

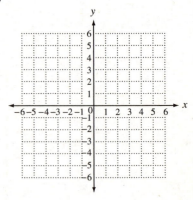

Objective 7 Use the midpoint formula.

For extra help, see Example 6 on pages 142–143 of your text, the Section Lecture video for Section 3.1, and Exercise Solution Clip 57.

Find the midpoint of each pair of points.

13. $(7, -1)$ and $(-4, 6)$

13. _____

14. $(1.5, -4.6)$ and $(-2.7, -8.2)$

14. _____

15. $\left(\dfrac{17}{4}, \dfrac{5}{3}\right)$ and $\left(\dfrac{5}{4}, -\dfrac{11}{3}\right)$

15. _____

Chapter 3 GRAPHS, LINEAR EQUATIONS, AND FUNCTIONS

3.2   The Slope of a Line

Learning Objectives
1   Find the slope of a line, given two points on the line.
2   Find the slope of a line, given the equation of a line.
3   Graph a line, given its slope and a point on the line.
4   Use slopes to determine whether two lines are parallel, perpendicular, or neither.
5   Solve problems involving average rate of change.

Key Terms
Use the vocabulary terms listed below to complete each statement in exercises 1–3.

    slope   **rise**   **run**

1.   _____ is the horizontal change between two points on a line, that is, the change in the x-values.

2.   _____ is the vertical change between two points on a line, that is, the change in the y-values.

3.   The ratio of the change in y to the change in x along a line is called the _____ of the line.

Guided Examples

Review this example for Objective 1:	Now Try:
1. Find the slope of the line through the points $(5, -2)$ and $(2, 7)$. Let $(5, -2) = (x_1, y_1)$ and $(2, 7) = (x_2, y_2)$ in the slope formula. $$m = \frac{y_2 - y_1}{x_2 - x_1} = \frac{7 - (-2)}{2 - 5} = \frac{9}{-3} = -3$$ Thus, the slope is -3.	1. Find the slope of the line through the points $(-3, -6)$ and $(-1, -7)$. _____
Review these examples for Objective 2:	**Now Try:**
2. Find the slope of the line $2x - y = 4$. The intercepts can be used as the two different points needed to find the slope. Let $y = 0$ to find that the x-intercept is $(2, 0)$. Then let $x = 0$ to find that the y-intercept is $(0, -4)$.	2. Find the slope of the line $7y - 4x = 28$. _____

Use these two points in the slope formula.

$$m = \frac{y_2 - y_1}{x_2 - x_1} = \frac{-4 - 0}{0 - 2} = \frac{-4}{-2} = 2$$

Thus, the slope is 2.

3. Find the slope of each line.

   a. $y = -4$ **b.** $x = 7$

   a. The graph of $y = -4$ is a horizontal line. Two points on the line are $(1, -4)$ and $(5, -4)$.

$$m = \frac{y_2 - y_1}{x_2 - x_1} = \frac{-4 - (-4)}{5 - 1} = \frac{0}{4} = 0$$

   Thus, the slope is 0.

   b. The graph of $x = 7$ is a vertical line. Two points on the line are $(7, -4)$ and $(7, 4)$.

$$m = \frac{y_2 - y_1}{x_2 - x_1} = \frac{4 - (-4)}{7 - 7} = \frac{8}{0}, \text{ which is}$$

   undefined. Thus, the slope is undefined.

4. Find the slope of the graph of $4x - 3y = 12$.

   Solve the equation for y.

$$4x - 3y = 12$$
$$-3y = -4x + 12 \quad \text{Subtract } 4x.$$
$$y = \frac{4}{3}x - 4 \quad \text{Divide each term by } -3.$$

   The slope is given by the coefficient of x, so the slope is $\frac{4}{3}$.

3. Find the slope of each line.

   a. $y = 8$ _____

   b. $x + 3 = 0$ _____

4. Find the slope of the line $4y - 9x = 11$.

   _____

Review this example for Objective 3:

5. Graph the line passing through $(-3, -2)$ that has slope $\frac{4}{3}$.

   Begin by plotting the point, then use the slope to find a second point.

$$m = \frac{\text{change in } y}{\text{change in } x} = \frac{4}{3}$$

   Move 4 units up from $(-3, -2)$ and then 3 units to the right to locate another point on the graph, $(0, 2)$. The line through the two points is the required line.

Now Try:

5. Graph the line passing through $(-1, 2)$ that has slope -1.

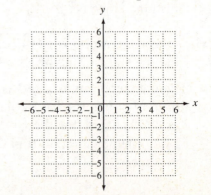

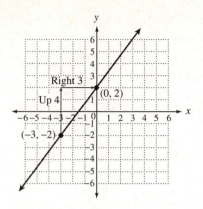

Review these examples for Objective 4:

6. Determine whether the lines L_1, through $(-1, 7)$ and $(2, 4)$, and L_2, through $(7, 10)$ and $(8, 9)$, are parallel.

   Find the slope of L_1.

   $$m_1 = \frac{4-7}{2-(-1)} = \frac{-3}{3} = -1$$

   Find the slope of L_2.

   $$m_2 = \frac{9-10}{8-7} = \frac{-1}{1} = -1$$

   The slopes are equal, so the lines are parallel.

7. Are the lines with equations $4x - 3y = 2$ and $3x + 4y = 12$ perpendicular?

   Find the slope of each line by solving each equation for y.
   $$4x - 3y = 2$$
   $$-3y = -4x + 2 \quad \text{Subtract } 4x.$$
   $$y = \frac{4}{3}x - \frac{2}{3} \quad \text{Divide each term by } -3.$$
   $$3x + 4y = 12$$
   $$4y = -3x + 12 \quad \text{Subtract } 3x.$$
   $$y = -\frac{3}{4}x + 3 \quad \text{Divide each term by } 4.$$

   Since the product of the slopes is $\left(\frac{4}{3}\right)\left(-\frac{3}{4}\right) = -1$, the lines are perpendicular.

Now Try:

6. Determine whether the lines L_1, through $(0, -2)$ and $(-3, 13)$, and L_2, through $(-2, 1)$ and $(-4, 9)$, are parallel.

   _____

7. Are the lines with equations $9x + 3y = 2$ and $x - 3y = 5$ perpendicular?

   _____

8. Determine whether the lines with equations $3x - 2y = 6$ and $3x + 2y = 12$ are *parallel*, *perpendicular*, or *neither*.

Find the slope of each line by solving each equation for y.

$3x - 2y = 6$

$\quad -2y = -3x + 6$ Subtract $3x$.

$\quad\quad y = \dfrac{3}{2}x - 3$ Divide each term by -2.

$3x + 2y = 12$

$\quad 2y = -3x + 12$ Subtract $3x$.

$\quad\quad y = -\dfrac{3}{2}x + 6$ Divide each term by 2.

The slopes, $\dfrac{3}{2}$ and $-\dfrac{3}{2}$, are not equal, and they are not negative reciprocals because their product is $-\dfrac{9}{4}$, not -1. Thus, the two lines are neither parallel nor perpendicular.

8. Determine whether the lines with equations $-5x - y = 8$ and $y = 5x + 1$ are *parallel*, *perpendicular*, or *neither*.

_____

Review this example for Objective 5:

9. A company had 44 employees during the first year of operation. During their eighth year, the company had 79 employees. What was the average rate of change in the number of employees per year?

Let one ordered pair be $(1, 44)$ and let the other ordered pair be $(8, 79)$.

$$\text{average rate of change} = \frac{79 - 44}{8 - 1} = \frac{35}{7} = 5$$

The company grew at an average rate of 5 employees per year.

Now Try:

9. On July 9, 2010, Verizon stock closed at \$26.65 per share. On October 8, 2010, the price was \$32.83 per share. What was the average change in price per month? (Source: USAToday.com)

_____

Objective 1 Find the slope of a line, given two points on the line.
For extra help, see Example 1 on page 149 of your text, the Section Lecture video for Section 3.2, and Exercise Solution Clip 27.

Find the slope of the line through the pair of points.

1. $(5, 7), (7, -9)$

1. _____

2. $(9, -4), (7, -7)$

2. _____

3. $(3,\ 6),\ (-2,\ 6)$ **3.** _____

Objective 2 Find the slope of a line, given the equation of a line.

For extra help, see Examples 2–4 on pages 149–151 of your text, the Section Lecture video for Section 3.2, and Exercise Solution Clips 45, 47, 51, and 53.

Find the slope of the line.

4. $9x + 5y = 18$ **4.**_____

5. $2x - 7y = 1$ **5.**_____

6. $x + 4 = 0$ **6.**_____

Objective 3 Graph a line, given its slope and a point on the line.

For extra help, see Example 5 on page 151 of your text, the Section Lecture video for Section 3.2, and Exercise Solution Clip 59.

Graph the line described.

7. Slope: $\dfrac{2}{3}$; through: $(-1, 4)$

7.

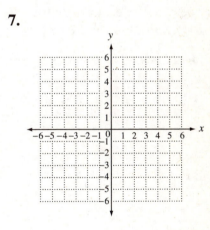

8. Slope: $-\dfrac{1}{2}$; through: $(2, -3)$

8.

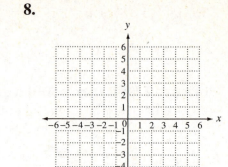

9. Slope: 0; through: $(2, 4)$

9.

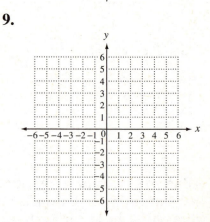

Objective 4 Use slopes to determine whether two lines are parallel, perpendicular, or neither.

For extra help, see Examples 6–8 on pages 152–153 of your text, the Section Lecture video for Section 3.2, and Exercise Solution Clips 69 and 71.

Decide whether the pair of lines is parallel, perpendicular, or neither.

10. $2x - y = 4$, $x + 2y = 9$ **10.** _____

11. $5x - y = 8$, $y = 5x + 1$ **11.** _____

12. The line through $(9, -2)$ and $(-1, 3)$ and the line through   **12.** _____
$(5, 7)$ and $(6, 5)$.

Objective 5 Solve problems involving average rate of change.
For extra help, see Examples 9 and 10 on pages 154–155 of your text and the Section Lecture video for Section 3.2.

Solve each problem.

13. Suppose a man's salary was $45,750 in 1995 and $60,000 in 2010. If the yearly salaries were plotted on a graph, what would be the slope of the line on which they approximately lie?

13. _____

14. A plane had an altitude of 8500 feet at 4:02 P.M. and 12,700 feet at 4:39 P.M. What was the average rate of change in the altitude in feet per minute?

14. _____

15. Enrollment in college was 11,500 two years ago, 10,975 last year, and 10,800 this year.
   (a) What is the average rate of change in enrollment per year for this 3-year period?
   (b) Explain why the rate of change is negative

15. a._____

    b. _____

Chapter 3   GRAPHS, LINEAR EQUATIONS, AND FUNCTIONS

3.3   Linear Equations in Two Variables

Learning Objectives
1.   Write an equation of a line, given its slope and y-intercept.
2.   Graph a line, using its slope and y-intercept.
3.   Write an equation of a line, given its slope and a point on the line.
4.   Write equations of horizontal and vertical lines.
5.   Write an equation of a line, given two points on the line.
6.   Write an equation of a line parallel or perpendicular to a given line.
7.   Write an equation of a line that models real data.

Key Terms

Use the vocabulary terms listed below to complete each statement in exercises 1–2.

    slope-intercept form   **point-slope form**

1.   A linear equation is written in _____ if it is in the form $y - y_1 = m(x - x_1)$, where m is the slope and (x_1, y_1) is a point on the line.

2.   A linear equation is written in _____ if it is in the form $y = mx + b$, where m is the slope and $(0, b)$ is the y-intercept.

Guided Examples

Review this example for Objective 1:

1.   Write an equation of the line with slope $-\frac{3}{2}$ and y-intercept $(0, 4)$.

    We have $m = -\frac{3}{2}$ and $b = 4$. Substitute these values into the slope-intercept form.
$$y = mx + b$$
$$y = -\frac{3}{2}x + 4$$

Now Try:

1.   Write an equation of the line with slope $\frac{4}{5}$ and y-intercept $(0, -2)$.

         _____

Review this example for Objective 2:

2. Graph the line $2x + y = 4$ using the slope and y-intercept.

First write the equation in slope intercept form: $y = -2x + 4$. Then $m = -2$ and $b = 4$. Plot the y-intercept $(0, 4)$.
The slope -2 can be interpreted as

$$m = \frac{\text{rise}}{\text{run}} = \frac{\text{change in } y}{\text{change in } x} = \frac{-2}{1}.$$

From $(0, 4)$, move two units down and one unit to the right, and plot a second point at $(1, 2)$. Join the two points with a line.

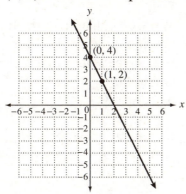

Now Try:

2. Graph the line $x - 2y = -4$ using the slope and y-intercept.

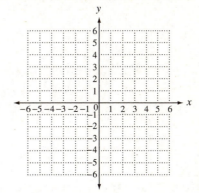

Review this example for Objective 3:

3. Write an equation of the line with slope $\frac{4}{3}$ and passing through the point $(-3, -7)$.

Method 1: Use the point-slope form of the equation of a line with $(x_1, y_1) = (-3, -7)$ and $m = \frac{4}{3}$.

$$y - y_1 = m(x - x_1)$$
$$y - (-7) = \frac{4}{3}(x - (-3)) \quad \text{Substitute}$$
$$y + 7 = \frac{4}{3}(x + 3)$$
$$y + 7 = \frac{4}{3}x + 4 \qquad \text{Distributive property}$$
$$y = \frac{4}{3}x - 3 \qquad \begin{array}{l}\text{Subtract 7.}\\ \text{Slope-intercept form.}\end{array}$$

Now Try:

3. Write an equation of the line with slope $-\frac{3}{2}$ and passing through the point $(-2, 2)$.

_____

Method 2: An alternative method for finding the equation uses slope-intercept form with $(x, y) = (-3, -7)$ and $m = \frac{4}{3}$.

$$y = mx + b$$
$$-7 = \frac{4}{3}(-3) + b \quad \text{Substitute.}$$
$$-7 = -4 + b \qquad \text{Multiply.}$$
$$-3 = b \qquad\qquad \text{Add 4.}$$

Since $m = \frac{4}{3}$ and $b = -3$, the equation is

$$y = \frac{4}{3}x - 3.$$

Review this example for Objective 4:

4. Write an equation of the line passing through the point $(4, -2)$ that satisfies the given condition.

   a. undefined slope   **b.** slope 0

   a. Since the slope is undefined, this is a vertical line. The x-coordinate of the point is 4, so the equation is $x = 4$.

   b. A line with slope 0 is a horizontal line. The y-coordinate of the point is -2, so the equation is $y = -2$.

Now Try:

4. Write an equation of the line passing through the point $(-4, 2)$ that satisfies the given condition.

   a. slope 0 _____

   b. undefined slope _____

Review this example for Objective 5:

5. Write an equation of the line passing through the points $(6, -2)$ and $(-2, 1)$. Give the final answer is standard form.

   First find the slope.
   $$m = \frac{y_2 - y_1}{x_2 - x_1} = \frac{1 - (-2)}{-2 - 6} = \frac{3}{-8} = -\frac{3}{8}$$

   Now use either point as (x_1, y_1) in the point-slope form of the equation of a line.
   $$y - y_1 = m(x - x_1)$$
   $$y - 1 = -\frac{3}{8}(x - (-2)) \quad \begin{array}{l}\text{Use } (-2, 1) \\ \text{as } (x_1, y_1).\end{array}$$
   $$y - 1 = -\frac{3}{8}(x + 2)$$
   $$8y - 8 = -3(x + 2) \qquad \text{Multiply by 8.}$$
   $$8y - 8 = -3x - 6 \qquad \text{Distributive property}$$
   $$3x + 8y = 2 \qquad\qquad \text{Add } 3x; \text{ add 8.}$$

   Verify that $(6, -2)$ also satisfies the equation.

Now Try:

5. Write an equation of the line passing through the points $(3, 7)$ and $(5, 4)$. Give the final answer is standard form.

   _____

Review this example for Objective 6:

6. Write an equation of the line passing through $(9, -3)$ and

   a. parallel to the line $2x + 3y = -12$

   b. perpendicular to the line $2x + 3y = -12$. Give final answers in slope-intercept form.

   a. Find the slope of the line $2x + 3y = -12$ by solving for y.

   $2x + 3y = -12$

   $\quad 3y = -2x - 12$ Subtract $2x$.

   $\quad\quad y = -\frac{2}{3}x - 4$ Divide by 3.

   The slope is $-\frac{2}{3}$. Since parallel lines have equal slopes, the slope of the required line is also $-\frac{2}{3}$. Using the point-slope form, we have

   $y - y_1 = m(x - x_1)$

   $y - (-3) = -\frac{2}{3}(x - 9)$ Substitute.

   $y + 3 = -\frac{2}{3}x + 6$ Simplify; distributive property

   $y = -\frac{2}{3}x + 3$ Subtract 3.

   The required equation is $y = -\frac{2}{3}x + 3$.

   b. To be perpendicular to the line $2x + 3y = -12$, a line must have a slope that is the negative reciprocal of $-\frac{2}{3}$, which is $\frac{3}{2}$. Using the point-slope form, we have

   $y - y_1 = m(x - x_1)$

   $y - (-3) = \frac{3}{2}(x - 9)$ Substitute.

   $y + 3 = \frac{3}{2}x - \frac{27}{2}$ Simplify; distributive property

   $y = \frac{3}{2}x - \frac{33}{2}$ Subtract 3.

   The required equation is $y = \frac{3}{2}x - \frac{33}{2}$.

Now Try:

6. Write an equation of the line passing through $(-3, 2)$ and

   a. parallel to the line $4x - 3y = 8$ _____

   b. perpendicular to the line $x + 2y = 4$. Give final answers in slope-intercept form.

   _____

Review this example for Objective 7:

7. The table shows the average annual telephone expenditures for cell phones from 2001 through 2007, where year 0 represents 2001. (Source: Bureau of Labor Statistics)

Year	Annual Cell Phone Expenditures
0	$210
1	$294
2	$316
3	$378
4	$455
5	$524
6	$608

Use the points (0, 210) and (4, 455) to write an equation in slope-intercept form to approximate the data. How well does this equation approximate the annual expenditure for 2007?

First, find the slope.

$$m = \frac{y_2 - y_1}{x_2 - x_1} = \frac{455 - 210}{4 - 0} = \frac{245}{4}$$

The y-intercept is (0, 210), so the equation is $y = \frac{245}{4}x + 210$.

The year 2007 is represented by $x = 6$.

$$y = \frac{245}{4}(6) + 210 = 577.5$$

The model underestimates the expenditure in 2007. This is illustrated in the graph below.

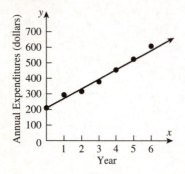

Now Try:

7. Using the table at the left, write an equation in slope-intercept form using the data for 2003 and 2006. Use this equation to approximate the annual expenditure for 2007.

_____

_____

Objective 1 Write an equation of a line, given its slope and *y*-intercept.

For extra help, see Example 1 on page 161 of your text, the Section Lecture video for Section 3.3, and Exercise Solution Clip 21.

Write an equation in standard form for the line.

1. Slope: 2; *y*-intercept: $(0, -5)$

   1. _____

2. Slope: $-\dfrac{2}{3}$; *y*-intercept: $(0, \ 2)$

   2. _____

Objective 2 Graph a line, using its slope and *y*-intercept.

For extra help, see Example 2 on page 162 of your text, the Section Lecture video for Section 3.3, and Exercise Solution Clip 29.

Graph each line using its slope and y-intercept.

3. $x + y = -4$

   3.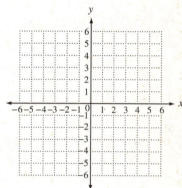

4. $3x - 2y = 5$

   4.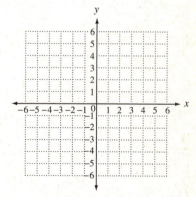

Objective 3 Write an equation of a line, given its slope and a point on the line.
For extra help, see Example 3 on page 163 of your text, the Section Lecture video for Section 3.3, and Exercise Solution Clip 37.

Write an equation in standard form for the line that satisfies the given conditions.

5. Slope: –4 through $(-2,\ 5)$

5. _____

6. Slope: $-\dfrac{3}{4}$ through: $(-1,\ -3)$

6. _____

Objective 4 Write equations of horizontal and vertical lines.
For extra help, see Example 4 on page 164 of your text, the Section Lecture video for Section 3.3, and Exercise Solution Clip 45.

Write an equation in standard form for the line that satisfies the given conditions.

7. Slope: undefined through $(3, 0)$

7. _____

8. Slope: 0 through $(3, -2)$

8. _____

Objective 5 Write an equation of a line, given two points on the line.
For extra help, see Example 5 on pages 164–165 of your text, the Section Lecture video for Section 3.3, and Exercise Solution Clip 55.

Write an equation for the line passing through each pair of points. Give the final answer in (a) slope-intercept form and (b) standard form.

9. $(4, 9)$ and $(3, 8)$

9. a._____

b. _____

10. $(-6, 2)$ and $(-4, 1)$

10. a._____

b. _____

11. $\left(\dfrac{1}{2}, \dfrac{2}{3}\right)$ and $\left(-\dfrac{3}{2}, 2\right)$

11. a._____

b. _____

Objective 6 Write an equation of a line parallel or perpendicular to a given line.
For extra help, see Example 6 on pages 165–166 of your text, the Section Lecture video for
Section 3.3, and Exercise Solution Clip 67.

Write an equation in standard form of the line.

12. Parallel to $x - y = 4$, through $(4, -7)$

12. _____

13. Perpendicular to $5x + y = 8$, through $(2, -1)$

13. _____

Objective 7 Write an equation of a line that models real data.
For extra help, see Examples 7–9 on pages 166–169 of your text, the Section Lecture video
for Section 3.3, and Exercise Solution Clip 75.

Solve.

14. Suppose that you are in charge of your office holiday
party. You call the local caterer who informs you that
the standard holiday party package costs $54.95 per
person plus a $150 setup fee.

14. a._____

    b. _____

  a. Write a linear equation in slope-intercept form that
represents the cost in dollars y, for catering the
office party.

  b. If 71 people attend, what will be the total cost in
dollars to cater the party?

15. The total expenditures (in millions of dollars) for the purchase of raw materials for Smith Manufacturing is given below.

15. a. _____

b. _____

Year	Millions of dollars (y)
2003	84
2004	101
2005	123
2006	136
2007	160
2008	181
2009	196

a. Use the data from 2004 and 2009, letting $x = 4$ represent 2004 and $x = 9$ represent 2009, to write an equation in slope-intercept form that models the data.

b. Use the equation to approximate the expenditures for raw materials in 2007. How does this result compare with the actual value, $160 million?

Chapter 3 GRAPHS, LINEAR EQUATIONS, AND FUNCTIONS

3.4   Linear Inequalities in Two Variables

Learning Objectives
1   Graph linear inequalities in two variables.
2   Graph the intersection of two linear inequalities.
3   Graph the union of two linear inequalities.

Key Terms

Use the vocabulary terms listed below to complete each statement in exercises 1–5.

linear inequality in two variables   boundary line   test point

intersection of two or more inequalities   union of two inequalities

1.   A(n) _____ can be written in the form $Ax + By < C$ or $Ax + By > C$
     (or with $\leq$ or $\geq$), where A, B, and C are real numbers and A and B are not both 0.

2.   A(n) _____ is used to determine the region that should be shaded
     when graphing a linear inequality.

3.   The graph of the _____ is the region of the plane where all points
     satisfy all of the inequalities at the same time.

4.   The graph of the _____ includes all of the points that satisfy either
     inequality.

5.   In the graph of a linear inequality, the _____ separates the region
     that satisfies the inequality from the region that does not satisfy the inequality.

Guided Examples

Review these examples for Objective 1:
 1.   Graph $3x - 2y \leq 6$.

       Begin by graphing the line $3x - 2y = 6$ with
       intercepts $(2, 0)$ and $(0, -3)$. This boundary
       line divides the plane into two regions, one
       of which satisfies the inequality. Choose a
       test point not on the boundary line to see
       whether the resulting statement is true or
       false. We will choose $(0, 0)$.

Now Try:
 1.   Graph $3x + 2y \leq -6$.

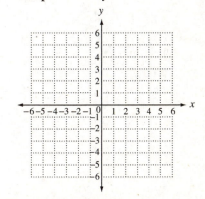

$$3x - 2y \leq 6 \quad \text{Original inequality}$$
$$\overset{?}{3(0) - 2(0) \leq 6} \quad \text{Let } x = 0 \text{ and } y = 0.$$
$$\overset{?}{0 - 0 \leq 6}$$
$$0 \leq 6 \quad \text{True}$$

Shade the region that includes the test point $(0, 0)$.

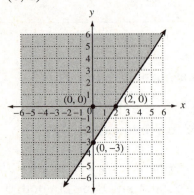

2. Graph $3x - 5y < -15$

To graph the inequality, first graph the equation $3x - 5y = -15$. Use a dashed line to show that the point on the line are not solution of the inequality. Then use $(0, 0)$ as a test point to see which side of the line satisfies the inequality.

$$3x - 5y < -15 \quad \text{Original inequality}$$
$$\overset{?}{3(0) - 5(0) < -15} \quad \text{Let } x = 0 \text{ and } y = 0.$$
$$\overset{?}{0 - 0 < -15}$$
$$0 < -15 \quad \text{False}$$

Since $0 < -15$ is false, the graph of the inequality is the region that does not contain the test point.

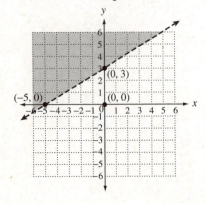

2. Graph $2x - 5y < 10$.

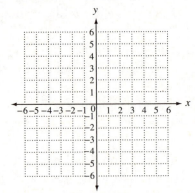

Review this example for Objective 2:

3. Graph $x + y \leq 4$ and $x > 2$.

A pair of inequalities joined with the word *and* is interpreted as the intersection of the solution sets of the inequalities. The graph of the intersection of two or more inequalities is the region of the plane where all points satisfy all of the inequalities at the same time.

Begin by graphing each of the two inequalities separately.

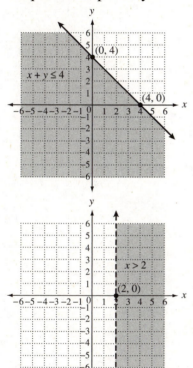

The intersection of the two graphs is the graph we are seeking.

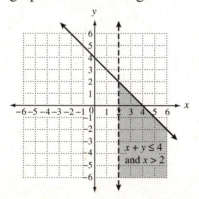

Now Try:

3. Graph $3x - 4y < 12$ and $y > -4$.

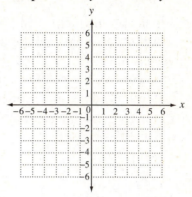

Review this example for Objective 3:

4. Graph $x + y \le 4$ or $x > 2$.

4. Graph $4x + 3y < 12$ or $x \ge 2$.

When two inequalities are joined by the word *or*, we must find the union of the graphs of the inequalities. The graph of the union of two inequalities includes all of the points that satisfy either inequality.

Begin by graphing each of the two inequalities separately.

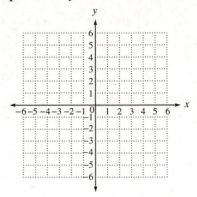

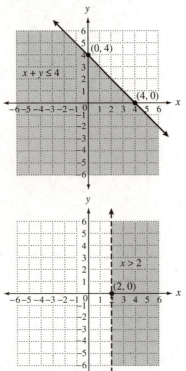

The union of the two graphs is the graph we are seeking.

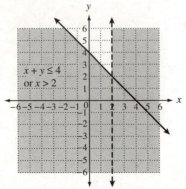

Name: Date:

Instructor: Section:

Objective 1 Graph linear inequalities in two variables.
For extra help, see Examples 1 and 2 on pages 176–177 of your text, the Section Lecture video for Section 3.4, and Exercise Solution Clips 7 and 9.

Graph the linear inequality.

1. $2x + 3y \geq 6$

    1.

    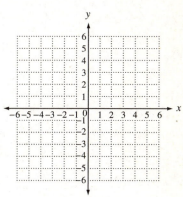

2. $3x - 2y \leq 12$

    2.

    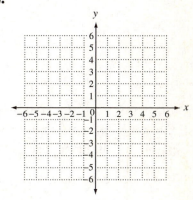

3. $3x - y > -3$

    3.

    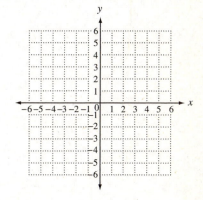

4. $5x + 2y < -10$

4.

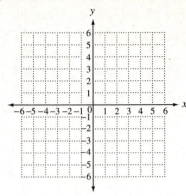

5. $x \leq 4y$

5.

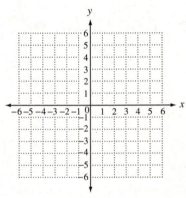

Objective 2 Graph the intersection of two linear inequalities.

For extra help, see Example 3 on page 177 of your text, the Section Lecture video for Section 3.4, and Exercise Solution Clip 23.

Graph each compound inequality.

6. $x - y < 3$ and $x > -2$

6.

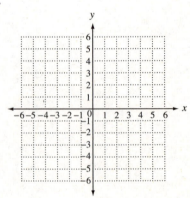

7. $3x - 4y > 12$ and $x \le 4$

7.

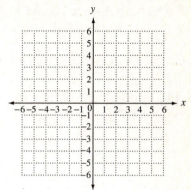

8. $5x + 2y > 10$ and $y < 3$

8.

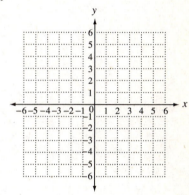

9. $2x + y \ge 6$ and $y \ge 3$

9.

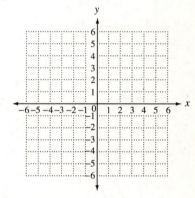

10. Use the method described in Section 2.7 to write $|y - 1| < 4$ as a compound inequality, and then graph the inequality.

10. _____

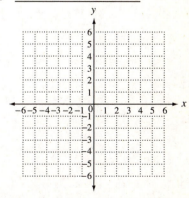

Name: _____ Date: _____

Instructor: _____ Section: _____

Objective 3 Graph the union of two linear inequalities.
For extra help, see Example 4 on page 178 of your text, the Section Lecture video for Section 3.4, and Exercise Solution Clip 33.

Graph each compound inequality.

11. $4x - 2y \geq -4$ or $x \geq 1$

11.

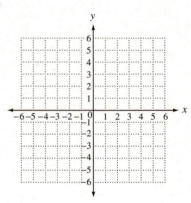

12. $x - 4y \leq -2$ or $x \leq 3$

12.

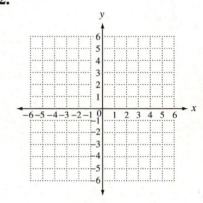

13. $4x - 2y \geq 8$ or $y \geq 2$

13.

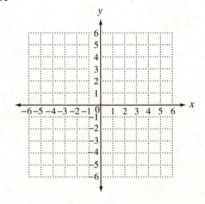

14. $x \geq 4$ or $y \leq -3$

14.

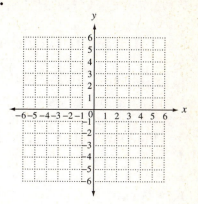

15. $2x + y < -1$ or $x - 2y > 1$

15.

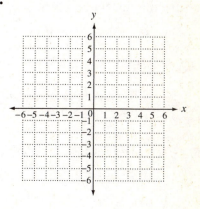

Chapter 3 GRAPHS, LINEAR EQUATIONS, AND FUNCTIONS

3.5   Introduction to Relations and Functions

Learning Objectives
1   Distinguish between independent and dependent variables.
2   Define and identify relations and functions.
3   Find the domain and range.
4   Identify functions defined by graphs and equations.

Key Terms

Use the vocabulary terms listed below to complete each statement in exercises 1–6.

dependent variable   independent variable   relation   function

domain   range

1.   In an equation relating x and y, if the value of the variable y depends on the variable x, then y is called the _____.

2.   The set of all second components (y-values) in the ordered pairs of a relation is the _____.

3.   A(n) _____ is a set of ordered pairs (relation) in which each value of the first component x corresponds to exactly one value of the second component y.

4.   In an equation relating x and y, if the value of the variable y depends on the variable x, then x is called the _____.

5.   The set of all first components (x-values) in the ordered pairs of a relation is the _____.

6.   A(n) _____ is a set of ordered pairs.

Guided Examples

Review this example for Objective 2:
1. Determine whether each relation defines a function.
    a. $F = \{(1, 3), (5, 7), (8, -2), (6, -7), (-4, -3)\}$
    b. $G = \{(2, 7), (5, -4), (-3, -1), (0, 8), (5, 2)\}$

Now Try:
1. Determine whether each relation defines a function.
    a. $\{(3, 5), (5, 2), (4, 3), (5, 3)\}$

    _____

    b. $\{(-3, 5), (-2, 5), (-1, 0), (1, 5)\}$

    _____

a. Each first component appears once and only once. The relation is a function.

b. The first component 5 appears in two ordered pairs and corresponds to two different second components. Therefore, this relation is not a function.

Review these examples for Objective 3:

2. Find the domain and range of each relation. Tell whether the relation defines a function.

a. $\{(3, 4), (5, 2), (4, 3), (5, 3), (-2, 2)\}$

b.

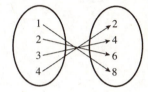

c.

x	y
4	−1
3	−2
2	1
1	1

a. The first component 5 appears in two ordered pairs and corresponds to two different second components, so this relation is not a function. The domain (the set of x-values) is $\{-2, 3, 4, 5\}$. The range (the set of y-values) is $\{2, 3, 4\}$.

b. The mapping defines a function since each x-value maps to exactly one y-value. The domain is $\{1, 2, 3, 4\}$, and the range is $\{2, 4, 6, 8\}$.

c. The table defines a function since each x-value corresponds to exactly one y-value. The domain is $\{1, 2, 3, 4\}$ and the range is $\{-2, -1, 1\}$.

Now Try:

2. Find the domain and range of each relation. Tell whether the relation defines a function.

a. $\{(1, 4), (2, 4), (3, 4), (4, 4)\}$

_____

domain: _____

range: _____

b.

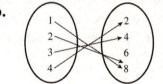

_____

domain: _____

range: _____

c.

x	y
4	−1
3	−2
2	1
4	1

_____

domain: _____

range: _____

3. Give the domain and range of the relation.

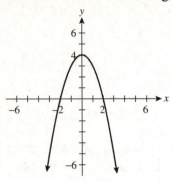

The graph extends indefinitely left and right, as well as downward. The domain is $(-\infty, \infty)$. Because there is a greatest y-value, 4, the range includes all numbers less than or equal to 4, written $(-\infty, 4]$.

3. Give the domain and range of the relation.

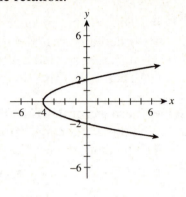

_____

Review these examples for Objective 4:

4. Find the domain and range of each relation. Tell whether the relation defines a function.

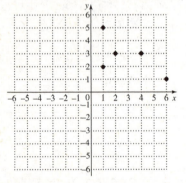

The vertical line test shows that this relation is not a function.

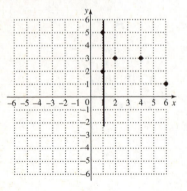

Now Try:

4. Use the vertical line test to decide whether the relation is a function.

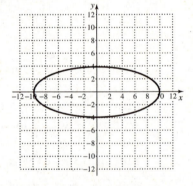

_____

5. Decide whether each relation defines y as a function of x, and give the domain.

   a. $y = 3x - 7$ **b.** $y = \sqrt{3x - 4}$

   c. $y^2 = x + 1$ **d.** $y = \dfrac{2}{x + 1}$

   e. $y < 3x - 2$

   a. In the defining equation, $y = 3x - 7$, y is always found by multiplying x by 3, then subtracting 7. Thus, each value of x corresponds to just one value of y, and the equation defines a function. Since x can be any real number, the domain is $\{x \mid x$ is a real number$\}$, or $(-\infty, \infty)$.

   b. For any choice of x in the domain, there is exactly one corresponding value for y. Thus, the equation defines a function. Since the equation involves a square root, the quantity under the radical symbol cannot be negative, that is, $3x - 4$ must be greater than or equal to 0.

$$3x - 4 \geq 0$$
$$3x \geq 4 \quad \text{Add 4.}$$
$$x \geq \tfrac{4}{3} \quad \text{Divide by 3.}$$

The domain of the function is $\left[\tfrac{4}{3}, \infty\right)$.

   c. The ordered pairs (8, 3) and (8, −3) both satisfy the equation. Since one value of x corresponds to two values of y, the equation does not define a function. Because $x + 1$ is equal to the square of y, the values of $x + 1$ must be nonnegative (greater than or equal to 0).

$$x + 1 \geq 0$$
$$x \geq -1$$

The domain of the relation is $[-1, \infty)$.

5. Decide whether each relation defines y as a function of x, and give the domain.

   a. $y = -5x + 9$ _____

   b. $y = \sqrt{-2x}$ _____

   c. $y^2 = x - 1$ _____

   d. $y = \dfrac{1}{3x + 1}$ _____

   e. $y \geq 2$ _____

   117

d. Given any value of x in the domain, we find y by adding 1 and then dividing 2 by the result. This process produces exactly one value of y for each value in the domain, so the equation defines a function. The domain includes all real numbers except those which make the denominator equal 0.

$x + 1 = 0$

$x = -1$

Thus, the domain includes all real numbers except -1. We write this as $(-\infty, -1) \cup (-1, \infty)$.

e. Because this is an inequality, a particular value of x corresponds to many values of y. Therefore, this is not a function. Any number can be used for x, so the domain of the relation is $(-\infty, \infty)$.

Objective 1 Distinguish between independent and dependent variables.

For extra help, see page 181–182 of your text and the Section Lecture video for Section 3.5.

1. Identify the dependent variable in the statement: A student's grades are related to the amount of time spent on homework.

1. _____

2. Identify the independent variable in the statement: The cost of mailing a package is related to the weight of the package.

2. _____

Objective 2 Define and identify relations and functions.

For extra help, see Example 1 on page 182 of your text, the Section Lecture video for Section 3.5, and Exercise Solution Clips 9, 11 and 15.

Decide whether each relation defines a function.

3. $\{(1, 3), (1, 4), (2, -1), (3, 7)\}$

3. _____

4. $\{(2, -2), (3, -3), (4, -4)\}$ 4. _____

Objective 3 Find the domain and range.

For extra help, see Examples 2 and 3 on pages 184–185 of your text, the Section Lecture video for Section 3.5, and Exercise Solution Clips 9, 11, and 15.

Decide whether the relation is a function, and give the domain and range of the relation.

5. $\{(-1, 1), (-2, 2), (0, 0)\}$ 5. _____

6. $\{(5, 2), (3, -1), (1, -3), (-1, -5)\}$ 6. _____

7.   7. _____

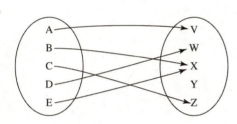

8.   8. _____

x	y
1	3
2	-1
-1	4
1	4

9.   9. _____

x	y
4	2
3	2
2	2
1	2

Objective 4 Identify functions defined by graphs and equations.
For extra help, see Examples 4 and 5 on pages 185–186 of your text, the Section Lecture video for Section 3.5, and Exercise Solution Clips 29, 31, and 35.

Does the graph represent a function?

10.

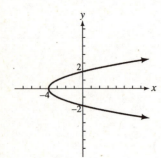

10._____

11.

11._____

12.

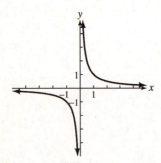

12._____

Determine if the relation represents a function, and give the domain.

13. $y = 3x + 4$

13._____

14. $y = \sqrt{x - 3}$

14._____

15. $y \geq 8x - 2$

15._____

Chapter 3 GRAPHS, LINEAR EQUATIONS, AND FUNCTIONS

3.6   Function Notation and Linear Functions

Learning Objectives
1   Use function notation.
2   Graph linear and constant functions.

Key Terms

Use the vocabulary terms listed below to complete each statement in exercises 1–3.

function notation   **linear function**   **constant function**

1.   A function defined by an equation of the form $f(x) = ax + b$, for real numbers a and b, is a _____.

2.   _____ $f(x)$ represents the value of the function at x, that is, the y-value that corresponds to x.

3.   A _____ is a linear function of the form $f(x) = b$, for a real number b.

Guided Examples

Review these examples for Objective 1:
1.   Let $f(x) = 3x + 5$. Find the value of the function f for $x = -1$.

$$f(x) = 3x + 5$$
$$f(-1) = 3(-1) + 5$$
$$= -3 + 5 = 2$$
Thus $f(-1) = 2$.

2.   Let $f(x) = x^2 - 3x + 2$. Find the following.
    a. $f(1)$ **b.** $f(r)$

    a. $f(x) = x^2 - 3x + 2$
    $$f(1) = 1^2 - 3(1) + 2$$
    $$= 1 - 3 + 2 = 0$$

    b. $f(x) = x^2 - 3x + 2$
    $$f(r) = r^2 - 3r + 2$$

Now Try:
1.   Let $f(x) = 4x - 7$. Find the value of the function f for $x = -3$.

    _____

2.   Let $f(x) = 2x^2 + x - 5$. Find the following.

    a. $f(-2)$ _____

    b. $f(s)$ _____

3. Let $g(x) = 6x - 5$. Find and simplify $g(a - 3)$.

$$g(x) = 6x - 5$$
$$g(a - 3) = 6(a - 3) - 5$$
$$= 6a - 18 - 5$$
$$= 6a - 23$$

3. Let $g(x) = 5x + 2$. Find and simplify $g(a + 3)$.

_____

4. Find $f(-2)$ for each function.
   a. $\{(-2, 1), (-1, -2), (0, -3)\}$

   b. $f(x) = \dfrac{2x + 1}{5}$

   a. $f(-2) = 1$

   b. $f(x) = \dfrac{2x + 1}{5}$

   $$f(-2) = \dfrac{2(-2) + 1}{5} = -\dfrac{3}{5}$$

4. Find $f(1)$ for each function.
   a. $\{(-2, 1), (-1, -2), (1, 2), (2, 3)\}$

   _____

   b. $f(x) = x^3 - 1$ _____

5. Refer to the function graphed below.

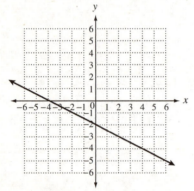

   a. Find $f(-2)$.
   b. For what value of x is $f(x) = -5$?

   a. $f(-2) = -1$

   b. Since $f(x) = y$, we want the value of x that corresponds to $y = -5$. Locate -5 on the y-axis. Moving across to the graph of f and up to the x-axis gives $x = 6$. Thus, $f(6) = -5$.

5. Refer to the function graphed below.

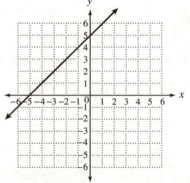

   a. Find $f(-2)$. _____

   b. For what value of x is $f(x) = 2$? _____

6. Rewrite the equation using function notation $f(x)$. Then find $f(-3)$ and $f(m)$.

$$\frac{1}{2}x + \frac{2}{3}y = -1$$

Solve the equation for y.

$$\frac{1}{2}x + \frac{2}{3}y = -1$$

$$6\left(\frac{1}{2}x + \frac{2}{3}y\right) = -1(6) \quad \text{Multiply by the LCD, 6.}$$

$$3x + 4y = -6 \qquad \text{Multiply.}$$

$$4y = -3x - 6 \quad \text{Subtract } 3x.$$

$$y = -\frac{3}{4}x - \frac{6}{4} \quad \text{Divide by 4.}$$

$$y = -\frac{3}{4}x - \frac{3}{2} \quad \text{Simplify.}$$

$$f(x) = -\frac{3}{4}x - \frac{3}{2} \quad \text{Replace } y \text{ with } f(x).$$

$$f(-3) = -\frac{3}{4}(-3) - \frac{3}{2} = \frac{9}{4} - \frac{3}{2} = \frac{9}{4} - \frac{6}{4} = \frac{3}{4}$$

$$f(m) = -\frac{3}{4}m - \frac{3}{2}$$

6. Rewrite the equation using function notation $f(x)$. Then find $f(3)$ and $f(w)$.

$$2y + \frac{1}{2}x = -2$$

_____

_____

_____

Review this example for Objective 2:

7. Graph the function $f(x) = -2x - 3$. Give the domain and range.

The graph of the function has slope -2 and y-intercept -3. To graph this function, plot the y-intercept $(0, -3)$ and use the definition of slope as $\frac{\text{rise}}{\text{run}}$ to find a second point on the line. Since the slope is -2, move down two units and right one unit to the point $(1, -5)$. Draw the straight line through the points to obtain the graph. The domain and range are both $(-\infty, \infty)$.

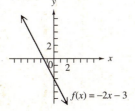

Now Try:

7. Graph the function $f(x) = \frac{1}{2}x + \frac{1}{2}$. Give the domain and range.

domain   _____

range   _____

Objective 1 Use function notation.

For extra help, see Examples 1–6 on pages 190–193 of your text, the Section Lecture video for Section 3.6, and Exercise Solution Clips 3, 15, 25, and 37.

For each function f, find (a) $f(2)$ and (b) $f(6)$.

1.

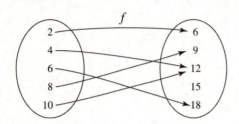

1. a. _____

   b. _____

2. $f = \{(-1, 4), (2, -6), (6, -9), (10, -15)\}$

2. a. _____

   b. _____

3.

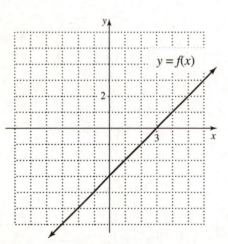

3. a. _____

   b. _____

4.

x	y
6	2
3	2
2	2
−1	2

4. a. _____

   b. _____

For each function f, find (a) $f(-2)$, (b) $f(0)$, and (c) $f(-x)$.

5. $f(x) = 3x - 7$

      5. **a.** _____

         b. _____

         c. _____

6. $f(x) = 2x^2 + x - 5$

      6. **a.** _____

         b. _____

         c. _____

7. $f(x) = |2x + 3|$

      7. **a.** _____

         b. _____

         c. _____

8. $f(x) = 9$

      8. **a.** _____

         b. _____

         c. _____

9. $f(x) = \dfrac{4}{x^2 + 1}$

      9. **a.** _____

         b. _____

         c. _____

10. Rewrite the equation $3x - \dfrac{2}{3}y = 1$ using function

notation $f(x)$. Then find $f(-2)$.

10. _____

_____

Objective 2 Graph linear and constant functions.

For extra help, see Example 7 on page 193 of your text, the Section Lecture video for Section 3.6, and Exercise Solution Clips 45 and 51.

Write each equation using function notation, then graph the function. Give the domain and range.

11. $2x - y = -2$

11. _____

domain_____

range _____

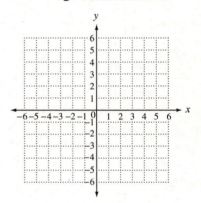

12. $y + \dfrac{1}{2}x = -2$

12. _____

domain_____

range _____

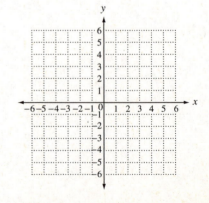

13. $\dfrac{1}{2}x + \dfrac{1}{3}y = -1$

13. _____

domain _____

range _____

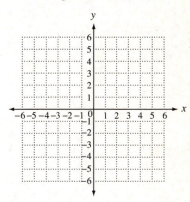

14. $y = 2$

14. _____

domain _____

range _____

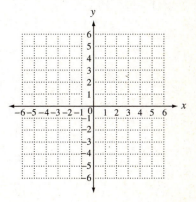

15. $f(x) = -\dfrac{1}{4}x + 2$

Chapter 4   SYSTEMS OF LINEAR EQUATIONS

4.1   Systems of Linear Equations in Two Variables

Learning Objectives
1   Decide whether an ordered pair is a solution of a linear system.
2   Solve linear systems by graphing.
3   Solve linear systems (with two equations and two variables) by substitution.
4   Solve linear systems (with two equations and two variables) by elimination.
5   Solve special systems.

Key Terms

Use the vocabulary terms listed below to complete each statement in exercises 1–9.

system of equations   **system of linear equations**

solution set of a linear system   **consistent system**

independent equations   **inconsistent system**

dependent equations   **elimination method**

substitution method

1.   Two or more linear equations form a(n) _____.

2.   A system of equations with a single solution is called a(n) _____.

3.   The _____ is an algebraic method used to solve a system of equations in which the equations of the system are combined so that one or more variables is eliminated.

4.   A system of equations with an infinite number of solutions is called a(n) _____.

5.   The _____ of a linear system of equations includes all ordered pairs that satisfy all the equations of the system at the same time.

6.   The _____ is an algebraic method for solving a system of equations in which one equation is solved for one of the variables and the result is substituted into the other equation.

7.   A(n) _____ of equations is a system with no solutions.

8.   _____ are different forms of the same equation.

9. Equations of a system that have different graphs are called

_____.

Guided Examples

Review this example for Objective 1:

1. Decide whether the ordered pair (4, 1) is a solution of each system.

   a. $2x + 3y = 11$
   $3x - 2y = 10$

   b. $4x + 3y = 16$
   $x - 4y = -4$

   Substitute 4 for x and 1 for y in each equation.

   a. $2(4) + 3(1) \overset{?}{=} 11$ $3(4) - 2(1) \overset{?}{=} 10$

   $8 + 3 \overset{?}{=} 11$ $12 - 2 \overset{?}{=} 10$

   $11 = 11$ ✓ $10 = 10$ ✓

   Because (4, 1) satisfies both equations, it is a solution of the system.

   b. $4(4) + 3(1) \overset{?}{=} 16$

   $16 + 3 \overset{?}{=} 16$

   $19 \neq 16$

   Because (4, 1) does not satisfy the first equation, it is not a solution of the system.

Now Try:

1. Decide whether the ordered pair $(-3, -1)$ is a solution of each system.

   a. $5x - 3y = -12$
   $2x + 3y = -9$

   _____

   b. $2x + 3y = 5$
   $3x - y = -8$

   _____

Review this example for Objective 2:

2. Solve the system by graphing
   $x - 2y = 6$
   $2x + y = 2$

   Graph each line by plotting three points for each line.

   $x - 2y = 6$

x	y
0	-3
6	0
4	-1

   $2x + y = 2$

x	y
0	2
1	0
-1	4

Now Try:

2. Solve the system by graphing.
   $6x - 5y = 4$
   $2x - 5y = 8$

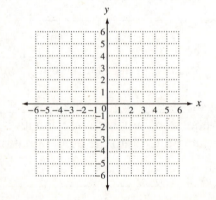

_____

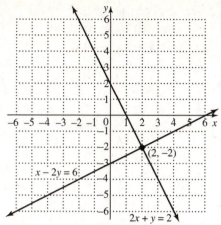

The lines intersect at (2, −2).
Check by substituting 2 for x and −2 for y in both equations.

$$2 - 2(-2) \overset{?}{=} 6 \qquad 2(2) + (-2) \overset{?}{=} 2$$
$$2 + 4 \overset{?}{=} 6 \qquad 4 - 2 \overset{?}{=} 2$$
$$6 = 6 \; \checkmark \qquad 2 = 2 \; \checkmark$$

Because (2, −2) satisfies both equations, it is a solution of the system.

Review these examples for Objective 3:	**Now Try:**
3. Solve the system by the substitution method.	3. Solve the system by the substitution method.

Review these examples for Objective 3:

3. Solve the system by the substitution method.

$$-8x + 5y = 11 \qquad (1)$$
$$x = y - 1 \qquad (2)$$

Equation (2) is already solved for x, so substitute $y - 1$ for x in equation (1).

$$-8x + 5y = 11 \quad \text{(1)}$$
$$-8(y - 1) + 5y = 11 \quad \text{Let } x = y - 1.$$
$$-8y + 8 + 5y = 11 \quad \text{Distributive property}$$
$$-3y + 8 = 11 \quad \text{Combine terms.}$$
$$-3y = 3 \quad \text{Subtract 8.}$$
$$y = -1 \quad \text{Divide by } -3.$$

Now find the value of x by substituting −1 for y in equation (2).

$$x = -1 - 1 = -2.$$

Check that the solution set of the equation is $\{(-2, -1)\}$.

Now Try:

3. Solve the system by the substitution method.

$$x = y + 6$$
$$-2x + 3y = -19$$

_____

4. Solve the system by the substitution method.

$x - 4y = 17$ (1)

$3x - 4y = 11$ (2)

Solve one of the equations for either x or y. Since the coefficient of x in equation (1) is 1, avoid fractions by solving this equation for x.

$x - 4y = 17$

    $x = 4y + 17$

Now substitute $4y + 17$ for x in equation (2) and solve for y.

$$3x - 4y = 11 \quad \text{(2)}$$
$$3(4y + 17) - 4y = 11 \quad \text{Let } x = 4y - 17.$$
$$12y + 51 - 4y = 11 \quad \text{Distributive property}$$
$$8y + 51 = 11 \quad \text{Combine terms.}$$
$$8y = -40 \quad \text{Subtract 51.}$$
$$y = -5 \quad \text{Divide by 8.}$$

Now find the value of x by substituting -5 for y in equation (1).

$$x - 4y = 17 \quad \text{(1)}$$
$$x - 4(-5) = 17 \quad \text{Let } y = -5.$$
$$x + 20 = 17 \quad \text{Multiply.}$$
$$x = -3 \quad \text{Subtract 20.}$$

Check that the solution set of the system is $\{(-3, -5)\}$.

5. Solve the system by the substitution method.

$$\frac{5}{4}x - \frac{1}{2}y = -\frac{1}{4} \quad \text{(1)}$$
$$-\frac{5}{4}x + \frac{1}{8}y = 1 \quad \text{(2)}$$

Clear the fractions in equation (1) by multiplying each side by 4.

$$4\left(\frac{5}{4}x - \frac{1}{2}y\right) = 4\left(-\frac{1}{4}\right) \quad \text{Multiply by 4.}$$
$$5x - 2y = -1 \quad \text{Distributive property}$$

Clear the fractions in equation (2) by multiplying each side by 8.

$$8\left(-\frac{5}{4}x + \frac{1}{8}y\right) = 8(1) \quad \text{(2)}$$
$$-10x + y = 8 \quad \text{Distributive property}$$

4. Solve the system by the substitution method.

$$x + 6y = -1$$
$$-2x - 9y = 0$$

_____

5. Solve the system by the substitution method.

$$\frac{1}{4}x + \frac{3}{8}y = -3$$
$$\frac{5}{6}x - \frac{3}{7}y = -10$$

_____

The original system of equations has been simplified to an equivalent system.

$$5x - 2y = -1 \quad (3)$$
$$-10x + y = 8 \quad (4)$$

Now solve equation (4) for y.

$$y = 10x + 8 \quad \text{Add } 10x.$$

Substitute $10x + 8$ for y in equation (3) and solve for x.

$$5x - 2y = -1 \quad (3)$$
$$5x - 2(10x + 8) = -1 \quad \text{Let } y = 10x + 8.$$
$$5x - 20x - 16 = -1 \quad \text{Distributive property}$$
$$-15x - 16 = -1 \quad \text{Combine like terms.}$$
$$-15x = 15 \quad \text{Add 16.}$$
$$x = -1 \quad \text{Divide by } -15.$$

Substitute -1 for x in equation (4) to find y.

$$y = 10(-1) + 8 = -10 + 8 = -2$$

Check that the solution set of the original system is $\{(-1, -2)\}$.

Review these examples for Objective 4:

6. Solve the system by the elimination method.

$$y = 3x - 5 \quad (1)$$
$$2x = y + 4 \quad (2)$$

Write both equations in standard form.

$$-3x + y = -5 \quad (1)$$
$$2x - y = 4 \quad (2)$$

Now add the two equations.

$$-3x + y = -5 \quad (1)$$
$$\underline{2x - y = 4 \quad (2)}$$
$$-x \quad\quad = -1 \quad \text{Add.}$$
$$x = 1 \quad \text{Multiply by } -1.$$

Now find the value of y by substituting 1 for x in equation (1).

$$y = 3(1) - 5 = 3 - 5 = -2.$$

Check the solution $(1, -2)$ in both equations.

$$
\begin{array}{c|c}
y = 3x - 5 & 2x = y + 4 \\
\quad ? & \quad ? \\
-2 = 3(1) - 5 & 2(1) = -2 + 4 \\
-2 = -2 \quad \checkmark & 2 = 2 \quad \checkmark
\end{array}
$$

Since $(1, -2)$ satisfies both equations, the solution set is $\{(1, -2)\}$.

Now Try:

6. Solve the system by the elimination method.

$$4x + 3y = -4$$
$$2x - 3y = 16$$

---

7. Solve the system by the elimination method.

$$5x - 3y = 13 \quad (1)$$
$$-3x + 2y = -10 \quad (2)$$

Eliminate x by multiplying (1) by 3 and (3) by 5 and then adding.

$$15x - 9y = 39 \qquad 3 \times (1)$$
$$-15x + 10y = -50 \qquad 5 \times (3)$$
$$y = -11 \qquad \text{Add.}$$

Find the value of x by substituting -11 for y in either equation. Using equation (1), we have

$$5x - 3(-11) = 13 \qquad \text{Let } y = -11.$$
$$5x + 33 = 13 \qquad \text{Multiply.}$$
$$5x = -20 \qquad \text{Subtract 33.}$$
$$x = -4 \qquad \text{Divide by 5.}$$

Check that the solution set of the system is $\{(-4, -11)\}$.

7. Solve the system by the elimination method.

$$5x + 3y = -4$$
$$3x + 5y = 2$$

_____

Review these examples for Objective 5:

8. Solve the system.

$$2x + 4y = -6 \quad (1)$$
$$-x - 2y = 3 \quad (2)$$

Multiply each side of equation (2) by 2, then add the equations.

$$2x + 4y = -6 \quad (1)$$
$$-2x - 4y = 6 \qquad 2 \times (2)$$
$$0 = 0 \qquad \text{True}$$

The true statement means that the equations are equivalent and that every solution of one equation is also a solution of the other. The solution set is $\left\{ (x,\ y) \,\middle|\, x + 2x = -3 \right\}$.

9. Solve the system.

$$6x - 7y = 32 \quad (1)$$
$$-18x + 21y = -12 \quad (2)$$

Multiply each side of equation (1) by 3, then add the equations.

$$18x - 21y = 96 \qquad 3 \times (1)$$
$$-18x + 21y = -12 \qquad (2)$$
$$0 = 84 \qquad \text{False}$$

The false statement means that the solution set of the system is $\varnothing$.

Now Try:

8. Solve the system.

$$12x - 10y = -2$$
$$25y - 30x = 5$$

_____

9. Solve the system.

$$6y = -15x + 3$$
$$10x + 4y = 6$$

_____

10. Write each equation in slope-intercept form and then tell how many solutions the system has.

a. $x - y = 1$ **b.** $x + y = 2$
$-2x + 2y = -2$ $x + y = 5$

a. Write each equation in slope-intercept form.

$$x - y = 1 \qquad\quad -2x + 2y = -2$$
$$-y = -x + 1 \qquad 2y = 2x - 2$$
$$y = x - 1 \qquad\quad y = x - 1$$

The equations are exactly the same, so the graphs are the same line. Thus, the system has an infinite number of solutions.

b. Write each equation in slope-intercept form.

$$x + y = 2 \qquad x + y = 5$$
$$y = -x + 2 \qquad y = -x + 5$$

Both equation have the same slope, but different *y*-intercepts. Thus, the equations have graphs that are parallel lines, and the system has no solution.

10. Write each equation in slope-intercept form and then tell how many solutions the system has.

a. $x - 2y = 5$ _____
$2x - 4y = 10$

b. $2x - y = -1$ _____
$-2x + y = 7$

Objective 1 Decide whether an ordered pair is a solution of a linear system.
For extra help, see Example 1 on pages 210–211 of your text, the Section Lecture video for Section 4.1, and Exercise Solution Clip 11.

1. Decide whether the ordered pair $(4, -2)$ is a solution of each system.

a. $x + 2y = 7$ **b.** $4x - 5y = 26$
$2x - y = 4$ $3x + 2y = 8$

1. a. _____

b. _____

Name: Date:

Instructor: Section:

Objective 2 Solve linear systems by graphing.

For extra help, see Example 2 on page 211 of your text, the Section Lecture video for Section 4.1, and Exercise Solution Clip 15.

Solve the system by graphing.

2. $x + 2y = 5$
$2x + y = 4$

2.

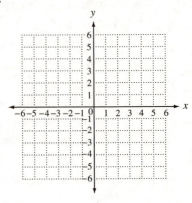

_____

3. $x + 2y = 0$
$2x + y = -6$

3.

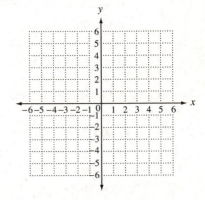

_____

4. $2x = y$
$5x + 3y = 0$

4.

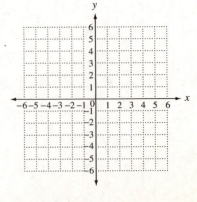

_____

Objective 3 Solve linear systems (with two equations and two variables) by substitution.
For extra help, see Examples 3–5 on pages 212–215 of your text, the Section Lecture video for Section 4.1, and Exercise Solution Clip 21, 27, and 31.

Solve the system by substitution.

5. $3x + 7y = 16$
$\qquad y = 2x - 5$

5. _____

6. $2x - 5y = 11$
$\qquad 3x = 2y$

6. _____

7. $\dfrac{x}{2} + \dfrac{y}{3} = \dfrac{3}{2}$
$\qquad\quad y = 3x$

7. _____

8. $y = 11 - 2x$
$\quad x = 18 - 3y$

8. _____

Objective 4 Solve linear systems (with two equations and two variables) by elimination.
For extra help, see Examples 6 and 7 on pages 215–216 of your text, the Section Lecture video for Section 4.1, and Exercise Solution Clips 41 and 45.

Use the elimination method to solve the system of linear equations.

9. $4x - 5y = -14$
$\quad 2x + 3y = \quad 4$

9. _____

10. $4x - 9y = 7$
$\quad\, 3x + 2y = 14$

10. _____

11. $\dfrac{x}{5} + \dfrac{y}{2} = 0$

      $\dfrac{3x}{5} - \dfrac{5y}{2} = -8$

11. _____

12. $3x + 4y = 3$

      $9x - 8y = 4$

12. _____

Objective 5 Solve special systems.

For extra help, see Examples 8–10 on pages 217–218 of your text, the Section Lecture video for Section 4.1, and Exercise Solution Clips 47, 51, and 59.

Solve.

13. $5x - 2y = 4$

      $10x - 4y = 8$

13. _____

14. $3x + 5y = 8$

      $-6x - 10y = 16$

14. _____

15.   Write each equation in slope-intercept form and then tell how many solutions the system has.

    a. $4x - 3y = 6$

         $8x - 6y = 10$

    b. $4x + 3y = 12$

         $-12x = -36 + 9y$

15. a. _____

    b. _____

Chapter 4 SYSTEMS OF LINEAR EQUATIONS

4.2   Systems of Linear Equations in Three Variables

Learning Objectives
1   Understand the geometry of systems of three equations in three variables.
2   Solve linear systems (with three equations and three variables) by elimination.
3   Solve linear systems (with three equations and three variables) in which some of the equations have missing terms.
4   Solve special systems.

Key Terms

Use the vocabulary terms listed below to complete each statement in exercises 1–6.

ordered triple   **common point**   **line in common**   **coincide**

no points common   **false statement**

1.   The three planes may have the points of a(n) _____, so that the infinite set of points that satisfy the equation of the line is the solution of the system.

2.   If you get a(n) _____ when adding two equations, you do not need to go any further with the solution. Since two of the three planes are parallel, it is not possible for the three planes to have any points in common.

3.   A solution of an equation in three variables, written (x, y, z), is called a(n) _____.

4.   The three planes may _____, so that the solution of the system is the set of all points on a plane.

5.   The three planes may meet at a single, _____ that is the solution of the system.

6.   The planes may have _____ to all three, so that there is no solution of the system.

Guided Examples

Review this example for Objective 2:

1. Solve the system.

$$x - 2y + 5z = -7 \quad (1)$$
$$2x + 3y - 4z = 14 \quad (2)$$
$$3x - 5y + z = 7 \quad (3)$$

Step 1: Since x in equation (1) has coefficient 1, choose x as the focus variable and (1) as the working equation.

Step 2: Multiply working equation (1) by -2 and add the result to equation (2) to eliminate focus variable x.

$$\begin{array}{ll} -2x + 4y - 10z = 14 & -2 \times (1) \\ \underline{2x + 3y - 4z = 14} & (2) \\ 7y - 14z = 28 & (4) \end{array}$$

Step 3: Multiply working equation (1) by -3 and add the result to equation (3) to eliminate focus variable x.

$$\begin{array}{ll} -3x + 6y - 15z = 21 & -3 \times (1) \\ \underline{3x - 5y + z = 7} & (3) \\ y - 14z = 28 & (5) \end{array}$$

Step 4: Write equations (4) and (5) as a system, then solve the system.

$$7y - 14z = 28 \quad (4)$$
$$y - 14z = 28 \quad (5)$$

We will eliminate z.

$$\begin{array}{ll} -7y + 14z = -28 & -1 \times (4) \\ \underline{y - 14z = 28} & (5) \\ -6y = 0 & \text{Add.} \\ y = 0 & \text{Divide by } -6. \end{array}$$

Substitute 0 for y in either equation to find z.

$$\begin{array}{ll} y - 14z = 28 & (5) \\ 0 - 14z = 28 & \text{Let } y = 0. \\ -14z = 28 & \\ z = -2 & \text{Divide by } -14. \end{array}$$

Step 5: Now substitute $y = 0$ and $z = -2$ in working equation (1) to find the value of the remaining variable, focus variable x.

Now Try:

1. Solve the system.

$$2x + y + 2z = -1$$
$$3x - y + 2z = -6$$
$$3x + y - z = -10$$

_____

$$x - 2y + 5z = -7 \quad (1)$$
$$x - 2(0) + 5(-2) = -7 \quad \text{Let } y = 0 \text{ and } z = -2.$$
$$x - 10 = -7 \quad \text{Multiply, then add.}$$
$$x = 3 \quad \text{Add 10.}$$

Step 6: It appears that the ordered triple (3, 0 −2) is the solution of the system. We must check that the solution satisfies all three original equations of the system.

$$x - 2y + 5z = -7 \quad (1)$$
$$3 - 2(0) + 5(-2) \overset{?}{=} -7 \quad x = 3,\ y = 0,\ z = -2$$
$$3 - 0 - 10 \overset{?}{=} -7$$
$$-7 = -7 \quad \checkmark$$

$$2x + 3y - 4z = 14 \quad (2)$$
$$2(3) + 3(0) - 4(-2) \overset{?}{=} 14 \quad x = 3,\ y = 0,\ z = -2$$
$$6 + 0 + 8 \overset{?}{=} 14$$
$$14 = 14 \quad \checkmark$$

$$3x - 5y + z = 7 \quad (3)$$
$$3(3) - 5(0) + (-2) \overset{?}{=} 7 \quad x = 3,\ y = 0,\ z = -2$$
$$9 - 0 - 2 \overset{?}{=} 7$$
$$7 = 7 \quad \checkmark$$

The solution set is $\{(3, 0, -2)\}$.

Review this example for Objective 3:

2. Solve the system.
$$3x \qquad - 4z = -23 \quad (1)$$
$$y + 5z = 24 \quad (2)$$
$$x - 3y \qquad = 2 \quad (3)$$

Since equation (1) is missing the variable y, one way to begin is to eliminate y again, using equations (2) and (3).

$$3y + 15z = 72 \quad 3 \times (2)$$
$$\underline{x - 3y \qquad = 2 \quad (3)}$$
$$x \qquad + 15z = 74 \quad (4)$$

Now Try:

2. Solve the system.
$$x + 5y \qquad = -23$$
$$4y - 3z = -29$$
$$2x \qquad + 5z = 19$$

_____

Now solve the system composed of equations (1) and (4).

$$3x - 4z = -23 \quad (1)$$
$$\underline{-3x - 45z = -222} \quad -3 \times (4)$$
$$-49z = -245$$
$$z = 5$$

Substitute 5 for z in (1) and solve for x.

$$3x - 4(5) = -23 \quad \text{(1); Let } z = 5.$$
$$3x - 20 = -23 \quad \text{Multiply.}$$
$$3x = -3 \quad \text{Add 20.}$$
$$x = -1 \quad \text{Divide by 3.}$$

Substitute 5 for z in (2) and solve for y.

$$y + 5(5) = 24 \quad \text{(2); Let } z = 5.$$
$$y + 25 = 24 \quad \text{Multiply.}$$
$$y = -1 \quad \text{Subtract 25.}$$

Check to verify that the solution set is $\{(-1, -1, 5)\}$.

Review these examples for Objective 4:

3. Solve the system.

$$x - y + z = 7 \quad (1)$$
$$2x + 5y - 4z = 2 \quad (2)$$
$$-x + y - z = 4 \quad (3)$$

Since x in equation (1) has coefficient 1, choose x as the focus variable and (1) as the working equation. Using equations (1) and (3), we have

$$x - y + z = 7 \quad (1)$$
$$\underline{-x + y - z = 4} \quad (3)$$
$$0 = 11 \quad \text{Add; false}$$

The resulting false statement indicates that equations (1) and (3) have no common solution. Thus, the system is inconsistent and the solution set is $\varnothing$. The graph of this system would show that three planes are parallel to each other as shown below.

Now Try:

3. Solve the system.

$$-4x - 2y + z = -19$$
$$-6x + 2y - 6z = -8$$
$$-4x + 2y - 5z = -6$$

_____

4. Solve the system.
$$3x - 2y + 5z = 4 \quad (1)$$
$$-6x + 4y - 10z = -8 \quad (2)$$
$$\frac{3}{2}x - y + \frac{5}{2}z = 2 \quad (3)$$

Multiplying each side of equation (1) by -2 gives equation (2). Multiplying each side of equation (1) by $\frac{1}{2}$ gives equation (3). Thus, the equations are dependent, and all three equations have the same graph as shown below. The solution set is written $\{(x, \ y, \ z) \,|\, 3x - 2y + 5z = 4\}.$

(1), (2), (3)

5. Solve the system.
$$3x + 2y + z = 3 \quad (1)$$
$$-6x - 4y - 2z = -6 \quad (2)$$
$$x + \frac{2}{3}y + \frac{1}{3}z = 4 \quad (3)$$

Multiplying each side of equation (1) by -2 gives equation (2), so these two equations are dependent. Multiplying each side of equation (1) by $\frac{1}{3}$ does not give equation (3). Instead, we obtain two equations with the same coefficients, but with different constant terms. The graphs of equations (1) and (3) have no points in common, that is the planes are parallel. Thus, the system is inconsistent and the solution set is $\varnothing$, as illustrated below.

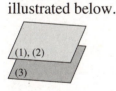
(1), (2)
(3)

4. Solve the system.
$$x - 5y + 2z = 0$$
$$-x + 5y - 2z = 0$$
$$\frac{1}{2}x - \frac{5}{2}y + z = 0$$

_____

5. Solve the system.
$$3x + 2y + z = 3$$
$$-6x - 4y - 2z = -6$$
$$9x + 6y + 3z = 1$$

_____

Objective 2 Solve linear systems (with three equations and three variables) by elimination.

For extra help, see Example 1 on pages 228–229 of your text, the Section Lecture video for Section 4.2, and Exercise Solution Clip 3.

Solve the system of equations.

1. $x + y + z = 2$
   $x - y + z = -2$
   $x - y - z = -4$

         1. _____

2. $x - y + z = 2$
   $x + y - z = 0$
   $x - y - z = 4$

         2. _____

3. $2x + y - z = 9$
   $x + 2y + z = 3$
   $3x + 3y - z = 14$

         3. _____

4. $2x - 5y + 2z = 30$
   $x + 4y + 5z = -7$
   $\dfrac{1}{2}x - \dfrac{1}{4}y + z = 4$

         4. _____

5. $5x - 2y + z = 28$
   $3x + 5y - 2z = -23$
   $\dfrac{2}{3}x + \dfrac{1}{3}y + z = 1$

         5. _____

Objective 3 Solve linear systems (with three equations and three variables) in which some of the equations have missing terms.

For extra help, see Example 2 on pages 229–230 of your text, the Section Lecture video for Section 4.2, and Exercise Solution Clip 23.

Solve the system of equations.

6.
$$x + 2y \quad = 1$$
$$y - z = -6$$
$$x \quad + z = 8$$

6. _____

7.
$$4x - 5y \quad = -13$$
$$3x \quad + z = 9$$
$$2y + 5z = 10$$

7. _____

8.
$$7x \quad + z = -1$$
$$3y - 2z = 8$$
$$5x + y \quad = 2$$

8. _____

9.
$$2x + 5y \quad = 18$$
$$3y + 2z = 4$$
$$\frac{1}{4}x - y \quad = -1$$

9. _____

10.
$$5x \quad - 2z = 8$$
$$4y + 3z = -9$$
$$\frac{1}{2}x + \frac{2}{3}y \quad = -1$$

10. _____

    145

Objective 4 Solve special systems.
For extra help, see Examples 3–5 on pages 230–231 of your text, the Section Lecture video for Section 4.2, and Exercise Solution Clips 31, 37, and 41.

Solve the system of equations. If the system is inconsistent or has dependent equations, say so.

11. $\begin{aligned} 8x - 7y + 2z &= 1 \\ 3x + 4y - z &= 6 \\ -8x + 7y - 2z &= 5 \end{aligned}$

11. _____

12. $\begin{aligned} 3x - 2y + 4z &= 5 \\ -3x + 2y - 4z &= -5 \\ \frac{3}{2}x - y + 2z &= \frac{5}{2} \end{aligned}$

12. _____

13. $\begin{aligned} -x + 5y - 2z &= 3 \\ 2x - 10y + 4z &= -6 \\ -3x + 15y - 6z &= 9 \end{aligned}$

13. _____

14. $\begin{aligned} 2x + 7y - 8z &= 3 \\ 5x - y - z &= 1 \\ x + \frac{7}{2}y - 4z &= 3 \end{aligned}$

14. _____

15. $\begin{aligned} 8x - 4y + 2z &= 12 \\ -4x + 2y - z &= 6 \\ x - \frac{1}{2}y + z &= 8 \end{aligned}$

15. _____

Chapter 4   SYSTEMS OF LINEAR EQUATIONS

4.3   Applications of Systems of Linear Equations

Learning Objectives

1   Solve geometry problems by using two variables.
2   Solve money problems by using two variables.
3   Solve mixture problems by using two variables.
4   Solve distance-rate-time problems by using two variables.
5   Solve problems with three variables by using a system of three equations.

Key Terms

Use the vocabulary terms listed below to complete each statement in exercises 1–2.

    elimination method   **substitution**

1.   Using the addition property to solve a system of equations is called the

     _____ .

2.   _____ is being used when one expression is replaced by another.

Guided Examples

Review this example for Objective 1:

1.   The perimeter of a rectangular room is 50 feet. The length is three feet greater than the width. Find the dimensions of the room.

    Step 1: Read the problem again. We are asked to find the dimensions of the room.

    Step 2: Assign variables. Let l = the length and w = the width.

    Step 3: Write a system of equations.

    $2l + 2w = 50$ (1)

       $l = 3 + w$ (2)

    Step 4: Solve the system of equations. Since equation (2) is solved for l, we can substitute $3 + w$ for l in equation (1) and solve for w.

Now Try:

1.   A rectangle is three times as long as it is wide. The perimeter is 75 inches. Find the length and width.

        length _____

        width _____

$$2l + 2w = 50 \quad \text{(1)}$$
$$2(3 + w) + 2w = 50 \quad \text{Let } l = 3 + w.$$
$$6 + 2w + 2w = 50 \quad \text{Distributive property}$$
$$6 + 4w = 50 \quad \text{Combine like terms.}$$
$$4w = 44 \quad \text{Subtract 6.}$$
$$w = 11 \quad \text{Divide by 4.}$$

Let $w = 11$ in equation (2) to find l.

$$l = 3 + 11 = 14$$

Step 5: State the answer. The length is 14 feet and the width is 11 feet.

Step 6: Check. The perimeter is $2(14) + 2(11) = 50$ feet, as required.

Review this example for Objective 2:

2. There were 411 tickets sold for a soccer game, some for students and some for nonstudents. Student tickets cost $4.25 and nonstudent tickets cost $8.50 each. The total receipts were $3021.75. How many of each type were sold?

Step 1: Read the problem again. There are two unknowns.

Step 2: Assign variables. Let s = the number of student tickets and n = the number of nonstudent tickets.

Step 3: Write a system of equations.

$$s + \quad n = 411 \quad \text{(1)}$$
$$4.25s + 8.50n = 3021.75 \quad \text{(2)}$$

Step 4: Solve the system of equations. We will eliminate n by multiplying equation (1) by -8.50 and then adding the equations.

$$-8.50s - 8.50n = -3493.50 \quad {-8.50 \times \text{(1)}}$$
$$\underline{4.25s + 8.50n = \ \ 3021.75 \quad \text{(2)}}$$
$$-4.25s = -471.75$$
$$s = 111 \qquad \text{Divide by } -4.25.$$

To find the value of n, let $s = 111$ in (1).

$$s + n = 411 \quad \text{(1)}$$
$$111 + n = 411 \quad \text{Let } s = 111.$$
$$n = 300 \quad \text{Subtract 111.}$$

Step 5: State the answer. 111 student tickets were sold and 300 nonstudent tickets were sold.

Now Try:

2. Norma has some $5-bills and some $10-bills. The total value of the money is $260, with a total of 32 bills. How many of each are there?

$5-bills _____

$10-bills _____

Step 6: Check.

The total amount for 111 student tickets and 300 nonstudent tickets is

$4.25(111) + 8.50(300) = 3021.75$, and

$111 + 300 = 411$, as required.

Review this example for Objective 3:

3. Milton needs 45 liters of 20% alcohol solution. He has only 15% alcohol solution and 30% alcohol solution on hand to make the mixture. How many liters of each solution should he combine to make the mixture?

Step 1 Read the problem. Note the percentage of each solution and of the mixture.

Step 2 Assign variables.

Let x = the number of liters of 15% solution and let y = the number of liters of 30% solution.

Summarize the information in a table.

	Number of liters	Percent (as a decimal)	Liters of pure alcohol
15% solution	x	0.15	$0.15x$
30% solution	y	0.30	$0.30y$
20% solution	45	0.20	$(0.20)(45) = 9$

Step 3 Write two equations.

$$x + \quad y = 45 \quad (1)$$
$$0.15x + 0.30y = \quad 9 \quad (2)$$

Step 4 Solve the system by eliminating x.

$$\begin{array}{ll} -0.15x - 0.15y = -6.75 & \text{Multipy (1) by } -0.15. \\ \underline{0.15x + 0.30y = 9} & (2) \\ \phantom{-0.15x + {}}0.15y = 2.25 & \text{Add.} \\ y = 15 & \text{Divide by 0.15.} \end{array}$$

Substitute 15 for y in equation (1) and solve for x.

$$x + 15 = 45$$
$$x = 30 \quad \text{Subtract 15.}$$

Now Try:

3. How many ounces each of a 20% acid solution and a 50% acid solution must be mixed together to get 120 ounces of a 30% solution acid?

20% solution _____

50% solution _____

Step 5 State the answer.
Milton needs 30 liters of 15% alcohol
solution and 15 liters of 30% alcohol
solution.

Step 6 Check.
Since 30 and 15 is 45 and
$0.15(30) + 0.30(15) = 9$,
this mixture will give 45 liters of the 20%
solution, as required.

Review this example for Objective 4:

4. Pablo left Somerset traveling to Akron 240
    miles away at the same time as Shawn left
    Akron traveling to Somerset. They met
    after 2 hours. If Shawn was traveling twice
    as fast as Pablo, what were their speeds?

    Step 1 Read the problem several times.

    Step 2 Assign variables.
    Let x = Pablo's rate and let y = Shawn's
    rate.
    Make a table and draw a sketch to
    summarize the information given in the
    problem.

	Rate	Time	Distance
Pablo	x	2	$2x$
Shawn	y	2	$2y$

    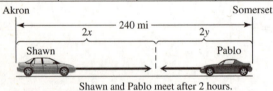

    Shawn and Pablo meet after 2 hours.

    Step 3 Write two equations.
    $$y = 2x \quad (1)$$
    $$2x + 2y = 240 \quad (2)$$

    Step 4 Solve the system using the
    substitution method.

    $$2x + 2(2x) = 240 \quad \text{Substitute } 2x \text{ for } y \text{ in (2).}$$
    $$6x = 240 \quad \text{Combine like terms}$$
    $$x = 40 \quad \text{Divide by 6.}$$

    Substitute 40 for x in equation (1) and solve
    for y.
    $$y = 2(40) = 80$$

Now Try:

4. Rick and Hilary drive from
    positions 378 miles apart
    toward each other. They meet
    after 3 hours. Find the average
    speed of each if Hilary travels
    30 miles per hour faster than
    Rick.

    Rick_____

    Hilary_____

Step 5 State the answer.

Pablo traveled at 40 miles per hour and Shawn traveled at 80 miles per hour.

Step 6 Check.

Since each man travels for 2 hours, the total distance traveled is $2(40) + 2(80) = 240$ miles, as required.

Review these examples for Objective 5:	**Now Try:**

Review these examples for Objective 5:

5. Lee has some $5, $10, and $20-bills. He has a total of 51 bills, worth $795. The number of $5-bills is 25 less than the number of $20-bills. Find the number of each type of bill he has.

Step 1: Read the problem again. There are three unknowns.

Step 2: Assign variables.

Let x = the number of $5 bills,

let y = the number of $10 bills,

let z = the number of $20 bills.

Step 3: Write a system of three equations. There are a total of 51 bills, so

$x + y + z = 51$ (1).

The bills amounted to $795, so

$5x + 10y + 20z = 795$ (2).

The number of $5-bills is 25 less than the number of $20-bills, so $x = z - 25$ or

$x - z = -25$ (3).

The system is

$$
\begin{aligned}
x + \quad y + \quad z &= 51 \quad (1)\\
5x + 10y + 20z &= 795 \quad (2)\\
x \qquad\qquad - \quad z &= -25 \quad (3)
\end{aligned}
$$

Step 4: Solve the system.

Eliminate y.

$$
\begin{array}{rll}
-10x - 10y - 10z &= -510 & -10 \times (1)\\
5x + 10y + 20z &= \underline{795} & (2)\\
-5x + \qquad\quad 10z &= 285 & (4)
\end{array}
$$

Solve the system consisting of equations (3) and (4).

$$
\begin{array}{rll}
5x - 5z &= -125 & 5 \times (3)\\
-5x + 10z &= \underline{285} & (4)\\
5z &= 160 & \text{Add.}\\
z &= 32 & \text{Divide by 5.}
\end{array}
$$

Now Try:

5. The manager of the Sweet Candy Shop wishes to mix candy worth $4 per pound, $6 per pound, and $10 per pound to get 100 pounds of a mixture worth $7.60 per pound. The amount of $10 candy must equal the total amounts of the $4 and the $6 candy. How many pounds of each must be used?

$4 candy _____

$6 candy _____

$10 candy _____

Substitute 32 for z in equation (3) and solve for x.

$x - 32 = -25$ (3); let $z = 32$.

$\quad\quad x = 7$ Add 32.

Substitute 7 for x and 32 for z in equation (1) and solve for y.

$7 + y + 32 = 51$ (1); let $x = 7$, $z = 32$.

$\quad\quad y + 39 = 51$ Add.

$\quad\quad\quad\quad y = 12$ Subtract 39.

Step 5: State the answer. Lee has 7 $5-bills, 12 $10-bills, and 32 $20-bills.

Step 6: Check that the total value of the bills is $795 and that the number of $5-bills is 25 less than the number of $20-bills.

6. A manufacturer produces three different products, A, B, and C. Product A requires 2 units of molded plastic, 3 fasteners, and 1 unit of paint. Product B requires 1 unit of molded plastic, 2 fasteners, and 2 units of paint. Product C requires 3 units of molded plastic, 2 fasteners, and 4 units of paint. If the manufacturer has 50 units of molded plastic, 60 fasteners, and 50 units of paint available, how many of each product can be manufactured?

Step 1: Read the problem again. There are three unknowns.

Step 2: Assign variables. Then organize the information in a table.

Let x = the number of product A produced, let y = the number of product B produced, let z = the number of product C produced.

	Product A	Product B	Product C	Totals
Molded plastic	2	1	3	50
Fasteners	3	2	2	60
Paint	1	2	4	50

Step 3: Write a system of three equations. Product A requires 2 units of molded plastic, product B requires 1 unit of molded plastic, and product C requires 3 units of molded plastic. Since 50 units are available, we have $2x + y + 3z = 50$ (1).

6. Janet takes a daily prescription composed of 8 units of drug A, 14 units of drug B, and 16 units of drug C. Generic capsule I contains 2 units of A, 3 units of B, and 2 units of C. Generic capsule II contains 1 unit of A, 2 units of B, and 4 units of C. Generic capsule III contains 3 units of A, 5 units of B, and 2 units of C. How many capsules of each drug should Janet take daily to obtain the prescribed dosage?

Generic capsule I _____

Generic capsule II _____

Generic capsule III _____

Product *A* requires 3 fasteners, product *B* requires 2 fasteners, and product *C* requires 2 fasteners. Since 60 fasteners are available, we have $3x + 2y + 2z = 60$ (2).

Product *A* requires 1 unit of paint, product *B* requires 2 units of paint, and product *C* requires 4 units of paint. Since 50 units are available, we have $x + 2y + 4z = 50$ (3).

Step 4: Solve the system.

$$2x + \ \ y + 3z = 50 \quad (1)$$
$$3x + 2y + 2z = 60 \quad (2)$$
$$\ \ x + 2y + 4z = 50 \quad (3)$$

We find that $x = 12$, $y = 5$, and $z = 7$.

Step 5: State the answer. The manufacturer should produce 12 product *A*, 5 product *B*, and 7 product *C*.

Step 6: Check that these values satisfy the conditions of the problem.

Objective 1 Solve geometry problems by using two variables.

For extra help, see Example 1 on pages 234–235 of your text, the Section Lecture video for Section 4.3, and Exercise Solution Clip 3.

Solve each problem.

1. The length of a rectangle is 5 feet more than the width. The perimeter of the rectangle is 58 feet. Find the length and width of the rectangle.

    1. length _____

       width _____

2. The measure of the largest angle of a triangle is 45° more than the measure of the smallest angle. The middle angle measures 65°. What are the measures of the largest and smallest angles?

    2. largest _____

       smallest _____

3. The side of a square is 4 centimeters longer than the side of an equilateral triangle. The perimeter of the square is 28 centimeters longer than the perimeter of the triangle. Find the length of a side of the triangle and of a side of a square.

3. square _____

   triangle _____

Objective 2 Solve money problems by using two variables.

For extra help, see Example 2 on pages 235–236 of your text and the Section Lecture video for Section 4.3.

Solve each problem.

4. The Garden Center ordered 6 ounces of marigold seed and 8 ounces of carnation seed paying $214.54. They later ordered another 12 ounces of marigold seed and 18 ounces of carnation seed, paying $464.28. Find the price per ounce for each type of seed.

4. marigold _____

   carnation _____

5. Guy has $15,000 to invest. He plans to invest part at 4%, with the remainder invested at 6%. Find the amount invested at each rate if the total annual interest income is $700.

5. 4% _____

   6% _____

6. A taxi charges a flat rate plus a certain charge per mile. A trip of 4 miles costs $2.05, while a trip of 8 miles costs $2.85. Find the flat rate and the charge per mile.

6. flat rate _____

   per mile _____

Objective 3 Solve mixture problems by using two variables.

For extra help, see Example 3 on pages 236–237 of your text, the Section Lecture video for Section 4.3, and Exercise Solution Clip 17.

Solve each problem.

7. A radiator holds 10 liters. How much pure antifreeze must be added to a mixture that is 10% antifreeze to make enough of a 20% mixture to fill the radiator?

7. _____

8. A candy mix is to be made by mixing candy worth $12 per kilogram with candy worth $15 per kilogram to get 120 kilograms of a mixture worth $13 per kilogram. How many kilograms of each should be used?

8. $12/kg _____

   $15/kg _____

9. A chemist needs to mix a 75% acid solution with a 55% acid solution in order to obtain 70 liters of a 63% acid solution. How many liters of each solution should be used?

9. 55% solution _____

   75% solution _____

Objective 4 Solve distance-rate-time problems by using two variables.

For extra help, see Example 4 on pages 237–238 of your text, the Section Lecture video for Section 4.3, and Exercise Solution Clip 27.

Solve each problem.

10. Two cars start at the same point and travel in opposite directions. One car travels at 65 miles per hour and the other at 60 miles per hour. How far has each traveled when they are 375 miles apart?

10. 65 mph car _____

    60 mph car _____

11. Two trains start at the same point going in the same direction on parallel tracks. One train travels at 70 miles per hour and the other at 42 miles per hour. In how many hours will they be 154 miles apart?

11. _____

12. A motorboat traveling with the current went 72 miles in 3 hours. Against the current, the boat could only go 48 miles in the same amount of time. Find the rate of the boat in still water if the rate of the current is 4 miles per hour.

12. _____

Objective 5 Solve problems with three variables by using a system of three equations. For extra help, see Examples 5 and 6 on pages 239–240 of your text, the Section Lecture video for Section 4.3, and Exercise Solution Clip 47.

Solve each problem involving three unknowns.

13. Julie has $80,000 to invest. She invests part at 5%, one fourth this amount at 6%, and the balance 7%. Her total annual income from interest is $4700. Find the amount invested at each rate.

13. 5% _____

6% _____

7% _____

14. A merchant wishes to mix gourmet coffee selling for $8 per pound, $10 per pound, and $15 per pound to get 50 pounds of a mixture that can be sold for $11.70 per pound. The amount of the $8 coffee must be 3 pounds more than the amount of the $10 coffee. Find the number of pounds of each that must be used.

14. $8/lb _____

$10/lb _____

$15/lb _____

15. A boy scout troop is selling popcorn. There are three different kinds of popcorn in three different arrangements. Arrangement I contains 1 bag of cheddar cheese popcorn, 2 bags of caramel popcorn, and 3 bags of microwave popcorn. Arrangement II contains 3 bags of cheddar cheese popcorn, 1 bag of caramel popcorn, and 2 bags of microwave popcorn. Arrangement III contains 2 bags of cheddar cheese popcorn, 3 bags of caramel popcorn, and 1 bag of microwave popcorn. Jim needs 28 bags of cheddar cheese popcorn, 22 bags of caramel popcorn, and 22 bags of microwave popcorn to give as stocking stuffers for Christmas. How many of each arrangement should he buy?

15. _____

Chapter 4 SYSTEMS OF LINEAR EQUATIONS

4.4   Solving Systems of Linear Equations by Matrix Methods

Learning Objectives
1   Define a matrix.
2   Write the augmented matrix of a system.
3   Use row operations to solve a system with two equations.
4   Use row operations to solve a system with three equations.
5   Use row operations to solve special systems.

Key Terms

Use the vocabulary terms listed below to complete each statement in exercises 1–8.

> **matrix**   **elements of a matrix**   **row**   **column**
>
> **square matrix**   **augmented matrix**   **row operations**
>
> **row echelon form**

1.   A _____ of a matrix is a group of elements that are read horizontally.

2.   A(n) _____ has a vertical bar that separates the columns of the matrix into two groups.

3.   A(n) _____ is a rectangular array of numbers, consisting of horizontal rows and vertical columns.

4.   _____ are operations on a matrix that produce equivalent matrices leading to systems that have the same solutions as the original system of equations.

5.   The numbers in a matrix are called the _____ of the matrix.

6.   A(n) _____ is a matrix that has the same number of rows as columns.

7.   If a matrix is written with 1s on the diagonal from upper left to lower right and 0s below the 1s, it is said to be in _____.

8.   A _____ of a matrix is a group of elements that are read vertically.

Guided Examples

Review this example for Objective 3:

1. Use row operations to solve the system.

$$x - 2y = 4$$

$$-2x + y = 1$$

Start by writing the augmented matrix of the system.

$$\begin{bmatrix} 1 & -2 & | & 4 \\ -2 & 1 & | & 1 \end{bmatrix}$$

There is already a 1 in the first row, first column. Next, we must get 0 row 2, column 1. This is obtained by multiplying each number in row 1 by 2 and then adding the results to each number in row 2. This is abbreviated as $2R_1 + R_2$.

$$\begin{bmatrix} 1 & -2 & | & 4 \\ -2 + 2(1) & 1 + 2(-2) & | & 1 + 2(4) \end{bmatrix}$$

$$= \begin{bmatrix} 1 & -2 & | & 4 \\ 0 & -3 & | & 9 \end{bmatrix} \quad 2R_1 + R_2$$

Now we need 1 in row 2, column 2. This is obtained by multiplying each term in row 2 by $-\frac{1}{3}$.

$$\begin{bmatrix} 1 & -2 & | & 4 \\ 0\left(-\frac{1}{3}\right) & -3\left(-\frac{1}{3}\right) & | & 9\left(-\frac{1}{3}\right) \end{bmatrix}$$

$$= \begin{bmatrix} 1 & -2 & | & 4 \\ 0 & 1 & | & -3 \end{bmatrix} \quad -\frac{1}{3}R_2$$

The augmented matrix leads to the system of equations

$$\begin{matrix} 1x - 2y = 4 \\ 0x + 1y = -3 \end{matrix} \text{ or } \begin{matrix} x - 2y = 4 \\ y = -3 \end{matrix}$$

From the second equation, substitute −3 for y in the first equation and solve for x.

$$x - 2(-3) = 4$$

$$x + 6 = 4 \quad \text{Multiply.}$$

$$x = -2 \quad \text{Subtract 6.}$$

The solution set of the system is $\{(-2, -3)\}$. Check this solution by substitution in both equations of the original system.

Now Try:

1. Use row operations to solve the system.

$$x - 3y = 10$$

$$2x + y = 6$$

_____

Review this example for Objective 4:

2. Use row operations to solve the system.

$$x - y - z = 6$$

$$-x + 3y + 2z = -11$$

$$3x + 2y + z = 1$$

Start by writing the augmented matrix of the system.

$$\begin{bmatrix} 1 & -1 & -1 & | & 6 \\ -1 & 3 & 2 & | & -11 \\ 3 & 2 & 1 & | & 1 \end{bmatrix}$$

There is already a 1 in the first row, first column. Next, we must get 0's in the rest of column 1. First, add row one to row two.

$$\begin{bmatrix} 1 & -1 & -1 & | & 6 \\ 0 & 2 & 1 & | & -5 \\ 3 & 2 & 1 & | & 1 \end{bmatrix} \quad R_1 + R_2$$

Now add to the number in row three, the results of multiplying each number of row one by −3.

$$\begin{bmatrix} 1 & -1 & -1 & | & 6 \\ 0 & 2 & 1 & | & -5 \\ 0 & 5 & 4 & | & -17 \end{bmatrix} \quad -3R_1 + R_3$$

Introduce 1 in row two, column two, by multiplying each number in row two by $\frac{1}{2}$.

$$\begin{bmatrix} 1 & -1 & -1 & | & 6 \\ 0 & 1 & \frac{1}{2} & | & -\frac{5}{2} \\ 0 & 5 & 4 & | & -17 \end{bmatrix} \quad \frac{1}{2}R_2$$

To obtain 0 in row three, column two, add to row three the results of multiplying each number in row two by −5.

$$\begin{bmatrix} 1 & -1 & -1 & | & 6 \\ 0 & 1 & \frac{1}{2} & | & -\frac{5}{2} \\ 0 & 0 & \frac{3}{2} & | & -\frac{9}{2} \end{bmatrix} \quad -5R_2 + R_3$$

Now Try:

2. Use row operations to solve the system.

$$x - 4y - z = 6$$

$$2x - y + 3z = 0$$

$$3x - 2y + z = 4$$

_____

Obtain 1 in row three, column three, by multiplying each number in row three by $\frac{2}{3}$.

$$\begin{bmatrix} 1 & -1 & -1 & | & 6 \\ 0 & 1 & \frac{1}{2} & | & -\frac{5}{2} \\ 0 & 0 & 1 & | & -3 \end{bmatrix} \begin{matrix} \\ \\ \frac{2}{3}R_3 \end{matrix}$$

The final matrix gives the system

$x - y - z = 6$

$y + \frac{1}{2}z = -\frac{5}{2}$

$z = -3$

Substitute -3 for z in the second equation and solve for y.

$y + \frac{1}{2}(-3) = -\frac{5}{2}$

$y - \frac{3}{2} = -\frac{5}{2}$ Multiply.

$y = -\frac{2}{2} = -1$ Add $\frac{3}{2}$; simplify.

Now substitute -3 for z and -1 for y in the first equation, and solve for x.

$x - (-1) - (-3) = 6$

$x + 1 + 3 = 6$ Definition of subtraction

$x + 4 = 6$ Add.

$x = 2$ Subtract 4.

The solution set is $\{(2, -1, -3)\}$.

Check this solution by substitution in the equations of the original system.

Review this example for Objective 5:

3. Use row operations to solve each system.

a. $x - 3y = 1$ **b.** $x - 2y = 3$

   $4x - 12y = 5$ $3x - 6y = 9$

a. $\begin{bmatrix} 1 & -3 & | & 1 \\ 4 & -12 & | & 5 \end{bmatrix}$ Write the augmented matrix.

$\begin{bmatrix} 1 & -3 & | & 1 \\ 0 & 0 & | & 1 \end{bmatrix} \begin{matrix} \\ -4R_1 + R_2 \end{matrix}$

Now Try:

3. Use row operations to solve each system.

a. $x + y = 3$ _____

   $3x + 3y = -2$

b. $2x + y = 10$ _____

   $-4x - 2y = -20$

The corresponding system of equations is

$$x - 3y = 1$$
$$0 = 1, \quad \text{False}$$

which is inconsistent and has no solution. The solution set is $\varnothing$.

b. $\begin{bmatrix} 1 & -2 & | & 3 \\ 3 & -6 & | & 9 \end{bmatrix}$ Write the augmented matrix.

$\begin{bmatrix} 1 & -2 & | & 3 \\ 0 & 0 & | & 0 \end{bmatrix}$ $-3R_1 + R_2$

The corresponding system of equations is

$$x - 2y = 3$$
$$0 = 0, \quad \text{True}$$

which has dependent equations. The solution set is $\{(x, y) \mid x - 3y = 1\}$.

Objective 1 Define a matrix.
For extra help, see page 247 of your text and the Section Lecture video for Section 4.4.

Give the number of rows and columns in each matrix.

1. a. $\begin{bmatrix} -2 & 5 & 3 \\ 1 & 0 & 1 \end{bmatrix}$ **b.** $\begin{bmatrix} -4 & 7 & 6 & 5 \\ 0 & 0 & 9 & 8 \\ 3 & 2 & 5 & 1 \end{bmatrix}$

1. a. _____

b. _____

c. $\begin{bmatrix} 5 & -8 \\ 1 & -9 \end{bmatrix}$ **d.** $\begin{bmatrix} -11 \\ 22 \\ -55 \end{bmatrix}$

c. _____

d. _____

Use the matrix $\begin{bmatrix} 3 & 2 & 6 \\ 6 & 0 & 1 \\ -2 & 10 & -11 \\ 1 & 5 & 2 \end{bmatrix}$ *to answer question 2.*

2. a. What are the elements of the second row?

b. What are the elements of the second column?

2. a. _____

b. _____

Objective 2 Write the augmented matrix of a system.

For extra help, see pages 247–248 of your text and the Section Lecture video for Section 4.4.

3. Write the augmented matrix for each system. Do not solve the system.

    a. $x - y + 4 = 0$
            $3x + y - 1 = 0$

    b. $x + 6 = 0$
           $y - 2 = 0$

    c. $2x + y - z = 3$
           $x - 2y + z = 5$
           $5x - y + 2z = 7$

    d. $4x - 7y + z - 1 = 0$
           $2x + 3y - 5z + 2 = 0$
           $6x - y + 8z - 5 = 0$

3. a._____

b. _____

c._____

d. _____

Objective 3 Use row operations to solve a system with two equations.

For extra help, see Example 1 on pages 248–249 of your text, the Section Lecture video for Section 4.4, and Exercise Solution Clip 9.

Use row operations to solve each system.

4. $x + 2y = 5$
     $2x + y = 4$

4. _____

5. $2x + y = -4$
     $x - y = -2$

5. _____

6. $2x + 3y = 2$
     $4x - 9y = -1$

6. _____

7. $4x + y = 1$
     $8x - 3y = 2$

7. _____

Objective 4 Use row operations to solve a system with three equations.
For extra help, see Example 2 on pages 250–251 of your text, the Section Lecture video for Section 4.4, and Exercise Solution Clip 19.

Use row operations to solve the system.

8. $\begin{aligned} x - y + z &= -1 \\ x - y + 2z &= 0 \\ x + y - z &= -3 \end{aligned}$

8. _____

9. $\begin{aligned} x + y + 2z &= -1 \\ x - y - z &= 3 \\ 2x + y + z &= 0 \end{aligned}$

9. _____

10. $\begin{aligned} x + 2y - z &= 0 \\ 2x - y + z &= 4 \\ 3x + y - z &= 1 \end{aligned}$

10. _____

11. $\begin{aligned} x - y \quad\;\; &= 1 \\ 2y + 3z &= 8 \\ 2x \quad\;\; + z &= 6 \end{aligned}$

11. _____

Objective 5 Use row operations to solve special systems.

For extra help, see Example 3 on page 251 of your text, the Section Lecture video for Section 4.4, and Exercise Solution Clip 13.

Use row operations to solve the system.

12. $5x - 10y = 8$
$-x + 2y = 3$

12. _____

13. $2x - 3y = 61$
$-4x + 6y = -122$

13. _____

14. $5x + y - 2z = 3$
$x + 6z = 13$
$-10x - 2y + 4z = 9$

14. _____

15. $-6x + 2y - 8z = -12$
$3x - y + 4z = 6$
$9x - 3y + 12z = 18$

15. _____

Chapter 5 EXPONENTS, POLYNOMIALS, AND POLYNOMIAL FUNCTIONS

5.1   Integer Exponents and Scientific Notation

Learning Objectives
1   Use the product rule for exponents.
2   Define 0 and negative exponents.
3   Use the quotient rule for exponents.
4   Use the power rules for exponents.
5   Simplify exponential expressions.
6   Use the rules for exponents with scientific notation.

Key Terms

Use the vocabulary terms listed below to complete each statement in exercises 1–6.

exponent   **base**   **exponential (power)**

product rule for exponents   **quotient rule for exponents**

power rule for exponents   **scientific notation**

1.   A number is written in _____ when it is expressed in the form $a \times 10^n$, where $1 \le |a| < 10$ and n is an integer.

2.   In the expression a^m, a is the _____ and m is the _____.

3.   The _____ states that, when dividing powers of like bases, keep the same base and subtract the exponent of the denominator from the exponent of the numerator.

4.   The statement "If m and n are any integers, then $\left(a^m\right)^n = a^{mn}$" is an example of the _____.

5.   The quantity 5^7 is called a(n) _____.

6.   The _____ states that, when multiplying powers of like bases, keep the same base and add the exponents.

Guided Examples

Review this example for Objective 1:

1. Apply the product rule, if possible, in each case.

   a. $7^5 \cdot 7^3$ b. $\left(3a^3b^4\right)\left(-3ab^2\right)$

   c. $x^2 \cdot y^4$

   a. $7^5 \cdot 7^3 = 7^{5+3} = 7^8$

   b. $\left(3a^3b^4\right)\left(-3ab^2\right) = 3(-3)a^3a^1b^4b^2$
   $$= -9a^{3+1}b^{4+2}$$
   $$= -9a^4b^6$$

   c. The product rule does not apply because the bases are not the same.

Now Try:

1. Apply the product rule, if possible, in each case.

   a. $6^4 \cdot 8^2$ _____

   b. $a^4 \cdot a^2$ _____

   c. $\left(-5x^3z^4\right)\left(-4x^3z\right)$

   _____

Review these examples for Objective 2:

2. Evaluate.

   a. 3^0 b. $(-3)^0$

   c. -3^0 d. $(-3x)^0$, $x \neq 0$

   e. $3^0 + 2^0$

   a. $3^0 = 1$ b. $(-3)^0 = 1$

   c. $-3^0 = -1$ d. $(-3x)^0 = 1$, $x \neq 0$

   e. $3^0 + 2^0 = 1+1 = 2$

3. Write with only positive exponents, then evaluate or simplify, if possible.

   a. 4^{-5} b. $(2x)^{-3}$

   c. $-5r^{-2}$ d. $3^{-1} + 9^{-1}$

   a. $4^{-5} = \dfrac{1}{4^5} = \dfrac{1}{1024}$

   b. $(2x)^{-3} = \dfrac{1}{(2x)^3} = \dfrac{1}{2^3 x^3} = \dfrac{1}{8x}$

Now Try:

2. Evaluate.

   a. 30^0 _____

   b. $(-30)^0$ _____

   c. -30^0 _____

   d. $(-30x)^0$, $x \neq 0$ _____

   e. $20^0 - 30^0$ _____

3. Write with only positive exponents, then evaluate or simplify, if possible.

   a. 3^{-4} _____

   b. $(4a)^{-2}$ _____

   c. $-3s^{-3}$ _____

   d. $3^{-1} - 6^{-1}$ _____

c. $-5r^{-2} = (-5)\dfrac{1}{r^2} = -\dfrac{5}{r^2}$

d. $3^{-1} + 9^{-1} = \dfrac{1}{3} + \dfrac{1}{9} = \dfrac{3}{9} + \dfrac{1}{9} = \dfrac{4}{9}$

4. Evaluate.

a. $\dfrac{1}{3^{-5}}$ **b.** $\dfrac{8^{-3}}{2^{-4}}$

a. $\dfrac{1}{3^{-5}} = \dfrac{1}{\frac{1}{3^5}} = 1 \div \dfrac{1}{3^5} = 1 \cdot 3^5 = 3^5 = 243$

b. $\dfrac{8^{-3}}{2^{-4}} = \dfrac{\frac{1}{8^3}}{\frac{1}{2^4}} = \dfrac{1}{8^3} \div \dfrac{1}{2^4} = \dfrac{1}{8^3} \cdot \dfrac{2^4}{1}$

$= \dfrac{2^4}{8^3} = \dfrac{16}{512} = \dfrac{1}{32}$

4. Evaluate.

a. $\dfrac{1}{4^{-2}}$ _____

b. $\dfrac{6^{-3}}{3^{-4}}$ _____

Review this example for Objective 3:

5. Apply the quotient rule, if possible, and write each result with only positive exponents.

a. $\dfrac{r^9}{r^{12}}$, $r \neq 0$ **b.** $\dfrac{12^{-7}}{12^{-6}}$

c. $\dfrac{a^5}{b^3}$

a. $\dfrac{r^9}{r^{12}} = r^{9-12} = r^{-3} = \dfrac{1}{r^3}$, $r \neq 0$

b. $\dfrac{12^{-7}}{12^{-6}} = 12^{-7-(-6)} = 12^{-7+6} = 12^{-1} = \dfrac{1}{12}$

c. $\dfrac{a^5}{b^3}$

This quotient rule does not apply because the bases are different. This expression cannot be simplified further.

Now Try:

5. Apply the quotient rule, if possible, and write each result with only positive exponents.

a. $\dfrac{k^7}{k^4}$ _____

b. $\dfrac{3}{3^{-4}}$ _____

c. $\dfrac{c^7}{b^9}$ _____

Review these examples for Objective 4:

6. Simplify using the power rules.

    a. $\left(5r^3\right)^4$ **b.** $\left(\dfrac{-2x}{5}\right)^3$

    a. $\left(5r^3\right)^4 = 5^4\left(r^3\right)^4 = 5^4 r^{3\cdot4} = 625r^{12}$

    b. $\left(\dfrac{-2x}{5}\right)^3 = \dfrac{(-2x)^3}{5^3} = \dfrac{(-2)^3 x^3}{5^3} = \dfrac{-8x^3}{125}$

7. Write with only positive exponents and then evaluate.

    a. $\left(\dfrac{4}{3}\right)^{-4}$ **b.** $\left(\dfrac{-2x}{5}\right)^{-3}$, $x \neq 0$

    a. $\left(\dfrac{4}{3}\right)^{-4} = \left(\dfrac{3}{4}\right)^4 = \dfrac{3^4}{4^4} = \dfrac{81}{256}$

    b. $\left(\dfrac{-2x}{5}\right)^{-3} = \left(\dfrac{5}{-2x}\right)^3 = \dfrac{5^3}{(-2x)^3}$

        $= \dfrac{125}{(-2)^3 x^3} = \dfrac{125}{-8x^3} = -\dfrac{125}{8x^3}$

Review this example for Objective 5:

8. Simplify. Assume that all variables represent nonzero real numbers.

    a. $\left(3a^{-2}b^{-4}\right)^2$ **b.** $\dfrac{9^{-3}}{9^7 \cdot 9^{-2}}$

    c. $\dfrac{3^{-2}x^{-4}\left(x^2\right)^{-3}}{7\left(x^2\right)^{-1}}$

    a. $\left(3a^{-2}b^{-4}\right)^2 = 3^2\left(a^{-2}\right)^2\left(b^{-4}\right)^2$

        $= 9a^{-4}b^{-8} = \dfrac{9}{a^4 b^8}$

Now Try:

6. Simplify using the power rules.

    a. $\left(3x^4\right)^3$ _____

    b. $\left(\dfrac{-2x^3}{z^4}\right)^6$ _____

7. Write with only positive exponents and then evaluate.

    a. $\left(\dfrac{3}{8}\right)^{-3}$ _____

    b. $\left(\dfrac{-2x^3}{z^4}\right)^{-4}$ _____

Now Try:

8. Simplify. Assume that all variables represent nonzero real numbers.

    a. $\dfrac{12w^7 w^{-3}}{20w^{-1}w^5}$ _____

    b. $\left(\dfrac{y}{2w}\right)^{-3}\left(\dfrac{w^2}{3y}\right)^2$ _____

    c. $\left(-5r^{-2}s^5 t^{-3}\right)^2\left(3r^2 s^{-3}t\right)^{-2}$

    _____

b. $\dfrac{9^{-3}}{9^7 \cdot 9^{-2}} = \dfrac{9^{-3}}{9^{7+(-2)}} = \dfrac{9^{-3}}{9^5} = 9^{-3-5} = 9^{-8}$

$\qquad = \dfrac{1}{9^8}$

c. $\dfrac{3^{-2}x^{-4}\left(x^2\right)^{-3}}{7\left(x^2\right)^{-1}} = \dfrac{3^{-2}x^{-4}x^{-6}}{7x^{-2}} = \dfrac{3^{-2}x^{-10}}{7x^{-2}}$

$\qquad = \dfrac{1}{3^2 \cdot 7}x^{-10-(-2)}$

$\qquad = \dfrac{1}{63}x^{-8} = \dfrac{1}{63x^8}$

Review these examples for Objective 6:

9. Write each number in scientific notation.

   a. 325,000   **b.** 0.0257

   a. Move the decimal point to follow the first nonzero digit (the 3). Count the number of places the decimal point was moved. This is the exponent.

   $325{,}000 = 3.25 \times 10^5$
   5 places

   b. Move the decimal point to the right of the first nonzero digit (the 2). Count the number of places the decimal point was moved. Since the decimal was moved to the right, the exponent is negative.

   $0.0257 = 2.57 \times 10^{-2}$
   2 places

10. Write each number in standard notation.

   a. 2.3×10^4 **b.** 7.24×10^{-4}

   a. Move the decimal four places to the right.

   $2.3 \times 10^4 = 23{,}000$

   b. Move the decimal four places to the left.

   $7.24 \times 10^{-4} = 0.000724$

Now Try:

9. Write each number in scientific notation.

   a. 23,651,000,000

   _____

   b. −0.00047

   _____

10. Write each number in standard notation.

   a. 7.2×10^7

   _____

   b. -4.5×10^{-5}

   _____

11. Evaluate: $\dfrac{210,000}{30 \times 0.007}$

$$\frac{210,000}{30 \times 0.007} = \frac{2.1 \times 10^5}{3.0 \times 10^1 \times 7 \times 10^{-3}}$$

Express all numbers in scientific notation.

$$= \frac{2.1 \times 10^5}{3.0 \times 7 \times 10^1 \times 10^{-3}}$$

Commutative property

$$= \frac{2.1 \times 10^5}{21.0 \times 10^{-2}}$$ Multiply; product rule

$$= \frac{2.1 \times 10^5}{2.1 \times 10^1 \times 10^{-2}}$$ Write in scientific notation.

$$= \frac{2.1 \times 10^5}{2.1 \times 10^{-1}}$$ Product rule

$$= \frac{2.1}{2.1} \times 10^6$$ Quotient rule

$$= 1 \times 10^6$$ Simplify.

$$= 1,000,000$$ Standard form

12. There are about 6×10^{23} atoms in a mole of atoms. About how many atoms are there in 81,000 moles?

First write 81,000 in scientific notation:

$$81,000 = 8.1 \times 10^4$$

Now multiply.

$$\left(6 \times 10^{23}\right) \times \left(8.1 \times 10^4\right)$$

$$= (6 \times 8.1) \times \left(10^{23} \times 10^4\right)$$

$$= 48.6 \times 10^{27} = \left(4.86 \times 10^1\right) \times 10^{27}$$

$$= 4.86 \times 10^{28}$$

There are about 4.86×10^{28} atoms in 81,000 moles.

11. Evaluate: $\dfrac{144,000}{0.016 \times 900}$

_____

12. The Milky Way galaxy is about 100,000 light-years in diameter. If one light-year is about 9.5×10^{15} meters, then about how many meters is the diameter of the Milky Way? (Source: NASA)

Name: _____   Date: _____

Instructor: _____   Section: _____

Objective 1 Use the product rule for exponents.

For extra help, see Example 1 on page 264 of your text, the Section Lecture video for Section 5.1, and Exercise Solution Clip 9.

Use the product rule to simplify each expression, if possible. Write each answer in exponential form.

1. $2^3 \cdot 2^5$

           1. _____

2. $7x^8 \cdot 2x^7 \cdot x^4$

           2. _____

Objective 2 Define 0 and negative exponents.

For extra help, see Examples 2–4 on pages 265–266 of your text, the Section Lecture video for Section 5.1, and Exercise Solution Clips 21, 37, and 53.

Evaluate or simplify each expression, and write it using only positive exponents. Assume that all variables represent nonzero real numbers.

3. $\left(-2p^2\right)(3q)^0\left(5r^2\right)$

           3. _____

4. $\left(6n^3\right)\left(2n^{-5}\right)$

           4. _____

Objective 3 Use the quotient rule for exponents.

For extra help, see Example 5 on page 267 of your text, the Section Lecture video for Section 5.1, and Exercise Solution Clip 63.

Use the quotient rule to simplify each expression, and write it using only positive exponents. Assume that all variables represent nonzero real numbers.

5. $\dfrac{r^4}{r^{-3}}$

           5. _____

6. $\dfrac{5^{-8}}{5^7}$

           6. _____

Objective 4 Use the power rules for exponents.
For extra help, see Examples 6 and 7 on pages 268–269 of your text, the Section Lecture
video for Section 5.1, and Exercise Solution Clips 57 and 79.

*Use the power rule to simplify each expression, and write it using only positive exponents.
Assume that all variables represent nonzero real numbers.*

7. $\left(5m^5\right)^3$

7. _____

8. $\left(-3x^{-4}\right)^{-3}$

8. _____

9. $\left(\dfrac{m^3}{3^{-4}}\right)^{-2}$

9. _____

Objective 5 Simplify exponential expressions.
For extra help, see Example 8 on page 270 of your text, the Section Lecture video for
Section 5.1, and Exercise Solution Clip 93.

*Simplify each expression, and write it using only positive exponents. Assume that all
variables represent nonzero real numbers.*

10. $\dfrac{8^{-7}}{8^3 \cdot 8^{-2}}$

10. _____

11. $\left(3x^2 y^{-2}\right)^{-2}\left(2x^{-2}y\right)^{-3}$

11. _____

12. $\left(\dfrac{3q}{4p^2}\right)^2\left(\dfrac{2p}{5q}\right)^{-2}$

12. _____

Name: Date:

Instructor: Section:

Objective 6 Use the rules for exponents with scientific notation.
For extra help, see Examples 9–12 on pages 272–273 of your text, the Section Lecture video for Section 5.1, and Exercise Solution Clips 133, 141, and 155.

13. Write 0.0000382 in scientific notation.   13. _____

14. Write 2.22×10^3 in standard notation.   14. _____

15. Use scientific notation and the rules for exponents to evaluate: 15. _____

$$\frac{9 \times 10^{-2}}{3 \times 10^4}$$

Chapter 5   EXPONENTS, POLYNOMIALS, AND POLYNOMIAL FUNCTIONS

5.2   Adding and Subtracting Polynomials

Learning Objectives
1   Know the basic definitions for polynomials.
2   Add and subtract polynomials.

Key Terms

Use the vocabulary terms listed below to complete each statement in exercises 1–14.

algebraic expression	**term**	**numerical coefficient (coefficient)**
polynomial	**degree of a term**	**polynomial in** x
descending powers	**leading term**	**leading coefficient**
trinomial	**binomial**	**monomial**
degree of a polynomial	**like terms**	**negative of a polynomial**

1.   The _____ is the sum of the exponents on the variables in that term.

2.   A polynomial in x is written in _____ if the exponents on x decrease from left to right.

3.   A(n) _____ is a number, a variable, or a product or quotient of a number and one or more variables raised to powers.

4.   A polynomial with exactly three terms is called a _____.

5.   A(n) _____ is a term, or a finite sum of terms, in which all variables have whole number exponents and no variables appear in denominators.

6.   The numerical factor in a term is its _____.

7.   A polynomial with exactly one term is called a _____.

8.   The _____ is the greatest degree of any term of the polynomial.

9.   A(n) _____ is a polynomial with exactly two terms.

10.   A(n) _____ is any combination of variables or constants joined by the basic operations of addition, subtraction, multiplication, and division (except by 0), or raising to powers or taking roots.

11. The _____ is obtained by changing the sign of every coefficient in the polynomial.

12. A polynomial containing only the variable x is a _____.

13. In the polynomial $7x^4 - 3x^3 + \frac{4}{3}x^2 + x - 15$, the term $7x^4$ is called the

_____ and 7 is the _____.

14. Terms with exactly the same variables raised to exactly the same powers are called

_____.

Guided Examples

Review these examples for Objective 1:

1. Write the polynomial $-3x^5 + 2x^2 - 4 - x^3$ in descending powers of the variable. Then, give the leading term and the leading coefficient.

$-3x^5 + 2x^2 - 4 - x^3$ is written as $-3x^5 - x^3 + 2x^2 - 4$. The leading term is $-3x^5$ and the leading coefficient is -3.

2. Identify the polynomial as a *monomial*, a *binomial*, a *trinomial*, or *none of these*. Also, give the degree.

$10y^4 - 12y^3 - 5y$

This is a trinomial of degree 4.

Now Try:

1. Write the polynomial

$-2x^3 - x + 8x^4 - 1 + x^2$ in descending powers of the variable. Then, give the leading term and the leading coefficient.

_____

2. Identify the polynomial as a *monomial*, a *binomial*, a *trinomial*, or *none of these*. Also, give the degree.

$3m^5 + 3m^4 - m^3 - 4m^2$

_____

Review these examples for Objective 2:

3. Combine like terms.

a. $7z^2 - 4z^3 + 5z^3 - 11z^2$

b. $2x^2y + 2xy - 5xy^2 + 6xy + 9xy^2$

a. $7z^2 - 4z^3 + 5z^3 - 11z^2$

$= -4z^3 + 5z^3 + 7z^2 - 11z^2$

Commutative property

$= z^3 - 4z^2$ Combine like terms.

Now Try:

3. Combine like terms.

a. $3c^2 - 6c + 9c^2 - 4c^3$

_____

b.

$3x^4 - 7x^2 + 5 - 4x^4 + 2x^3 - 9$

_____

b. $2x^2y + 2xy - 5xy^2 + 6xy + 9xy^2$

$= 2x^2y + 2xy + 6xy - 5xy^2 + 9xy^2$

Commutative property

$= 2x^2y + 8xy + 4xy^2$ Combine like terms

4. Add.

  a. $5m^4 + 2m^3 - 4$

  $\underline{-3m^4 + 5m^3 - 3}$

  b. $\left(3x^2 + 2x^4 - 3\right) + \left(8x^3 - 5x^4 - 6x^2\right)$

  a. Add column by column.

  $5m^4 + 2m^3 - 4$

  $\underline{-3m^4 + 5m^3 - 3}$

  $2m^4 + 7m^3 - 7$

  b. Combine like terms.

  $\left(3x^2 + 2x^4 - 3\right) + \left(8x^3 - 5x^4 - 6x^2\right)$

  $= -3x^4 + 8x^3 - 3x^2 - 3$

5. Subtract.

  a. $\left(4ab + 2bc - 9ac\right) - \left(3ca - 2cb - 9ba\right)$

  b. $5m^4 + 2m^3 - 4$

  $\underline{-3m^4 + 5m^3 - 3}$

  a. $\left(4ab + 2bc - 9ac\right) - \left(3ca - 2cb - 9ba\right)$

  $= \left(4ab + 2bc - 9ac\right) - \left(3ac - 2bc - 9ab\right)$

  $= 4ab + 2bc - 9ac - 3ac + 2bc + 9ab$

  $= 4ab + 9ab + 2bc + 2bc - 9ac - 3ac$

  $= 13ab + 4bc - 12ac$

  b. Change all of the signs in the subtrahend (the second row), then add.

  $5m^4 + 2m^3 - 4$

  $\underline{3m^4 - 5m^3 + 3}$

  $8m^4 - 3m^3 - 1$

4. Add.

  a. $9m^3 + 4m^2 - 2m + 3$

  $\underline{-4m^3 - 6m^2 - 2m + 1}$

  $\underline{}$

  b. $\left(x^2 + 6x - 8\right) + \left(3x^2 - 10\right)$

  $\underline{}$

5. Subtract.

  a.

  $\left(-4m^2n + 3mn - 6m\right) - \left(2n + 7mn\right)$

  $\underline{}$

  b. $2m^2 - 5m + 1$

  $\underline{-2m^2 - 5m + 3}$

  $\underline{}$

Objective 1 Know the basic definitions for polynomials.
For extra help, see Examples 1 and 2 on pages 279–280 of your text, the Section Lecture video for Section 5.2, and Exercise Solution Clips 13 and 23.

Write the polynomial in descending powers of the variable. Then give the leading term and the leading coefficient.

1. $-3x^5 + 2x^2 - x^3 - 4$

    1. _____

2. $6 + 2y - 2y^2 + y^3$

    2. _____

*Identify the polynomial as a **trinomial**, **binomial**, **monomial**, or **none of these**. Also, give the degree.*

3. $5a^3 + 4a^2 + 3a + 5$

    3. _____

4. $2x^{-3}$

    4. _____

5. -8

    5. _____

6. $xy^2 - y^2 + xy$

    6. _____

Objective 2 Add and subtract polynomials.
For extra help, see Examples 3–5 on pages 280–281 of your text, the Section Lecture video for Section 5.2, and Exercise Solution Clips 35, 53 and 63.

Add.

7. $-11z^2 + 3z + 1$

    $\underline{-7z^2 + 6z - 4}$

    7. _____

8. $\left(5a^3 - 4a^2 + 7\right) + \left(3a^3 + 2a^2 + 7a\right)$

    8. _____

9. $\left(3x^2 + 4x + 2\right) + \left(7x^2 - x + 2\right)$

    9. _____

10. $\left(2r^3 - 6r\right) + \left(4r - 3r^2\right)$

10. _____

Subtract.

11. $\left(11p + 14\right) - \left(6p - 5\right)$

11. _____

12. $\left(-4n^2 + 3n - 1\right) - \left(-2n^2 + 7n - 5\right)$

12. _____

13. $\quad -3y^3 + 7y^2$

$\underline{\quad y^4 - \quad y^3 \qquad\quad + 3y}$

13. _____

14. $\quad 8r^3 + 5r - 8$

$\underline{-5r^3 - 7r + 3}$

14. _____

15. $\left(3a^4 - 7a^2 + 5\right) - \left(4a^4 + 2a^2 - 9\right)$

15. _____

Chapter 5 EXPONENTS, POLYNOMIALS, AND POLYNOMIAL FUNCTIONS

5.3 Polynomial Functions, Graphs, and Composition

Learning Objectives
1 Recognize and evaluate polynomial functions.
2 Use a polynomial function to model data.
3 Add and subtract polynomial functions.
4 Find the composition of functions.
5 Graph basic polynomial functions.

Key Terms

Use the vocabulary terms listed below to complete each statement in exercises 1–5.

polynomial function   **composition of functions**   **identity function**

squaring function   **cubing function**

1. A function defined by a polynomial in one variable consisting of one or more terms, is called a(n) _____.

2. Replacing a variable with an algebraic expression is called _____.

3. The _____ pairs every real number with its cube.

4. The _____ is defined as $f(x) = x$.

5. The graph of the _____ is a parabola.

Guided Examples

Review this example for Objective 1:

1. Let $f(x) = 6x^3 - 6x + 1$. Find $f(-3)$.

$f(-3) = 6(-3)^3 - 6(-3) + 1$

   Substitute -3 for x.

$= 6(-27) - 6(-3) + 1$

   Apply the exponent.

$= -162 + 18 + 1$ Multiply.

$= -143$ Add.

Now Try:

1. Let $p(x) = -x^4 + 3x^2 - x + 7$. Find $p(2)$.

_____

Review this example for Objective 2:

2. The average undergraduate tuition, room, and board at public four-year colleges for the years 1989–2009 can be modeled by the function

$$f(x) = 112.38x^2 - 2949.32x + 29,488.08,$$

where $x = 1$ corresponds to 1989, $x = 2$ corresponds to 1990, etc. Use this model to estimate the tuition, room, and board in 2005. (Source: National Center for Education Statistics)

Since $x = 16$ corresponds to 2005, we must find $f(16)$.

$$f(16) = 112.38(16)^2 - 2949.32(16)$$
$$+ 29,488.08$$
$$= 11,068.24$$

According to the model, the average cost of undergraduate tuition, room, and board at four-year public colleges was $11,068.24 in 2005.

Now Try:

2. Use the function at the left to estimate Use this model to estimate the tuition, room, and board in 2000.

_____

Review these examples for Objective 3:

3. For $f(x) = 2x^2 + 4x - 5$ and

$g(x) = -x^2 + 3x - 8,$ find each of the following.

a. $(f + g)(x)$ b. $(f - g)(x)$

a. $(f + g)(x)$
$$= f(x) + g(x)$$
$$= \left(2x^2 + 4x - 5\right) + \left(-x^2 + 3x - 8\right)$$
$$= x^2 + 7x - 13$$

b. $(f - g)(x)$
$$= f(x) - g(x)$$
$$= \left(2x^2 + 4x - 5\right) - \left(-x^2 + 3x - 8\right)$$
$$= \left(2x^2 + 4x - 5\right) + \left(x^2 - 3x + 8\right)$$
$$= 3x^2 + x + 3$$

Now Try:

3. For $f(x) = 6x^2 - 7x + 12$ and

$g(x) = -3x^2 + x + 9,$ find each of the following.

a. $(f + g)(x)$

_____

b. $(f - g)(x)$

_____

4. For $f(x) = -6x^2 - 5x + 4$ and $g(x) = -3x^2 - 5x + 2$, find each of the following.

  a. $(f+g)(2)$ **b.** $(f-g)(-3)$

a. First find $(f+g)(x)$

$$(f+g)(x)$$
$$= f(x) + g(x)$$
$$= \left(-6x^2 - 5x + 4\right) + \left(-3x^2 - 5x + 2\right)$$
$$= -9x^2 - 10x + 6$$

Then, $(f+g)(2) = -9(2)^2 - 10(2) + 6$
$$= -36 - 20 + 6$$
$$= -50$$

Alternatively, we could find $f(2)$ and $g(2)$, and then add.

b. $(f-g)(-3)$
$$= f(-3) - g(-3)$$
$$= \left[-6(-3)^2 - 5(-3) + 4\right]$$
$$\quad - \left[-3(-3)^2 - 5(-3) + 2\right]$$
$$= \left[-6(9) + 15 + 4\right] - \left[-3(9) + 15 + 2\right]$$
$$= -35 - (-10) = -35 + 10 = -25$$

Alternatively, we could find $(f-g)(x)$ first, and then substitute -3 for x.

4. For $f(x) = 12x^2 + 3x - 22$ and $g(x) = -11x^2 + 3x + 4$, find each of the following.

  a. $(f+g)(-1)$

_____

  b. $(f-g)(2)$

_____

Review these examples for Objective 4:

5. Let $f(x) = 3x - 1$ and $g(x) = x^2 + 2$. Find $(f \circ g)(5)$.

$$(f \circ g)(5) = f\big(g(5)\big)$$
$$= f\left(5^2 + 2\right) = f(27)$$
$$= 3(27) - 1 = 80$$

Now Try:

5. Let $f(x) = -3x - 3$ and $g(x) = x^2 - 5$. Find $(f \circ g)(-3)$.

_____

Name: Date:
Instructor: Section:

6. Let $f(x) = 3x - 1$ and $g(x) = x^2 + 2$. Find $(f \circ g)(x)$ and $(g \circ f)(x)$.

$(f \circ g)(x) = f(g(x))$ Use $g(x)$ as the input for $f(x)$.

$\quad = 3(g(x)) - 1$ Use the rule for $f(x)$.

$\quad = 3(x^2 + 2) - 1$ $g(x) = x^2 + 2$

$\quad = 3x^2 + 6 - 1$ Distributive property

$\quad = 3x^2 + 5$ Combine like terms.

$(g \circ f)(x) = g(f(x))$ Use $f(x)$ as the input for $g(x)$.

$\quad = (f(x))^2 + 2$ Use the rule for $g(x)$.

$\quad = (3x - 1)^2 + 2$ $f(x) = 3x - 1$

$\quad = 3x^2 - 6x + 3$ Expand.

6. Let $f(x) = -3x - 3$ and $g(x) = x^2 - 5$. Find $(f \circ g)(x)$ and $(g \circ f)(x)$.

_____

_____

Review this example for Objective 5:

7. Graph $f(x) = x^2 + 2$. Give the domain and range.

For each input, square it and then add 2.

x	$F(x) = x^2 + 2$
-2	6
-1	3
0	2
1	3
2	6

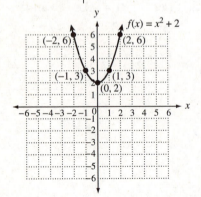

The domain is $(-\infty, \infty)$ and the range is $[2, \infty)$.

Now Try:

7. Graph $f(x) = -x^3 + 1$. Give the domain and range.

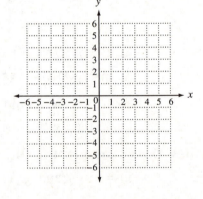

_____

Objective 1 Recognize and evaluate polynomial functions.
For extra help, see Example 1 on page 284 of your text, the Section Lecture video for Section 5.3, and Exercise Solution Clip 7.

Find the indicated values for each polynomial function.

1. If $P(x) = 2x^2 + 3x - 5,$ find each of the following.

   a. $P(3)$ b. $P(-3)$

   c. $P(-1)$ d. $P(1)$

   1. a. _____

   b. _____

   c. _____

   d. _____

2. If $P(x) = -x^2 - x - 5,$ find each of the following.

   a. $P(-2)$ b. $P(2)$

   c. $P(0)$ d. $P(-1)$

   2. a. _____

   b. _____

   c. _____

   d. _____

Objective 2 Use a polynomial function to model data.
For extra help, see Example 2 on page 285 of your text and the Section Lecture video for Section 5.3.

Solve each problem

3. If an object is propelled upward from a height of 16 feet with an initial velocity of 48 feet per second, its height h (in feet) t seconds later is given by the equation

   $h = -16t^2 + 48t + 16$.

   a. Find the height of the object after 1.5 seconds.

   b. Find the height of the object after 3 seconds.

   3. a._____

   b. _____

4. A widget manufacturer estimates that her monthly
   revenue can be modeled by the function
   $$f(x) = -.006x^2 + 32x - 10,000.$$

   a. Find the revenue if 2000 widgets are sold.

   b. Find the revenue if 3500 widgets are sold.

4. a._____

   b. _____

5. The unemployment rate in a certain community can be
   modeled by the equation

   $y = 0.0248x^2 - 0.4810x + 7.8543,$ where y is the
   unemployment rate (percent) and x is the month
   ($x = 1$ represents January, $x = 2$ represents February,
   etc.) Use the model to find the unemployment rate in
   August. Round your answer to the nearest tenth.

5. _____

Objective 3 Add and subtract polynomial functions.
For extra help, see Examples 3 and 4 on pages 285–286 of your text, the Section Lecture
video for Section 5.3, and Exercise Solution Clips 15 and 17.

Find (a) $(f + g)(x)$ and (b) $(f - g)(x)$ for each of the following pairs of functions.

6. $f(x) = 2x^2 + 7x - 4$, $g(x) = 3x^2 - 3x + 2$

6. a._____

   b. _____

7. $f(x) = 3x^2 - 4$, $g(x) = 8x^2 - 5x + 4$

7. a._____

   b. _____

Find (a) $(f+g)(-2)$ and (b) $(f-g)(1)$ for each of the following pairs of functions.

8. $f(x) = 3x^2 + 2x - 6$, $g(x) = 4x^2 - 5x - 7$

8. a._____

b. _____

9. $f(x) = 8x^2 - 4$, $g(x) = -2x^2 - 9x$

9. a._____

b. _____

Objective 4 Find the composition of functions.
For extra help, see Examples 5 and 6 on pages 287–288 of your text and the Section Lecture video for Section 5.3.

Let $f(x) = 4x - 3$ and $g(x) = 2x^2 - 1$. Find the following.

10. Let $f(x) = 4x - 3$ and $g(x) = 2x^2 - 1$. Find the following.

    a. $(f \circ g)(2)$ b. $(g \circ f)(-1)$

    c. $(f \circ g)(x)$

10. a._____

b. _____

c._____

11. Let $f(x) = \dfrac{1}{x}$ and $g(x) = 3x^2 - 4x + 1$. Find the following.

    a. $(f \circ g)(2)$ b. $(g \circ f)\left(\dfrac{1}{3}\right)$

    c. $(g \circ f)(x)$

11. a._____

b. _____

c._____

Objective 5 Graph basic polynomial functions.

For extra help, see Example 7 on pages 289–290 of your text, the Section Lecture video for Section 5.3, and Exercise Solution Clip 55.

Graph each function. Give the domain and the range.

12. $f(x) = -3x + 2$

12.

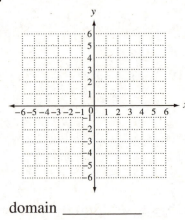

domain _____

range _____

13. $f(x) = -2x^2$

13.

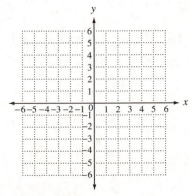

domain _____

range _____

14. $f(x) = \dfrac{1}{3}x^2$

14.

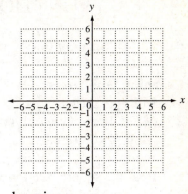

domain _____

range _____

15. $f(x) = x^3 + 2$

15.

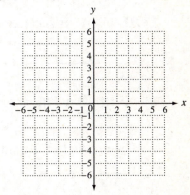

domain _____

range _____

Chapter 5 EXPONENTS, POLYNOMIALS, AND POLYNOMIAL FUNCTIONS

5.4   Multiplying Polynomials

Learning Objectives
1.   Multiply terms.
2.   Multiply any two polynomials.
3.   Multiply binomials.
4.   Find the product of the sum and difference of two terms.
5.   Find the square of a binomial.
6.   Multiply polynomial functions.

Key Terms

Use the vocabulary terms listed below to complete each statement in exercises 1–4.

     FOIL   **difference of the squares**   **square a binomial**   **intersection**

1.   _____ is a method for multiplying two binomials.

2.   To _____, take the sum of the first term, twice the product of the two terms, and the square of the last term.

3.   The product of the sum and difference of the two terms x and y is the _____ of the terms.

4.   When multiplying functions, the domain is found by taking the _____ of the domains of the original functions.

Guided Examples

Review this example for Objective 1:

1.   Find the product: $\left(-11xy^2\right)\left(2x^3y\right)$

$$\left(-11xy^2\right)\left(2x^3y\right) = -11\cdot 2\cdot x\cdot x^3\cdot y^2\cdot y$$

           Commutative and associative properties

$$= -22x^{1+3}y^{2+1}$$

           Multiply; product rule for exponents

$$= -22x^4y^3$$

Now Try:

1.   Find the product:

$$\left(-12r^4s^7\right)(-9r)$$

_____

Name:

Instructor:

Date:

Section:

Review these examples for Objective 2:

2. Find each product.

   a. $2m\left(3m^3 + 7m^2 + 3\right)$

   b. $-y^2\left(3y + 1\right)\left(2y - 5\right)$

   a. $2m\left(3m^3 + 7m^2 + 3\right)$

      $= 2m\left(3m^3\right) + 2m\left(7m^2\right) + 2m(3)$

                    Distributive property

      $= 6m^4 + 14m^3 + 6m$ Multiply monomials

   b. $-y^2\left(3y + 1\right)\left(2y - 5\right)$

      $= -y^2\left[\left(3y + 1\right)2y + \left(3y + 1\right)\left(-5\right)\right]$

                  Distributive property

      $= -y^2\left[\begin{array}{l}\left(3y\right)\left(2y\right) + 1\left(2y\right) \\ \quad + \left(3y\right)\left(-5\right) + 1\left(-5\right)\end{array}\right]$

                  Distributive property

      $= -y^2\left(6y^2 + 2y - 15y - 5\right)$

                  Multiply.

      $= -y^2\left(6y^2 - 13y - 5\right)$

                  Combine like terms.

      $= -6y^4 + 13y^3 + 5y^2$

                  Distributive property

3. Find the product vertically.

   $\left(2r^2 + r - 1\right)\left(r - 3\right)$

$$
\begin{array}{r}
2r^2 + r - 1 \\
r - 3 \\
\hline
-6r^2 - 3r + 3 \\
2r^3 + r^2 - r \\
\hline
2r^3 - 5r^2 - 4r + 3
\end{array}
$$

Begin by multiplying each of the terms in the top row by -3.
Then multiply each of the terms in the top row by r.
Finally, add like terms.

Now Try:

2. Find each product.

   a. $-7z^3\left(5z^3 - 4z^2 + 2\right)$

      _____

   b. $a^2\left(3 - 4a\right)\left(1 - 2a\right)$

      _____

3. Multiply

   $\left(3x^2 + \frac{1}{2}x\right)\left(2x^2 + 6x - 4\right)$

   vertically.

      _____

Review this example for Objective 3:

4.   Use the FOIL method to find each product.

    a. $(4m + 3)(m - 7)$

    b. $(3x + 2y)(2x - 3y)$

    a. $(4m + 3)(m - 7)$

        First terms:

        $(4m + 3)(m - 7)$ $4m(m) = 4m^2$

        Outer terms

        $(4m + 3)(m - 7)$ $4m(-7) = -28m$

        Inner terms

        $(4m + 3)(m - 7)$ $3(m) = 3m$

        Last terms

        $(4m + 3)(m - 7)$ $3(-7) = -21$

$$
\begin{array}{cccc}
\text{F} & \text{O} & \text{I} & \text{L}
\end{array}
$$
$$(4m + 3)(m - 7) = 4m^2 - 28m + 3m - 21$$
$$= 4m^2 - 25m - 21$$
                       Combine like terms.

    b. $(3x + 2y)(2x - 3y)$

$$
\begin{array}{cccc}
\text{F} & \text{O} & \text{I} & \text{L}
\end{array}
$$
$$= 6x^2 - 9xy + 4xy - 6y^2$$
$$= 6x^2 - 5xy - 6y^2 \quad \text{Combine like terms.}$$

Review this example for Objective 4:

5.   Find each product.

    a. $(2k + 5)(2k - 5)$

    b. $y(5y - 3)(5y + 3)$

    a. $(2k + 5)(2k - 5) = (2k)^2 - 5^2$
$$= 4k^2 - 25$$

    b. $y(5y - 3)(5y + 3) = y\left[(5y)^2 - 3^2\right]$
$$= y\left(25y^2 - 9\right)$$
$$= 25y^3 - 9y$$

Now Try:

4.   Use the FOIL method to find each product.

    a. $(x - 3)(2x - 9)$

    _____

    b. $(2m + 3n)(-3m + 4n)$

    _____

Now Try:

5.   Find each product.

    a. $(5m - 4n)(5m + 4n)$

    _____

    b. $x^2(8x - 3)(8x + 3)$

    _____

Name:

Date:

Instructor:

Section:

Review these examples for Objective 5:

6. Find each product.

   a. $(8k + 5p)^2$

   b. $(5y - 3)^2$

   a. $(8k + 5p)^2 = (8k)^2 + 2(40kp) + (5p)^2$
   $= 64k^2 + 80kp + 25p^2$

   b. $(5y - 3)^2 = (5y)^2 - 2(15y) + 3^2$
   $= 25y^2 - 30y + 9$

7. Find each product.

   a. $\left[(6x + 5) + 3y\right]\left[(6x + 5) - 3y\right]$

   b. $(3x + y)^3$

   c. $(2r - 3)^4$

   a. $\left[(6x + 5) + 3y\right]\left[(6x + 5) - 3y\right]$
   $= (6x + 5)^2 - (3y)^2$

   Product of sum and difference
   of terms

   $= 36x^2 + 60x + 25 - 9y^2$

   Square both quantities.

   b. $(3x + y)^3 = (3x + y)^2(3x + y)$ $a^3 = a^2 \cdot a$
   $= \left(9x^2 + 6xy + y^2\right)(3x + y)$

   Square the binomial.

   $= 27x^3 + 18x^2y + 3xy^2 + 9x^2y$
   $+ 6xy^2 + y^3$

   Multiply polynomials.

   $= 27x^3 + 27x^2y + 9xy^2 + y^3$

   Combine like terms.

Now Try:

6. Find each product.

   a. $(a + 2b)^2$

   _____

   b. $(2m - 3p)^2$

   _____

7. Find each product.

   a. $\left[(6x + 5) + 3y\right]^2$

   _____

   b. $(5m - 4n)^3$

   _____

   c. $(3a - 2)^4$

   _____

c. $(2r-3)^4$

$$= (2r-3)^2 (2r-3)^2 \qquad a^4 = a^2 \cdot a^2$$

$$= \left(4r^2 - 12r + 9\right)\left(4r^2 - 12r + 9\right)$$

Square each binomial.

$$= 16r^4 - 48r^3 + 36r^2 - 48r^3$$

$$\quad + 144r^2 - 108r + 36r^2 - 108r + 81$$

Multiply polynomials.

$$= 16r^4 - 96r^3 + 216r^2 - 216r + 81$$

Combine like terms.

Review this example for Objective 6:

8. For $f(x) = x^2 - 5$ and $g(x) = 3x + 4$, find $(fg)(x)$ and $(fg)(-3)$.

$$(fg)(x) = \left(x^2 - 5\right)(3x + 4)$$

$$= 3x^3 + 4x^2 - 15x - 20 \quad \text{FOIL}$$

$$(fg)(-3) = 3(-3)^3 + 4(-3)^2 - 15(-3) - 20$$

$$= -81 + 36 + 45 - 20$$

$$= -20$$

Now Try:

8. For $f(x) = 3x - 8$ and $g(x) = 2x^2 + 5x + 3$, find $(fg)(x)$ and $(fg)(3)$.

_____

_____

Objective 1 Multiply terms.

For extra help, see Example 1 on page 293 of your text, the Section Lecture video for Section 5.4, and Exercise Solution Clip 5.

Find each product.

1. $8x^2 y\left(9xy^3\right)$

1. _____

2. $3y^2 z^3 \left(6yz^4\right)$

2. _____

Objective 2 Multiply any two polynomials.
For extra help, see Examples 2 and 3 on pages 293–294 of your text, the Section Lecture video for Section 5.4, and Exercise Solution Clips 11 and 23.

Find each product.

3. $-8z^3\left(2z^3 + 4z\right)$

3. _____

4. $(3a - 2)\left(a^2 - a + 1\right)$

4. _____

5. $(8s + 1)\left(3s^2 + s - 5\right)$

5. _____

Objective 3 Multiply binomials.
For extra help, see Example 4 on pages 295–296 of your text, the Section Lecture video for Section 5.4, and Exercise Solution Clip 35.

Use the FOIL method to find each product.

6. $(4r - 2)(6r + 1)$

6. _____

7. $(3z - 2p)(2z - p)$

7. _____

Objective 4 Find the product of the sum and difference of two terms.
For extra help, see Example 5 on page 296 of your text, the Section Lecture video for Section 5.4, and Exercise Solution Clip 45.

Find each product.

8. $\left(\frac{3}{4}s + \frac{7}{5}t\right)\left(\frac{3}{4}s - \frac{7}{5}t\right)$

8. _____

9. $\left(4t^3 + 5\right)\left(4t^3 - 5\right)$ 9. _____

Objective 5 Find the square of a binomial.

For extra help, see Examples 6 and 7 on pages 297–298 of your text, the Section Lecture video for Section 5.4, and Exercise Solution Clips 55 and 75.

Find each square.

10. $(x + 2y)^2$ 10. _____

11. $(6r - 4s)^2$ 11. _____

12. $\left[(2a - 7b) + 1\right]^2$ 12. _____

Objective 6 Multiply polynomial functions.

For extra help, see Example 8 on page 298 of your text, the Section Lecture video for Section 5.4, and Exercise Solution Clip 119.

Find $(fg)(x)$ and $(fg)(-1)$ for each of the following pairs of functions.

13. $f(x) = x - 11$ and $g(x) = 2x^2 - 3x$ 13. _____

14. $f(x) = 2x + 3$ and $g(x) = 3x^2 - 4x$ 14. _____

15. $f(x) = 8x + 9$ and $g(x) = x^2 - 7x + 4$ 15. _____

Chapter 5 EXPONENTS, POLYNOMIALS, AND POLYNOMIAL FUNCTIONS

5.5   Dividing Polynomials

Learning Objectives

1   Divide a polynomial by a monomial.
2   Divide a polynomial by a polynomial of two or more terms.
3   Divide polynomial functions.

Key Terms

Use the vocabulary terms listed below to complete each statement in exercises 1–3.

     quotient   **dividend**   **divisor**

1.   In the division $\dfrac{5x^5 - 10x^3}{5x^2} = x^3 - 2x$, the expression $5x^5 - 10x^3$ is the

     _____.

2.   In the division $\dfrac{5x^5 - 10x^3}{5x^2} = x^3 - 2x$, the expression $x^3 - 2x$ is the

     _____.

3.   In the division $\dfrac{5x^5 - 10x^3}{5x^2} = x^3 - 2x$, the expression $5x^2$ is the

     _____.

Guided Examples

Review this example for Objective 1:

1. Divide.

   a. $\dfrac{6x^5 - 4x^3 + 12x}{-4x}$

   b. $\dfrac{-54m^3 + 30m^2 + 6m}{3m^2}$

   c. $\dfrac{30x^3 y^3 - 45x^2 y^2 + 15xy^2}{-15xy^2}$

Now Try:

1. Divide.

   a. $\dfrac{-24w^8 + 12w^6 - 18w^4}{-6w^3}$

     _____

   b. $\dfrac{63b^5 - 27b^3 - b}{9b^3}$

     _____

a. $\dfrac{6x^5 - 4x^3 + 12x}{-4x} = \dfrac{6x^5}{-4x} - \dfrac{-4x^3}{-4x} + \dfrac{12x}{-4x}$

Divide each term
by $-4x$.

$= -\dfrac{3}{2}x^4 + x^2 - 3$

b. $\dfrac{-54m^3 + 30m^2 + 6m}{3m^2}$

$= -\dfrac{54m^3}{3m^2} + \dfrac{30m^2}{3m^2} + \dfrac{6m}{3m^2}$

$= -18m + 10 + \dfrac{2}{m}$

c. $\dfrac{30x^3y^3 - 45x^2y^2 + 15xy^2}{-15xy^2}$

$= -2x^2y + 3x - 1$

c. $\dfrac{5x^2y^4 - 30x^4y^3 + 30x^5y^2}{-5x^2y^2}$

_____

Review these examples for Objective 2:

2. Divide $\dfrac{3x^2 - 3x - 6}{x + 1}$.

$$\begin{array}{r} 3x - 6 \\ x+1\overline{\smash{\big)}\,3x^2 - 3x - 6} \\ \underline{-(3x^2 + 3x)} \\ -6x - 6 \\ \underline{-(6x - 6)} \\ 0 \end{array}$$

$\dfrac{3x^2 - 3x - 6}{x + 1} = 3x - 6$

3. Divide $\dfrac{y^3 + 1}{y + 1}$.

Note that the dividend is missing the y^2 and y terms. Use 0 as the coefficient for each missing term. Thus,

$y^3 + 1 = y^3 + 0y^2 + 0y + 1$.

Now Try:

2. Divide $\dfrac{2r^2 + 3r - 20}{r + 4}$.

_____

3. Divide $\dfrac{3x^4 - 6x^2 + 3x + 6}{x + 1}$.

_____

$$\begin{array}{r} y^2 - y + 1 \\ y+1\overline{)y^3 + 0y^2 + 0y + 1} \end{array}$$

$$\begin{array}{r} y^3 + \ y^2 \\ \hline -y^2 + 0y \\ -y^2 - \ y \\ \hline y + 1 \\ y + 1 \\ \hline 0 \end{array}$$

$$\frac{y^3 + 1}{y + 1} = y^2 - y + 1$$

4. Divide $\dfrac{3x^4 - 2x^3 - 2x^2 - 2x - 1}{x^2 - 1}$.

Note that the divisor is missing the x term. Use 0 as the coefficient for each the term. Thus, $x^2 - 1 = x^2 + 0x - 1$.

$$\begin{array}{r} 3x^2 - 2x + 1 \\ x^2 + 0x - 1\overline{)3x^4 - 2x^3 - 2x^2 - 2x - 1} \end{array}$$

$$\begin{array}{r} 3x^4 + 0x^3 - 3x^2 \\ \hline -2x^3 + \ x^2 - 2x \\ -2x^3 + 0x^2 + 2x \\ \hline x^2 - 4x - 1 \\ x^2 + 0x - 1 \\ \hline -4x \end{array}$$

$$\frac{3x^4 - 2x^3 - 2x^2 - 2x - 1}{x^2 - 1}$$
$$= 3x^2 - 2x + 1 - \frac{4x}{x^2 - 1}$$

4. Divide
$$\frac{6x^4 - 12x^3 + 13x^2 - 5x - 1}{2x^2 + 3}.$$

_____

5. Divide $\dfrac{4x^3 - 5x^2 + x - 16}{2x - 4}$.

$$2x - 4 \overline{\smash{\big)}\, 4x^3 - 5x^2 + x - 16} \quad \begin{array}{l} 2x^2 + \frac{3}{2}x + \frac{7}{2} \end{array}$$

$$\underline{4x^3 - 8x^2}$$
$$3x^2 + x$$
$$\underline{3x^2 - 6x}$$
$$7x - 16$$
$$\underline{7x - 14}$$
$$-2$$

$$\dfrac{4x^3 - 5x^2 + x - 16}{2x - 4} = 2x^2 + \dfrac{3}{2}x + \dfrac{7}{2} - \dfrac{2}{2x - 4}$$

5. Divide $\dfrac{24x^3 + 16x^2 - 6x - 9}{6x + 3}$.

_____

Review this example for Objective 3:

6. For $f(x) = 6x^2 + 13x - 5$ and $g(x) = 3x - 1$, find $\left(\dfrac{f}{g}\right)(x)$ and $\left(\dfrac{f}{g}\right)(-1)$. Find any x-values that are not in the domain of the quotient function.

$$\left(\dfrac{f}{g}\right)(x) = \dfrac{6x^2 + 13x - 5}{3x - 1}$$

$$3x - 1 \overline{\smash{\big)}\, 6x^2 + 13x - 5} \quad \begin{array}{l} 2x + 5 \end{array}$$

$$\underline{-(6x^2 - 2x)}$$
$$15x - 5$$
$$\underline{-(15x - 5)}$$
$$0$$

Thus, $\left(\dfrac{f}{g}\right)(x) = 2x + 5$, $x \neq \frac{1}{3}$. The number $\frac{1}{3}$ is not in the domain because it causes the denominator $g(x) = 3x - 1$ to equal 0.

$$\left(\dfrac{f}{g}\right)(-1) = 2(-1) + 5 = 3$$

Alternatively, we could evaluate $\dfrac{f(-1)}{g(-1)}$.

Now Try:

6. For $f(x) = 8x^2 + 2x - 3$ and $g(x) = 2x - 1$, find $\left(\dfrac{f}{g}\right)(x)$ and $\left(\dfrac{f}{g}\right)(-2)$. Find any x-values that are not in the domain of the quotient function.

_____

Name: Date:

Instructor: Section:

Objective 1 Divide a polynomial by a monomial.

For extra help, see Example 1 on page 302 of your text, the Section Lecture video for Section 5.5, and Exercise Solution Clip 5.

Divide.

1. $\dfrac{45q^3 + 15q^2 + 9q^5}{5q^3}$

1. _____

2. $\dfrac{-12x^6 + 6x^5 + 3x^4 - 9x^3 + 3x}{3x}$

2. _____

3. $\dfrac{2n^5 - 6n^4 + 8n^2}{-2n^3}$

3. _____

Objective 2 Divide a polynomial by a polynomial of two or more terms.

For extra help, see Examples 2–5 on pages 303–305 of your text, the Section Lecture video for Section 5.5, and Exercise Solution Clips 17, 39, 45, and 47.

Divide.

4. $\dfrac{18a^2 - 9a - 5}{3a + 1}$

4. _____

5. $\dfrac{4q^2 - 4q + 5}{2q - 1}$

5. _____

6. $\left(2a^2 - 11a + 16\right) \div \left(2a + 3\right)$

6. _____

7. $\dfrac{12x^3 - 17x^2 + 30x - 10}{3x^2 - 2x + 5}$

7. _____

8. $\dfrac{6x^4 - 12x^3 + 13x^2 - 5x - 1}{2x^2 + 3}$

8. _____

9. $\dfrac{y^4 - 2y^2 + 5}{y^2 - 1}$

9. _____

10. $\dfrac{y^3 + 1}{y + 1}$

10. _____

Name: Date:

Instructor: Section:

Objective 3 Divide polynomial functions.

For extra help, see Example 6 on page 306 of your text, the Section Lecture video for Section 5.5, and Exercise Solution Clip 67.

For each pair of functions, find the quotient $\left(\dfrac{f}{g}\right)(x)$ and give any x-values that are not in the domain of the quotient function.

11. $f(x) = 27x^3 - 8$, $g(x) = 3x - 2$ **11.** _____

12. $f(x) = 2x^3 + 9x^2 + 8x - 7$, $g(x) = 2x + 5$ **12.** _____

13. $f(x) = 2x^4 + 4x^2 - 2x - 7$, $g(x) = x^2 - 2$ **13.** _____

Let $f(x) = 4x^2 - 81$, $g(x) = 3x$, and $h(x) = 2x + 9$. Find each of the following.

14. $\left(\dfrac{g}{h}\right)(-2)$ **14.** _____

15. $\left(\dfrac{f}{g}\right)(0)$ **15.** _____

Chapter 6 FACTORING

6.1   Greatest Common Factors and Factoring by Grouping

Learning Objectives
1   Factor out the greatest common factor.
2   Factor by grouping.

Key Terms

Use the vocabulary terms listed below to complete each statement in exercises 1–3.

factoring   **greatest common factor (GCF)**   **factoring by grouping**

1. _____ is a method of grouping the terms of a polynomial in such a way that the polynomial can be factored even though the greatest common factor of its individual terms is 1.

2. The _____ of a polynomial is the largest term that is a factor of all terms in the polynomial.

3. Writing a polynomial as the product of two or more simpler polynomials is called _____.

Guided Examples

Review these examples for Objective 1:

1. Factor out the greatest common factor.

   a. $42 - 6x$ **b.** $72r + 48$

   c. $14s - 9$

   a. Since 6 is the GCF of 42 and 6,
   $42 - 6x = 6(7 - x)$.

   b. Since 24 is the GCF of 72 and 48,
   $72r + 48 = 24(3r + 2)$.

   c. There is no common factor other than 1.

2. Factor out the greatest common factor.

   a. $16x^2 + 40x^4$

   b. $26x^8 - 13x^{12} - 52x^{10}$

   c. $t^8 - t^5$

   d. $24ab^3 - 8a^2b + 40a^3b^2$

Now Try:

1. Factor out the greatest common factor.

   a. $9x - 36$ _____

   b. $12r + 15$ _____

   c. $6a - 35$

   _____

2. Factor out the greatest common factor.

   a. $-15y^2 + 18y$

   _____

   b. $96a^3 - 48a^2 + 60a$

   _____

a. $16x^2 + 40x^4$

$\quad = 8x^2(2) + 8x^2(5x^2) \quad \text{GCF} = 8x^2$

$\quad = 8x^2(2 + 5x^2)$

b. $26x^8 - 13x^{12} - 52x^{10}$

$\quad = 13x^8(2) + 13x^8(-x^4) + 13x^8(-4x^2)$

$\quad\quad\quad\quad\quad\quad\quad\quad \text{GCF} = 13x^8$

$\quad = 13x^8(2 - x^4 - 4x^2)$

c. $t^8 - t^5 = t^5(t^3) + t^5(-1) \quad \text{GCF} = t^5$

$\quad\quad\quad = t^5(t^3 - 1)$

d. $24ab^3 - 8a^2b + 40a^3b^2$

$\quad = 8ab(3b^2) + 8ab(-a) + 8ab(5a^2b)$

$\quad\quad\quad\quad\quad\quad\quad\quad \text{GCF} = 8ab$

$\quad = 8ab(3b^2 - a + 5a^2b)$

3. Factor out the greatest common factor.

a. $2a(x - 2y) + 9b(x - 2y)$

b. $(x - 1)(3x - 1) + x(3x - 1)$

c. $x^2(2a - 5) + y^2(2a - 5)^2$

a. $2a(x - 2y) + 9b(x - 2y) \quad \text{GCF} = (x - 2y)$

$\quad = (x - 2y)(2a + 9b)$

b. $(x - 1)(3x - 1) + x(3x - 1) \quad \text{GCF} = (3x - 1)$

$\quad = (3x - 1)[(x - 1) + x]$

$\quad = (3x - 1)(2x - 1)$

c. $x^2(2a - 5) + y^2(2a - 5)^2 \quad \text{GCF} = (2a - 5)$

$\quad = (2a - 5)\left[x^2 + y^2(2a - 5)\right]$

$\quad = (2a - 5)\left(x^2 + 2ay^2 - 5y^2\right)$

4. Factor $-2s^5 + 4s^6 + 8s^3$ in two ways.

First, $2s^3$ could be used as the common factor.

$-2s^5 + 4s^6 + 8s^3 = 2s^3\left(-s^2 + 2s^3 + 4\right)$

c. $x^4z^2 + x^6z^2$

_____

d. $14x^3y^2 + 7x^2y - 21x^5y^3$

_____

3. Write in factored form by factoring out the greatest common factor.

a. $x^2(r - 4s) + z^2(r - 4s)$

_____

b. $(b + 5)(3x + 1) - (3x + 1)$

_____

c. $x^2(3r + 4)^3 - y^2(3r + 4)^2$

_____

4. Factor $-12t^5 - 6t^3 + 8t^2$ in two ways.

_____

_____

Alternatively, $-2s^3$ could be used as the common factor.

$$-2s^5 + 4s^6 + 8s^3 = -2s^3\left(s^2 - 2s^3 - 4\right)$$

Review these examples for Objective 2:

5. Factor. $4ax + 4ay + 3bx + 3by$

Group the first two terms and the last two terms since the first two terms have a common factor of $4a$ and the last two terms have a common factor of $3b$.

$4ax + 4ay + 3bx + 3by$
$= (4ax + 4ay) + (3bx + 3by)$
$= 4a(x + y) + 3b(x + y)$
            Factor each group.
$= (x + y)(4a + 3b)$
            Factor out $(x + y)$.

6. Factor. $5rs - 5rt - 2qs + 2qt$

$5rs - 5rt - 2qs + 2qt$
$= (5rs - 5rt) + (-2qs + 2qt)$
$= 5r(s - t) - 2q(s - t)$
            Factor each group.
$= (s - t)(5r - 2q)$ Factor out $(s - t)$.

7. Factor. $1 + p - q - pq$

$1 + p - q - pq$
$= (1 + p) + (-q - pq)$
$= 1(1 + p) - q(1 + p)$ Factor each group.
$= (1 + p)(1 - q)$ Factor out $(1 - q)$.

8. Factor. $21xy - 12b^2 + 14xb - 18by$

We must rearrange the terms in order to find a common factor in each group of two terms.

$21xy - 12b^2 + 14xb - 18by$
$= (21xy + 14xb) + \left(-12b^2 - 18by\right)$
$= 7x(3y + 2b) - 6b(2b + 3y)$
            Factor each group.
$= 7x(2b + 3y) - 6b(2b + 3y)$
            Commutative property
$= (2b + 3y)(7x - 6b)$
            Factor out $(2b + 3y)$.

Now Try:

5. Factor. $4xy - x + 24y - 6$

_____

6. Factor.
$12bc - 4bd - 15xc + 5xd$

_____

7. Factor. $x - 8y^2 + 2xy^2 - 4$

_____

8. Factor.
$9mz - 4nc + 3mc - 12nz$

_____

Objective 1 Factor out the greatest common factor.
For extra help, see Examples 1–4 on pages 320–322 of your text, the Section Lecture video for Section 6.1, and Exercise Solution Clips 1, 7, 21, and 33.

Factor out the greatest common factor. Simplify the factors, if possible.

1. $14x^3y^2 + 7x^2y - 21x^5y^3$

    1. _____

2. $6m^2n^3 + 11p^4r^5$

    2. _____

3. $10a^5 - 15a^4 + 25a^3$

    3. _____

4. $(x+2)(x+y) - (x+3)(x+y)$

    4. _____

5. $(x-1)(3x-1) + x(3x-1)$

    5. _____

6. $(x-1)(x+1) + (x-1)$

    6. _____

7. $9x^3y^2 - 12x^5y + 15x^6y^3$

    7. _____

8. Factor $-14x^3y^2 + 7x^2y - 21x^5y^3$ in two ways.

    8. _____

    _____

Name: Date:

Instructor: Section:

Objective 2 Factor by grouping.
For extra help, see Examples 5−8 on pages 322−324 of your text, the Section Lecture video for Section 6.1, and Exercise Solution Clips 39, 43, 47, and 53.

Factor by grouping.

9. $1 - x + xy - y$

9. _____

10. $x^3 + x^3y^2 - 3y^2 - 3$

10. _____

11. $5rs - 5rt - 2qs + 2qt$

11. _____

12. $14w^2 + 6wx - 35wx - 15x^2$

12. _____

13. $3x^3 + 3xy^2 + 4x^2y + 4y^3$

13. _____

14. $16x^3 + y^3 - 4x^2y^2 - 4xy$

14. _____

15. $x^2 + xy - 2x - 2y$

15. _____

Chapter 6 FACTORING

6.2 Factoring Trinomials

Learning Objectives
1 Factor trinomials when the coefficient of the quadratic term is 1.
2 Factor trinomials when the coefficient of the quadratic term is not 1.
3 Use an alternative method for factoring trinomials.
4 Factor by substitution.

Key Terms

Use the vocabulary terms listed below to complete each statement in exercises 1–5.

> **FOIL**   **product**   **sum**   **prime polynomial**   **common factor**

1. A _____ is a polynomial that cannot be factored using only integer coefficients.

2. When factoring a trinomial of the form $x^2 + bx + c$, find all pairs of integers whose _____ is c, the third term of the trinomial.

3. When factoring, always look for a _____ first.

4. After factoring a trinomial, you can use _____ to check your results.

5. When factoring a trinomial of the form $x^2 + bx + c$, find all pairs of integers whose _____ is b, the coefficient of the middle term.

Guided Examples

Review these examples for Objective 1:
1. Factor each trinomial.

   a. $x^2 + 8x + 15$ **b.** $x^2 - 13x + 30$

   a. We must find two integers whose product is 15 and whose sum is 8.

Factors of 15	Sum of Factors
1, 15	16
3, 5	8

   From the list, 3 and 5 are the required integers since $3 \cdot 5 = 15$ and $3 + 5 = 8$.

   Thus, $x^2 + 8x + 15 = (x + 3)(x + 5)$.

Now Try:
1. Factor each trinomial.

   a. $x^2 + 14x + 24$ _____

   b. $x^2 + 3x - 28$ _____

b. We must find two integers whose product is 30 and whose sum is -13. These integers have a product that is positive and a sum that is negative, so we consider only pairs of negative integers.

Factors of 15	Sum of Factors
$-1, -30$	-31
$-2, -15$	-17
$-3, -10$	-13
$-5, -6$	-11

From the list, -3 and -10 are the required integers since $(-3)(-10) = 30$ and $-3 + (-10) = -13$.

Thus, $x^2 - 13x + 30 = (x - 3)(x - 10)$.

2. Factor $r^2 - r - 5$.

We must find two integers whose product is -5 and whose sum is -1. These integers have a product that is negative, so the pairs of integers must have different signs.

Factors of -5	Sum of Factors
$-1, 5$	4
$1, -5$	-4

Neither of the pairs of integers has a sum of -1. Therefore, $r^2 - r - 5$ cannot be factored by using only integers. It is a *prime polynomial*.

3. Factor $r^2 + 4rs - 21s^2$.
We must find two expressions whose product is $-21s^2$ and whose sum is $4s$.

Factors of $-21s^2$	Sum of Factors
$-s, 21s$	$20s$
$s, -21s$	$-20s$
$-3s, 7s$	$4s$
$3s, -7s$	$-4s$

Thus, $r^2 + 4rs - 21s^2 = (r - 3s)(r + 7s)$.

2. Factor $x^2 + 14x - 49$.

_____

3. Factor $p^2 - 7mp + 12m^2$.

_____

4. Factor $3p^6 - 18p^5 + 24p^4$

First factor out the greatest common factor, $3p^4$.

$$3p^6 - 18p^5 + 24p^4 = 3p^4\left(p^2 - 6p + 8\right)$$

Now factor $p^2 - 6p + 8$. The integers -2 and -4 have a product of 8 and a sum of -6.

$$3p^6 - 18p^5 + 24p^4 = 3p^4\left(p^2 - 6p + 8\right)$$
$$= 3p^4(p - 2)(p - 4)$$

4. Factor $2s^2t - 16st - 40t$.

_____

Review this example for Objective 2:

5. Factor $2x^2 + 5x - 3$.

We must find two integers whose product is $2(-3) = -6$ and whose sum is 5.

Integers	Product	Sum
$-1, 6$	-6	5
$1, -6$	-6	-5
$-2, 3$	-6	1
$2, -3$	-6	-1

The integers are -1 and 6. Write the middle term as $-x + 6x$, then factor by grouping.

$$2x^2 + 5x - 3 = 2x^2 - x + 6x - 3$$
$$= \left(2x^2 - x\right) + (6x - 3)$$
$$= x(2x - 1) + 3(2x - 1)$$
$$= (2x - 1)(x + 3)$$

Now Try:

5. Factor $3m^2 - 5m - 12$.

_____

Review these examples for Objective 3:

6. Factor each trinomial.

a. $2x^2 + 11x + 5$.   **b.** $4x^2 - 12x + 5$

a. We use the FOIL method in reverse. We want to write $2x^2 + 11x + 5$ as the product of two binomials.

$$2x^2 + 11x + 5 = (\underline{\quad})(\underline{\quad})$$

The product of the first two terms of the binomials is $2x^2$.

Now Try:

6. Factor each trinomial.

a. $3y^2 + 13y + 4$

_____

b. $5x^2 - 18x + 9$

_____

Since all coefficients in the trinomial are positive, consider only positive factors. The factors of $2x^2$ are $2x$ and x.

$$2x^2 + 11x + 5 = (2x + \underline{\quad})(x + \underline{\quad})$$

The factors of the last term, 5, are 1 and 5 or 5 and 1. Try each pair to find the pair that gives the correct middle term, $11x$.

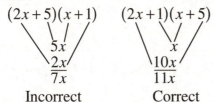

Thus, $2x^2 + 11x + 5 = (2x+1)(x+5)$.

b. The only factors of 5 are 1 and 5 or -1 and -5, so it is easier to begin by factoring 5. The middle term has a negative coefficient and the last term is positive, so consider only negative factors.

$$4x^2 - 12x + 5 = (\underline{\quad} - 1)(\underline{\quad} - 5)$$

The factors of the first term, $4x^2$, are x and $4x$ or $2x$ and $2x$. Try each pair to find the pair that gives the correct middle term, $11x$.

$(x-1)(4x-5)$ $(x-5)(4x-1)$

$-4x$ $-20x$

$\underline{-5x}$ $\underline{-x}$

$-9x$ $-21x$

Incorrect   Incorrect

$(2x-1)(2x-5)$

$-2x$

$\underline{-10x}$

$-12x$

Correct

Thus, $4x^2 - 12x + 5 = (2x-1)(x-5)$.

7. Factor $10x^2 - 29xy + 21y^2$.

The factors of $10x^2$ are x and $10x$ or $2x$ and $5x$. The coefficient of the middle term is negative and the coefficient of the third term is positive, so we need the factors of the third term that are both negative. The factors of $21y^2$ are $-y$ and $-21y$ or $-3y$ and $-7y$.

$10x^2 - 29xy + 21y^2 = (5x - 7y)(2x - 3y)$

7. Factor $40z^2 + 18zy - 9y^2$.

_____

8. Factor $-3x^2 + 11x + 4$.

First, factor out -1 so that the coefficient of the x^2-term is positive.

$$-3x^2 + 11x + 4 = -1\left(3x^2 - 11x - 4\right)$$
$$= -1(3x + 1)(x - 4)$$
$$= -(3x + 1)(x - 4)$$

8. Factor $-15r^2 + 2r + 24$.

_____

Review these examples for Objective 4:

9. Factor $9x^3 - 30x^2 + 24x$.

First, factor out the common factor, $3x$.

$$9x^3 - 30x^2 + 24x = 3x\left(3x^2 - 10x + 8\right)$$

Now factor $3x^2 - 10x + 8$.

$$9x^3 - 30x^2 + 24x = 3x\left(3x^2 - 10x + 8\right)$$
$$= 3x(3x - 4)(x - 2)$$

Now Try:

9. Factor $30a^4b - 3a^3b - 6a^2b$.

_____

10. Factor $8(p + 5)^2 + 2(p + 5) - 15$.

Since the binomial $p + 5$ appears to powers 2 and 1, let the substitution variable t represent $p + 5$

$8(p + 5)^2 + 2(p + 5) - 15$

$\quad = 8t^2 + 2t - 15 \qquad$ Let $t = p + 5$.

$\quad = (2t + 3)(4t - 5) \quad$ Factor.

$\quad = \left[2(p + 5) + 3\right]\left[4(p + 5) - 5\right]$

$\qquad\qquad\qquad$ Replace t with $p + 5$.

$\quad = (2p + 10 + 3)(4p + 20 - 5)$

$\qquad\qquad\qquad$ Simplify.

$\quad = (2p + 13)(4p + 15)$

10. Factor $8(5 - z)^2 - 14(5 - z) + 3$.

_____

11. Factor $18x^4 - x^2 - 5$.

The variable y appears to powers in which the greater exponent is twice the lesser exponent. We can let a substitution variable represent the variable to the lesser power.

Here, we let $t = x^2$.

$18x^4 - x^2 - 5$

$$= 18\left(x^2\right)^2 - x^2 - 5 \qquad x^4 = \left(x^2\right)^2$$

$$= 18t^2 - t - 5 \qquad \text{Substitute } t \text{ for } x^2.$$

$$= (2t + 1)(9t - 5) \qquad \text{Factor.}$$

$$= \left(2x^2 + 1\right)\left(9x^2 - 5\right) \qquad \text{Replace } t \text{ with } x^2.$$

11. Factor $4y^4 + 8y^2 - 45$.

_____

Objective 1 Factor trinomials when the coefficient of the quadratic term is 1.
For extra help, see Examples 1–4 on pages 327–328 of your text, the Section Lecture video for Section 6.2, and Exercise Solution Clips 5, 9, and 11.

Factor.

1. $x^2 + 3x - 28$

1. _____

2. $r^2 - 6r - 16$

2. _____

3. $x^2 + 8xy + 15y^2$

3. _____

4. $m^2 p^2 - 7mp + 12$

4. _____

    215

Objectives 2 and 3 Factor trinomials when the coefficient of the quadratic term is not 1. Use an alternative method for factoring trinomials.

For extra help, see Examples 5–9 on pages 328–331 of your text, the Section Lecture video for Section 6.2, and Exercise Solution Clips 19, 21, 23, 31, 33, and 39.

Factor.

5. $3z^2 + 2z - 8$

5. _____

6. $7x^2 + 8x + 3$

6. _____

7. $15q^2 - 7q - 4$

7. _____

8. $20x^2 + 39x - 11$

8. _____

9. $8x^2 + 2x - 15$

9. _____

10. $6p^2 - p - 15$

10. _____

11. $6y^2 - 5yz - 6z^2$

11. _____

Objective 4 Factor by substitution.

For extra help, see Examples 10 and 11 on page 331 of your text, the Section Lecture video for Section 6.2, and Exercise Solution Clips 49, and 59.

Factor completely using substitution.

12. $5(a-1)^2 - 7(a-1) - 6$ **12.** _____

13. $3(p-q)^2 + 10(p-q) + 7$ **13.** _____

14. $10(r+s)^2 - (r+s) - 24$ **14.** _____

15. $24m^4 - 37m^2 - 5$ **15.** _____

Chapter 6 FACTORING

6.3   Special Factoring

Learning Objectives
1   Factor a difference of squares.
2   Factor a perfect square trinomial.
3   Factor a difference of cubes.
4   Factor a sum of cubes.

Key Terms

Use the vocabulary terms listed below to complete each statement in exercises 1–4.

   difference of squares   **perfect square trinomial**   **difference of cubes**

   sum of cubes

1.   The _____, $x^2 - y^2$, can be factored as the product of the sum and
     difference of two terms, or $x^2 - y^2 = (x + y)(x - y)$.

2.   The _____, $x^3 + y^3$, can be factored as
     $x^3 + y^3 = (x + y)(x^2 - xy + y^2)$.

3.   A _____ is a trinomial that can be factored as the square of a
     binomial.

4.   The _____, $x^3 - y^3$, can be factored as
     $x^3 - y^3 = (x - y)(x^2 + xy + y^2)$.

Guided Examples

Review this example for Objective 1:
1.   Factor each polynomial.

   a. $x^2 - 25$ **b.** $98r^2 - 200$

   c. $121m^2 - 9n^2$ **d.** $(r - s)^2 - 9z^2$

   e. $m^4 - 625$

   The pattern to factor the difference of
   squares is $x^2 - y^2 = (x + y)(x - y)$.

Now Try:
1.   Factor out the greatest common
     factor.

   a. $y^2 - 16$ _____

   b. $25a^2 - 36$ _____

   c. $81x^4 - 900x^2$

   _____

a. $x^2 - 25 = (x+5)(x-5)$

b. $98r^2 - 200 = 2(49r^2 - 100)$

Factor out the GCF, 2.

$$= 2(7r+10)(7r-10)$$

c. $121m^2 - 9n^2 = (11m+3n)(11m-3n)$

d. $(r-s)^2 - 9z^2$

$$= \left[(r-s)-3z\right]\left[(r-s)+3z\right]$$
$$= (r-s-3z)(r-s+3z)$$

We could have used the method of substitution here.

e. $m^4 - 625 = (m^2+25)(m^2-25)$

Factor the difference of squares.

$$= (m^2+25)(m+5)(m-5)$$

Factor the difference of squares.

d. $(a+b)^2 - 36b^2$

_____

e. $100x^4 y^2 - 81z^2$

_____

Review this example for Objective 2:

2. Factor each polynomial.

   a. $100x^2 + 180x + 81$

   b. $9x^2 - 42xy + 49y^2$

   c. $(p-q)^2 - 20(p-q) + 100$

   d. $4x^2 + 12x + 9 - 4y^2$

   There are two patterns to factor perfect square trinomials:

   $$x^2 + 2xy + y^2 = (x+y)^2$$
   $$x^2 - 2xy + y^2 = (x-y)^2$$

   a. $100x^2 + 180x + 81 = (10x+9)^2$

   b. $9x^2 - 42xy + 49y^2 = (3x-7y)^2$

   c. $(p-q)^2 - 20(p-q) + 100$

   $$= \left[(p-q)-10\right]^2$$
   $$= (p-q-10)^2$$

Now Try:

2. Factor each polynomial.

   a. $16x^2 - 40x + 25$

   _____

   b. $16a^2 + 72ab + 81b^2$

   _____

   c. $(m-n)^2 - 12(m-n) + 36$

   _____

   d. $x^2 - 4x + 4 - 9y^2$

   _____

d. $4x^2 + 12x + 9 - 4y^2$

$\quad = \left(4x^2 + 12x + 9\right) - 4y^2$

$\quad = (2x+3)^2 - 4y^2$ Factor the perfect
square trinomial.

$\quad = (2x + 3 + 2y)(2x + 3 - 2y)$
Factor the difference of squares.

Review this example for Objective 3:	**Now Try:**

3. Factor each polynomial.

a. $m^3 - 27$

b. $8p^3 - 216$

c. $2x^3 - 128y^3$

The pattern to factor the difference of cubes

is $x^3 - y^3 = (x - y)\left(x^2 + xy + y^2\right)$.

a. $m^3 - 27 = m^3 - 3^2$

$\quad = (m-3)\left(m^2 + 3m + 9\right)$

b. $8p^3 - 216 = 8\left(p^3 - 27\right)$

$\qquad$ Factor out the GCF, 8.

$\quad = 8(p-3)\left(p^2 + 3p + 9\right)$

$\qquad$ Factor the difference of cubes.

c. $2x^3 - 128y^3 = 2\left(x^3 - 64y^3\right)$

$\qquad$ Factor out the GCF, 2.

$\quad = 2\left(x^3 - 4^3 y^3\right)$

$\quad = 2(x - 4y) \cdot$

$\qquad\qquad \left(x^2 + 8xy + 16y\right)$

3. Factor each polynomial.

a. $x^3 - 343$

_____

b. $375 - 81y^3$

_____

c. $250x^4 - 128xy^3$

_____

Review this example for Objective 4:	**Now Try:**

4. Factor each polynomial.

a. $q^3 + 8$

b. $250p^3 + 128r^3$

c. $1029x^4 + 648x$

d. $(x + 4)^3 + y^3$

4. Factor each polynomial.

a. $27x^3 + 125$

_____

b. $8x^3 + 64b^3$

_____

c. $u^3 v^6 + 216$

_____

The pattern to factor the sum of cubes is
$$x^3 + y^3 = (x+y)\left(x^2 - xy + y^2\right).$$

a. $q^3 + 8 = q^3 + 2^2$
$$= (q+2)\left(q^2 - 2q + 4\right)$$

b. $250p^3 + 128r^3$
$$= 2\left(125p^3 + 64r^3\right)$$

Factor out the GCF, 2.
$$= 2\left[(5p)^3 + (4r)^3\right]$$
$$= 2(5p + 4r)\left(25p^2 - 20pr + 16r^2\right)$$

c. $1029x^4 + 648x$
$$= 3x\left(343x^3 + 216\right)$$

Factor out the GCF, $3x$.
$$= 3x\left[(7x)^3 + 6^3\right]$$
$$= 3x(7x + 6)\left(49x^2 - 42x + 36\right)$$

d. $(x+4)^3 + y^3$
$$= \left[(x+4) + y\right] \cdot$$
$$\left[(x+4)^2 - (x+4)y + y^2\right]$$

Sum of cubes
$$= (x+4+y) \cdot$$
$$\left(x^2 + 8x + 16 - xy - 4y + y^2\right)$$

Multiply.

d. $(a-1)^3 + a^3$

_____

Objective 1 Factor a difference of squares.
For extra help, see Example 1 on pages 333–334 of your text, the Section Lecture video for
Section 6.3, and Exercise Solution Clip 7.

Factor.

1. $36z^2 - 121$

1. _____

2. $144x^2 - 25y^2$

2. _____

3. $s^4 - 16$

3. _____

4. $q^2 - (2r+3)^2$

4. _____

Objective 2 Factor a perfect square trinomial.

For extra help, see Example 2 on pages 334–335 of your text, the Section Lecture video for Section 6.3, and Exercise Solution Clip 23.

Factor the square trinomial.

5. $16t^2 + 56t + 49$

5. _____

6. $36r^2 - 60rs + 25s^2$

6. _____

7. $(y+z)^2 + 14(y+z) + 49$

7. _____

Objective 3 Factor a difference of cubes.

For extra help, see Example 3 on pages 335–336 of your text, the Section Lecture video for Section 6.3, and Exercise Solution Clip 37.

Factor.

8. $64a^3 - 343b^3$

8. _____

9. $8a^3 - 125b^3$

9. _____

10. $(r+s)^3 - 1$

10. _____

11. $(m+n)^3 - (m-n)^3$

11. _____

Objective 4 Factor a sum of cubes.

For extra help, see Example 4 on pages 336–337 of your text, the Section Lecture video for Section 6.3, and Exercise Solution Clip 41.

Factor.

12. $8a^3 + 64b^3$

12. _____

13. $125p^3 + q^3$

13. _____

14. $1 + (y+z)^3$

14. _____

15. $t^3 + (t+2)^3$

15. _____

Chapter 6 FACTORING

6.4   A General Approach to Factoring

Learning Objectives
1   Factor out any common factor.
2   Factor binomials.
3   Factor trinomials.
4   Factor polynomials of more than three terms.

Key Terms

Use the vocabulary terms listed below to complete each statement in exercises 1–3.

factor out a common factor   difference of squares   perfect square trinomial

1.   When factoring a binomial, check to see if it is a _____.

2.   The first step when factoring any polynomial is to _____ if there is one.

3.   When factoring a trinomial, check to see if it is a _____.

Guided Examples

Review this example for Objective 1:
1.   Factor each polynomial.

   a. $12a^2b^2 + 3a^2b - 9ab^2$

   b. $2m(m-n) - (m+n)(m-n)$

   a. $12a^2b^2 + 3a^2b - 9ab^2$
        $= 3ab(4ab + a - 3b)$ The GCF is $3ab$.

   b. $2m(m-n) - (m+n)(m-n)$
        $= (m-n)\left[2m - (m+n)\right]$
                        The GCF is $(m-n)$.
        $= (m-n)(m-n)$ Simplify.
        $= (m-n)^2$

Now Try:
1.   Factor each polynomial.

   a. $5r^2t - 10rt + 5rt^2$

        _____

   b. $(x+1)(2x+3) - (x+1)$

        _____

Review this example for Objective 2:
2.   Factor each binomial, if possible.

   a. $16s^4 - r^4$ b. $x^9 + 27$

   c. $25x^2 + 49$

Now Try:
2.   Factor each binomial.

   a. $128x^3 - 2y^3$

        _____

a. $16s^4 - r^4 = \left(4s^2 + r^2\right)\left(4s^2 - r^2\right)$

          Difference of squares

        $= \left(4s^2 + r^2\right)(2s + r)(2s - r)$

          Difference of squares

b. $x^9 + 27 = \left(x^3 + 3\right)\left(x^6 - 3x^3 + 9\right)$

          Sum of cubes

c. $25x^2 + 49$ is prime. It is a *sum* of squares.

b. $(a - b)^2 - (a + b)^2$

_____

c. $25x^2 + 169$

_____

Review this example for Objective 3:

3. Factor each trinomial.

  a. $12m^2 + 11m - 5$

  b. $25x^2 - 5xy - 2y^2$

  c. $(a + 1)^2 - 6(a + 1) + 9$

  a. $12m^2 + 11m - 5 = (4m + 5)(3m - 1)$

  b. $25x^2 - 5xy - 2y^2 = (5x - 2y)(5x + y)$

  c. Let $t = a + 1$.

    $(a + 1)^2 - 6(a + 1) + 9$

    $= t^2 - 6t + 9$ Replace $(a + 1)$ with t.

    $= (t - 3)^2$ Perfect square trinomial

    $= \left[(a + 1) - 3\right]^2$ Replace t with $(a + 1)$.

    $= (a - 2)^2$ Simplify.

Now Try:

3. Factor each trinomial.

  a. $2a^2 - 17a + 30$

  _____

  b. $4x^2 + 12xy + 9y^2$

  _____

  c.

  $4b^2(b + 2) - 3b(b + 2) - (b + 2)$

  _____

Review this example for Objective 4:

4. Factor each polynomial.

  a. $bx^2 - by^2 - ax + ay$

  b. $x^2 + 7xy + 2x + 14y$

  c. $x^3 - 64y^3 - x^2 + 16y^2$

Now Try:

4. Factor each polynomial.

  a. $st + 2s + 3t + 6$

  _____

  b. $r^3 - s^3 - sr^2 + rs^2$

  _____

  c. $8 - 6k^3 - 12k + 9k^4$

  _____

a. $bx^2 - by^2 - ax + ay$

$= \left(bx^2 - by^2\right) - \left(ax - ay\right)$

Group the terms.

$= b\left(x^2 - y^2\right) - a\left(x - y\right)$

Factor each group.

$= b\left(x + y\right)\left(x - y\right) - a\left(x - y\right)$

Difference of squares

$= \left(x - y\right)\left[b\left(x + y\right) - a\right]$

$x - y$ is a common factor.

b. $x^2 + 7xy + 2x + 14y$

$= x^2 + 2x + 7xy + 14y$

Rearrange the terms.

$= \left(x^2 + 2x\right) + \left(7xy + 14y\right)$

Group the terms.

$= x\left(x + 2\right) + 7y\left(x + 2\right)$

Factor each group.

$= \left(x + 2\right)\left(x + 7y\right)$

$x + 2$ is a common factor.

c. $x^3 - 64y^3 - x^2 + 16y^2$

$= \left(x^3 - 64y^3\right) - \left(x^2 - 16y^2\right)$

$= \left(x - 4y\right)\left(x^2 + 4xy + 16y^2\right)$

$\quad\quad\quad - \left(x - 4y\right)\left(x + 4y\right)$

$= \left(x - 4y\right) \cdot$

$\quad \left[\left(x^2 + 4xy + 16y^2\right) - \left(x + 4y\right)\right]$

$= \left(x - 4y\right)\left(x^2 + 4xy + 16y^2 - x - 4y\right)$

Objectives 1 and 2 Factor out any common factor. Factor binomials.
For extra help, see Example 1 on page 340 of your text, the Section Lecture video for Section 6.4, and Exercise Solution Clip 13.

Factor each polynomial.

1. $12a^3b^3 + 6a^2b^5$

1. _____

2. $3a^2 + 6a(x-y)$

2. _____

3. $(2q+1)(q-6) - (2q+1)(q+5)$

3. _____

Objective 2 Factor binomials.
For extra help, see Example 2 on page 340 of your text, the Section Lecture video for Section 6.4, and Exercise Solution Clip 7.

Factor each binomial.

4. $27y^3 - 10z^3$

4. _____

5. $a^6 - a^4$

5. _____

6. $2x^3 + 16y^3$

6. _____

7. $(s+2)^2 - (s-2)^2$

7. _____

Objective 3 Factor trinomials.
For extra help, see Example 3 on page 341 of your text, the Section Lecture video for Section 6.4, and Exercise Solution Clip 9.

Factor each trinomial.

8. $3n^4 - 11n^2 - 4$

8. _____

9. $3s^2 + 7st + 2t^2$

9. _____

10. $4a^3 + 24a^2 - 64a$

10. _____

11. $6(p-1)^2 + (p-1) - 2$

11. _____

Objective 4 Factor polynomials of more than three terms.
For extra help, see Example 4 on pages 341–342 of your text, the Section Lecture video for Section 6.4, and Exercise Solution Clip 21.

Factor each polynomial.

12. $2r^3 - r^2 - 2r + 1$

12. _____

13. $3ax + 3ay + 5ax + 5ay$

13. _____

14. $4s^3 - 8s^2 - 9s + 18$

14. _____

15. $xy - 9y + xz - 9z$

15. _____

Chapter 6 FACTORING

6.5   Solving Equations by Factoring

> **Learning Objectives**
> 1   Learn and use the zero-factor property.
> 2   Solve applied problems that require the zero-factor property.
> 3   Solve a formula for a specified variable, where factoring is necessary.

Key Terms

Use the vocabulary terms listed below to complete each statement in exercises 1–3.

> **zero-factor property**   **quadratic equation**   **standard form**

1.   The _____ states that if two numbers have a product of 0, then at least one of the numbers must be 0.

2.   An equation written in the form $ax^2 + bx + c = 0$ is written in the _____ of a quadratic equation.

3.   A _____ is an equation that can be written in the form $ax^2 + bx + c = 0$, for real numbers a, b, and c, with $a \neq 0$.

Guided Examples

Review these examples for Objective 1:

1.   Solve: $(3x + 7)(x - 4) = 0$

Since $(3x + 7)(x - 4) = 0$, the zero-factor property tells us that at least one of the factors equals 0. Therefore, either $3x + 7 = 0$ or $x - 4 = 0$.

$$3x + 7 = 0 \quad \text{or} \quad x - 4 = 0$$
$$3x = -7 \qquad\qquad x = 4$$
$$x = -\frac{7}{3}$$

Check each proposed solution in the original equation.

$$(3x + 7)(x - 4) = 0$$
$$\left[3\left(-\tfrac{7}{3}\right) + 7\right]\left(-\tfrac{7}{3} - 4\right) = 0$$
$$0\left(-\tfrac{19}{3}\right) = 0$$
$$0 = 0 \ \checkmark$$

Now Try:

1.   Solve: $p(5p + 25) = 0$

_____

$$(3x+7)(x-4)=0$$
$$[3(4)+7](4-4)=0$$
$$19(0)=0$$
$$0=0 \checkmark$$

The solution set is $\left\{-\frac{7}{3}, 4\right\}$.

2. Solve each equation.

   a. $z^2 = 7z - 12$

   b. $12x^2 - 12 = -7x$

   a. First, rewrite the equation in standard form.

   $$z^2 - 7z + 12 = 0$$

   Now factor $z^2 - 7z + 12$ and then apply the zero-factor property to solve the equation.

   $$z^2 - 7z + 12 = 0$$
   $$(z-3)(z-4)=0$$

   $$z - 3 = 0 \quad \text{or} \quad z - 4 = 0$$
   $$z = 3 \qquad\qquad z = 4$$

   Be sure to check each proposed solution in the original equation.
   The solution set is $\{3, 4\}$.

   b. $$12x^2 - 12 = -7x$$
   $$12x^2 + 7x - 12 = 0$$
   $$(3x+4)(4x-3)=0$$

   $$3x + 4 = 0 \quad \text{or} \quad 4x - 3 = 0$$
   $$3x = -4 \qquad\qquad 4x = 3$$
   $$x = -\frac{4}{3} \qquad\qquad x = \frac{3}{4}$$

   Be sure to check each proposed solution in the original equation.

   The solution set is $\left\{-\frac{4}{3}, \frac{3}{4}\right\}$.

2. Solve each equation.

   a. $a^2 = -18 - 9a$

   _____

   b. $14x^2 - 6 = 17x$

   _____

3. Solve: $25x^2 - 20x = 0$

This quadratic equation has a missing term. Comparing it with the standard form

$ax^2 + bx + c$ shows that $c = 0$. The zero-factor property can still be used.

$25x^2 - 20x = 0$

$5x(5x - 4) = 0$ Factor.

$5x = 0$ or $5x - 4 = 0$ Zero factor property

$x = 0$ $5x = 4$ Solve.

$x = \frac{4}{5}$

Be sure to check each proposed solution in the original equation.

The solution set is $\left\{0, \frac{4}{5}\right\}$.

3. Solve: $24r = -8r^2$

_____

4. Solve: $36z^2 - 25 = 0$

$36z^2 - 25 = 0$

$(6z - 5)(6z + 5) = 0$ Factor the difference of squares.

$6z - 5 = 0$ or $6z + 5 = 0$ Zero factor property

$6z = 5$ $6z = -5$ Solve.

$z = \frac{5}{6}$ $z = -\frac{5}{6}$

Be sure to check each proposed solution in the original equation.

The solution set is $\left\{\frac{5}{6}, -\frac{5}{6}\right\}$.

4. Solve: $9a^2 = 49$

_____

5. Solve: $(2c + 1)(3c - 5) = c(c - 24) + 7$

$(2c + 1)(3c - 5) = c(c - 24) + 7$

$6c^2 - 7c - 5 = c^2 - 24c + 7$

Multiply on each side.

$5c^2 + 17c - 12 = 0$ Subtract c^2; add 24c; subtract 7.

$(5c - 3)(c + 4) = 0$ Factor.

$5c - 3 = 0$ or $c + 4 = 0$ Zero factor property

$5c = 3$ $c = -4$ Solve.

$c = \frac{3}{5}$

5. Solve: $3x(x + 3) = (x + 2)^2 - 1$

_____

Be sure to check each proposed solution in the original equation.

The solution set is $\left\{\frac{3}{5}, -4\right\}$.

6. Solve: $4x^3 - 64x = 0$

$$4x^3 - 64x = 0$$

$4x\left(x^2 - 16\right) = 0$ Factor out $4x$.

$4x(x+4)(x-4) = 0$ Factor the difference of squares.

Now apply the zero-factor property.

$4x = 0$ or $x + 4 = 0$ or $x - 4 = 0$

   $x = 0$ $x = -4$ $x = 4$

Check each solution to verify that the solution set is $\{0, -4, 4\}$.

6. Solve: $3x^3 + 15x^2 - 42x = 0$

_____

Review these examples for Objective 2:

7. The top of a rectangular table has an area of 63 square feet. It has a length that is 2 feet more than the width. Find the width of the table top.

Step 1   Read the problem. We need to find the width of the table top.

Step 2   Assign a variable.
Let w = the width of the table top.
Then $w + 2$ = the length of the table top.

Step 3   Write an equation. The area of a rectangle is given by
$\mathcal{A} = $ length $\times$ width, so for this problem, we have $\mathcal{A} = 63 = (w + 2)w$.

Step 4   Solve.

$$63 = (w + 2)w$$

$63 = w^2 + 2w$ Multiply.

$w^2 + 2w - 63 = 0$ Standard form

$(w + 9)(w - 7) = 0$ Factor.

$w + 9 = 0$ or $w - 7 = 0$ Zero factor property

   $w = -9$ $w = 7$ Solve.

Now Try:

7. The area of a triangle is 42 square centimeters. The base is 2 centimeters less than twice the height. Find the base and height of the triangle.

   base   _____

   height _____

Step 5 State the answer. The solutions are −9 and 7. Because a rectangle cannot have a side of negative length, we discard the solution −9. The width is 7 feet.

Step 6 Check. The width is 7 feet, so the length is 2 feet more or 9 feet. The area is 7(9) = 63 sq ft, as required.

8. A projectile is thrown with an upward velocity of 160 feet per second from the top of a building that is 500 feet tall. The height of the baseball after t seconds is modeled by the equation $h = -16t^2 + 160t + 500$. How long will it take for the projectile to reach the ground?

When the ball reaches the ground, $h = 0$, so substitute 0 for h in the equation and solve for t.

$$0 = -16t^2 + 160t + 500$$

$-16t^2 + 160t + 500 = 0$ Interchange sides.

$-4\left(4t^2 - 40t - 125\right) = 0$ Factor out − 4.

$4t^2 - 40t - 125 = 0$ Divide by − 4.

$(2t + 5)(2t - 25) = 0$ Factor.

$2t + 5 = 0$ or $2t - 25 = 0$ Zero factor property

$2t = -5$ $2t = 25$ Solve.

$t = -\dfrac{5}{2}$ $t = \dfrac{25}{2}$

Since time cannot be negative, we discard the solution $t = -\dfrac{5}{2}$. Thus, the projectile will reach the ground after $\dfrac{25}{2} = 12.5$ seconds.

8. Jeff threw a stone straight upward at 46 feet per second from a dock 6 feet above a lake. The height of the stone above the lake t seconds after it is thrown is given by $h = -16t^2 + 46t + 6$. How long will it take for the stone to reach a height of 39 feet?

_____

Review this example for Objective 3:

9. Solve $A = \frac{1}{2}b_1h + \frac{1}{2}b_2h$ for b_1.

To solve for b_1, treat b_1 as the only variable and treat all other variables as constants.

$$A = \frac{1}{2}b_1h + \frac{1}{2}b_2h$$

$$A = \frac{1}{2}h(b_1 + b_2) \quad \text{Factor.}$$

$$\frac{2A}{h} = b_1 + b_2 \quad \begin{array}{l}\text{Multiply by 2.}\\ \text{Divide by } h.\end{array}$$

$$\frac{2A}{h} - b_2 = b_1 \quad \text{Subtract } b_2.$$

$$\frac{2A - b_2h}{h} = b_1 \quad \text{Combine terms.}$$

Now Try:

9. Solve $3 + \dfrac{a+b}{a-b} = 6a$ for b

_____

Objective 1 Learn and use the zero-factor property.
For extra help, see Examples 1–6 on pages 344–347 of your text, the Section Lecture video for Section 6.5, and Exercise Solution Clips 5, 11, 19, 23, 35, and 39.

Solve each equation.

1. $9s^2 - 18s = 0$

1. _____

2. $6z^2 = -19z - 10$

2. _____

3. $24a^2 + 22a - 10 = 0$

3. _____

4. $r(6r + 11) = -3$

4. _____

5. $8x^2 + 2x = 15$

5. _____

6. $4x^3 - 9x = 8x^2 - 18$

6. _____

7. $(3x+1)(4x+1) = (7x+1)(x-1) - 4$

7. _____

8. $15x^2 = x^3 + 56x$

8. _____

Objective 2 Solve applied problems that require the zero-factor property.
For extra help, see Examples 7 and 8 on pages 347–348 of your text, the Section Lecture video for Section 6.5, and Exercise Solution Clips 57 and 65.

Solve each problem.

9. If the square of the sum of two consecutive integers is reduced by twice their product, the result is 25. Find the integers.

9. _____

10. The Browns installed 96 feet of fencing around a rectangular play yard. If the yard covers 540 square feet, what are its dimensions?

10. _____

11. A company determines that its daily revenue R (in dollars) for selling x items is modeled by the equation $R = x(150 - x)$. How many items must be sold for its revenue to be $4400?

11. _____

12.   A ball is dropped from the roof of a 19.6 meter high building. Its height h (in meters) t seconds later is given by the equation $h = -4.9t^2 + 19.6$.

     (a) After how many seconds is the height 18.375 meters?

     (b) After how many seconds is the height 14.7 meters?

     (c) After how many seconds does the ball hit the ground?

12. a._____

    b. _____

    c._____

Objective 3 Solve a formula for a specified variable, where factoring is necessary.
For extra help, see Example 9 on page 349 of your text, the Section Lecture video for Section 6.5, and Exercise Solution Clip 69.

Solve each equation for the specified variable.

13. $axy + 25 = -bx - 20y$ for y

13. _____

14. $d - 7ac = 3a + 5d$ for a

14. _____

15. $\dfrac{2y + 5}{8 - 2y} = x$ for y

15. _____

Chapter 7 RATIONAL EXPRESSIONS AND FUNCTIONS

7.1   Rational Expressions and Functions; Multiplying and Dividing

---

Learning Objectives
1   Define rational expressions.
2   Define rational functions and describe their domains.
3   Write rational expressions in lowest terms.
4   Multiply rational expressions.
5   Find reciprocals of rational expressions.
6   Divide rational expressions.

Key Terms

Use the vocabulary terms listed below to complete each statement in exercises 1–5.

rational expression   rational function   opposites   reciprocal

fundamental property of rational numbers

1.   To divide two rational expressions, multiply the first (the dividend) by the
      _____ of the second (the divisor).

2.   In general, if the numerator and the denominator of a rational expression are
      _____, then the expression equals –1.

3.   A function that is defined by a quotient of polynomials is called a(n)
      _____.

4.   The _____ states that the numerator and denominator of a rational
      number may either be multiplied or divided by the same nonzero number without
      changing the value of the rational number.

5.   The quotient of two polynomials with denominator not 0 is called a
      _____, or algebraic fraction.

Guided Examples

Review this example for Objective 2:

1. For each rational function, find all numbers that are not in the domain. Then give the domain, using set-builder notation.

   a. $f(x) = \dfrac{2x-3}{4x-7}$

   b. $f(x) = \dfrac{x}{x^2 - 3x + 2}$

   c. $f(x) = \dfrac{x-6}{x^2 + 1}$

   The only values that cannot be used for x are those that make the denominator 0.

   a. Set the denominator equal to 0, then solve.
   $$4x - 7 = 0$$
   $$4x = 7$$
   $$x = \frac{7}{4}$$

   The given expression is undefined for $\frac{7}{4}$, so the domain of f includes all real numbers except $\frac{7}{4}$, written using set-builder notation as $\left\{ x \mid x \neq \frac{7}{4} \right\}$.

   b. $x^2 - 3x + 2 = 0$ Set the denominator equal to 0.

   $(x-2)(x-1) = 0$ Factor.

   $x - 2 = 0$ or $x - 1 = 0$ Zero-factor property

       $x = 2$ $x = 1$ Solve.

   The denominator equals 0 when $x = 2$ or $x = 1$, so the domain is $\{ x \mid x \neq 1,\, 2 \}$.

   c. $\dfrac{x-6}{x^2 + 1}$

   The denominator will not equal 0 for any value of x, so there are no values for which this expression is undefined. The domain is $\{ x \mid x \text{ is a real number} \}$.

Now Try:

1. For each rational function, find all numbers that are not in the domain. Then give the domain, using set-builder notation.

   a. $f(x) = \dfrac{9}{4x}$

   _____

   b. $f(x) = \dfrac{x-3}{x^2 + 9}$

   _____

   c. $f(x) = \dfrac{x+2}{x^4 - 16}$

   _____

Review these examples for Objective 3:

2. Write each rational expression in lowest terms.

  a. $\dfrac{x^2 - 25}{x^2 + 10x + 25}$

  b. $\dfrac{3y^2 - 13y - 10}{2y^2 - 9y - 5}$

  c. $\dfrac{12xy - 28x - 15y + 35}{6xy - 14x + 3y - 7}$

First, factor both numerator and denominator to find their greatest common factor. Next, apply the fundamental property and divide out common factors.

  a. $\dfrac{x^2 - 25}{x^2 + 10x + 25} = \dfrac{(x-5)(x+5)}{(x+5)(x+5)}$ Factor.

  $\qquad = \dfrac{x-5}{x+5}$ Fundamental property

  b. $\dfrac{3y^2 - 13y - 10}{2y^2 - 9y - 5} = \dfrac{(3y+2)(y-5)}{(2y+1)(y-5)}$

  $\qquad = \dfrac{3y+2}{2y+1}$

  c. $\dfrac{12xy - 28x - 15y + 35}{6xy - 14x + 3y - 7}$

  $\qquad = \dfrac{(12xy - 15y) + (-28x + 35)}{(6xy + 3y) + (-14x - 7)}$

  $\qquad\qquad$ Group the terms.

  $\qquad = \dfrac{3y(4x-5) - 7(4x-5)}{3y(2x+1) - 7(2x+1)}$

  $\qquad\qquad$ Factor within the groups.

  $\qquad = \dfrac{(3y-7)(4x-5)}{(3y-7)(2x+1)}$

  $\qquad\qquad$ Factor by grouping.

  $\qquad = \dfrac{4x-5}{2x+1}$ Lowest terms

Now Try:

2. Write each rational expression in lowest terms.

  a. $\dfrac{3y^2 - 13y - 10}{2y^2 - 9y - 5}$

  _____

  b. $\dfrac{x^2 + y^2}{x + y}$

  _____

  c. $\dfrac{r^3 - s^3}{r^2 - s^2}$

  _____

Name: Date:
Instructor: Section:

3. Write each rational expression in lowest terms.

a. $\dfrac{x^2 - 25}{25 - x^2}$ b. $\dfrac{16 - x^2}{2x - 8}$

a. $\dfrac{x^2 - 25}{25 - x^2} = \dfrac{x^2 - 25}{(-1)(x^2 - 25)} = -1$

b. $\dfrac{16 - x^2}{2x - 8} = \dfrac{(4 - x)(4 + x)}{2(x - 4)}$

Factor the numerator and denominator.

$= \dfrac{(4 - x)(4 + x)}{2(-1)(4 - x)}$

$x - 4 = (-1)(4 - x)$

$= \dfrac{4 + x}{-2} = -\dfrac{4 + x}{2}$

3. Write each rational expression in lowest terms.

a. $\dfrac{4r - s}{s - 4r}$ _____

b. $\dfrac{4x^2 - 100}{20 - 4x}$

_____

Review this example for Objective 4:

4. Multiply.

a. $\dfrac{x + 4}{4x^2} \cdot \dfrac{8x^5}{4x^2 + 32x + 64}$

b. $\dfrac{3x^2 - 12}{x^2 - x - 6} \cdot \dfrac{x^2 - 6x + 9}{2x - 4}$

In order to multiply rational expression, first factor all numerators and denominators, then apply the fundamental property. Next, multiply the numerators and multiply the denominators and check to be sure that the product is in lowest terms.

Now Try:

4. Multiply.

a. $\dfrac{4r + 4p}{8z^2} \cdot \dfrac{36z^6}{r^2 + rp}$

_____

b. $\dfrac{3x + 12}{6x - 30} \cdot \dfrac{x^2 - x - 20}{x^2 - 16}$

_____

a. $\dfrac{x+4}{4x^2}\cdot\dfrac{8x^5}{4x^2+32x+64}$

$=\dfrac{x+4}{4x^2}\cdot\dfrac{8x^5}{4(x+4)^2}$

$=\dfrac{(x+4)}{(x+4)^2}\cdot\dfrac{8x^5}{4\cdot 4x^2}$ Commutative property

$=\dfrac{1}{(x+4)}\cdot\dfrac{x^3}{2}$ Fundamental property

$=\dfrac{x^3}{2(x+4)}$ Multiply; lowest terms

b. $\dfrac{3x^2-12}{x^2-x-6}\cdot\dfrac{x^2-6x+9}{2x-4}$

$=\dfrac{3(x-2)(x+2)}{(x-3)(x+2)}\cdot\dfrac{(x-3)^2}{2(x-2)}$

$=\dfrac{3(x-3)}{2}$

Review this example for Objectives 5 and 6:

5. Divide.

a. $\dfrac{3x+7}{8x^6}\div\dfrac{3x+7}{2x^3}$

b. $\dfrac{2x+2y}{8x}\div\dfrac{x^2\left(x^2-y^2\right)}{24}$

To divide two rational expressions, multiply the first expression (the *dividend*) by the reciprocal of the second expression (the *divisor*).

a. $\dfrac{3x+7}{8x^6}\div\dfrac{3x+7}{2x^3}=\dfrac{3x+7}{8x^6}\cdot\dfrac{2x^3}{3x+7}$

$=\dfrac{1}{4x^3}$

Now Try:

5. Divide.

a. $\dfrac{4m-12}{2m+10}\div\dfrac{m^2-9}{m^2-25}$

_____

b. $\dfrac{6a^2+18a}{a-1}\div\dfrac{a^3+3a^2}{1-a^2}$

_____

b. $\dfrac{2x+2y}{8x} \div \dfrac{x^2\left(x^2 - y^2\right)}{24}$

$= \dfrac{2x+2y}{8x} \cdot \dfrac{24}{x^2\left(x^2 - y^2\right)}$ Multiply by the reciprocal.

$= \dfrac{2(x+y)}{8x} \cdot \dfrac{3 \cdot 8}{x^2(x+y)(x-y)}$ Factor.

$= \dfrac{6}{x^3(x-y)}$ Multiply.

Objectives 1 and 2 Define rational expressions. Define rational functions and describe their domains.

For extra help, see Example 1 on pages 362–363 of your text, the Section Lecture video for Section 7.1, and Exercise Solution Clip 3.

Find all real numbers that are not in the domain of the function.

1. $f(a) = \dfrac{2a - 3}{4a - 7}$

1. _____

2. $f(r) = \dfrac{r + 7}{r^2 - 25}$

2. _____

3. $p(q) = \dfrac{q + 7}{q^2 - 3q + 2}$

3. _____

Objective 3 Write rational expressions in lowest terms.

For extra help, see Examples 2 and 3 on pages 364–365 of your text, the Section Lecture video for Section 7.1, and Exercise Solution Clips 27 and 51.

Write in lowest terms.

4. $\dfrac{9 - x^2}{x^2 + 6x + 9}$

4. _____

5. $\dfrac{22r^3 - 11r^2}{6 - 12r}$

5. _____

6. $\dfrac{s^2 - s - 6}{s^2 + s - 12}$

6. _____

7. $\dfrac{r^3 - s^3}{r^2 - s^2}$

7. _____

Objective 4 Multiply rational expressions.
For extra help, see Example 4 on pages 366–367 of your text, the Section Lecture video for Section 7.1, and Exercise Solution Clip 63.

Multiply.

8. $\dfrac{9k - 18}{6k + 12} \cdot \dfrac{3k + 6}{15k - 30}$

8. _____

9. $\dfrac{3x + 12}{6x - 30} \cdot \dfrac{x^2 - x - 20}{x^2 - 16}$

9. _____

10. $\dfrac{x^2 + 10x + 21}{x^2 + 14x + 49} \cdot \dfrac{x^2 + 12x + 35}{x^2 - 6x - 27}$

10. _____

11. $\dfrac{x^2 - x - 6}{x^2 - 2x - 8} \cdot \dfrac{x^2 + 7x + 12}{9 - x^2}$

11. _____

Objective 5 Find reciprocals of rational expressions.
Objective 6 Divide rational expressions.
For extra help, see Example 5 on page 368 of your text, the Section Lecture video for Section 7.1, and Exercise Solution Clip 63.

Divide.

12. $\dfrac{27 - 3k^2}{3k^2 + 8k - 3} \div \dfrac{k^2 - 6k + 9}{6k^2 - 19k + 3}$

12. _____

13. $\dfrac{2k^2 + 5k - 12}{2k^2 + k - 3} \div \dfrac{k^2 + 8k + 16}{2k^2 + 11k + 12}$

13. _____

14. $\dfrac{z^4 + 2z^3 + z^2}{z^5 - 4z^3} \div \dfrac{9z + 9}{6z + 12}$

14. _____

15. $\dfrac{2a^2 - 5a - 12}{a^2 - 10a + 24} \div \dfrac{4a^2 - 9}{a^2 - 9a + 18}$

15. _____

Chapter 7 RATIONAL EXPRESSIONS AND FUNCTIONS

7.2   Adding and Subtracting Rational Expressions

Learning Objectives
1   Add and subtract rational expressions with the same denominator.
2   Find a least common denominator.
3   Add and subtract rational expressions with different denominators.

Key Terms

Use the vocabulary terms listed below to complete each statement in exercises 1–2.

    least common denominator (LCD)   **equivalent expressions**

1.   Given several denominators, the smallest expression that is divisible by all the denominators is called the _____.

2. $\dfrac{24x-8}{9x^2-1}$ and $\dfrac{8}{3x+1}$ are _____

Guided Examples

Review this example for Objective 1:

1.   Add or subtract as indicated.

    a. $\dfrac{2}{3x^2}+\dfrac{4}{3x^2}$

    b. $\dfrac{-4x+3}{x-7}+\dfrac{2x+11}{x-7}$

    c. $\dfrac{16}{x-8}-\dfrac{x+8}{x-8}$

The only values that cannot be used for x are those that make the denominator 0.

    a. $\dfrac{2}{3x^2}+\dfrac{4}{3x^2}=\dfrac{2+4}{3x^2}=\dfrac{6}{3x^2}=\dfrac{2}{x^2}$

    b. $\dfrac{-4x+3}{x-7}+\dfrac{2x+11}{x-7}=\dfrac{-2x+14}{x-7}$

                   $=\dfrac{-2(x-7)}{x-7}$ Factor.

                   $=-2$

Now Try:

1.   For each rational function, find all numbers that are not in the domain. Then give the domain, using set-builder notation.

    a. $\dfrac{b}{b^2-4}+\dfrac{2}{b^2-4}$

    _____

    b.

    $\dfrac{3m+4}{2m^2-7m-15}+\dfrac{m+2}{2m^2-7m-15}$

    _____

    c. $\dfrac{6p}{p-4}-\dfrac{p+20}{p-4}$

    _____

c. $\dfrac{16}{x-8} - \dfrac{x+8}{x-8} = \dfrac{16-(x+8)}{x-8}$

          Subtract numerators.
          Keep the same denominator.

        $= \dfrac{16-x-8}{x-8}$

          Distributive property

        $= \dfrac{8-x}{x-8} = -1$

          Combine like terms.

            $\dfrac{x-y}{y-x} = -1$

Review this example for Objective 2:

2. Find the LCD for each group of denominators.

  a. $9x^3y^2,\ 15x^2y$

  b. $x-y,\ y^2-x^2$

  c. $x^2+5x+6,\ x^2+4x+4,\ x^2-9$

  a. $9x^3 = 3^2 \cdot x^3$
      $15x^2y = 3 \cdot 5 \cdot x^2 \cdot y$

  Find the LCD by taking each different factor the greatest number of time it appears as a factor in any of the denominators.

  $\text{LCD} = 3^2 \cdot 5 \cdot x^3 \cdot y = 45x^3y$

  b. $y^2 - x^2 = (y-x)(y+x)$

  Since $y - x$ and $x - y$ are opposites of each other, we can multiply one of the factors by -1 to obtain the other.
  $\text{LCD} = -(x-y)(y+x)$ or
          $-(y-x)(y+x)$

  c. $x^2+5x+6 = (x+2)(x+3)$
     $x^2+4x+4 = (x+2)^2$
     $x^2-9 = (x+3)(x-3)$
     $\text{LCD} = (x+2)^2(x+3)(x-3)$

Now Try:

2. Find the LCD for each group of denominators.

  a. $8x^2y^5,\ 12x^3y^2$

  _____

  b. $a^2 - b^2,\ b^2 - a^2$

  _____

  c. $x^2 - 9,\ x^2 - x - 12,$
     $x^2 - 7x + 12$

  _____

Review these examples for Objective 3:

3. Add or subtract as indicated.

   a. $\dfrac{6}{5r} + \dfrac{3}{4r}$ b. $\dfrac{3}{x+4} - \dfrac{4}{x}$

   a. $\dfrac{6}{5r} + \dfrac{3}{4r} = \dfrac{6 \cdot 4}{5r \cdot 4} + \dfrac{3 \cdot 5}{4r \cdot 5}$ LCD $= 20r$
      Fundamental property

   $ = \dfrac{24}{20r} + \dfrac{15}{20r}$ Multiply.

   $ = \dfrac{39}{20r}$ Add numerators. Keep the common denominator.

   b. $\dfrac{3}{x+4} - \dfrac{4}{x} = \dfrac{3x}{x(x+4)} - \dfrac{4(x+4)}{x(x+4)}$

    LCD $= x(x+4)$
   Fundamental property

   $ = \dfrac{3x}{x(x+4)} - \dfrac{4x+16}{x(x+4)}$

    Distributive property

   $ = \dfrac{3x - 4x - 16}{x(x+4)}$

    Subtract numerators.

   $ = \dfrac{-x - 16}{x(x+4)}$

    Combine like terms in numerator.

4. Subtract.

   a. $\dfrac{x}{x^2 - 7x + 10} - \dfrac{2}{x^2 - 7x + 10}$

   b. $\dfrac{8}{x+8} - \dfrac{4}{x}$

Now Try:

3. Add or subtract as indicated.

   a. $\dfrac{2}{9z} + \dfrac{4}{5z}$

   _____

   b. $\dfrac{5}{s-6} - \dfrac{5}{s}$

   _____

4. Subtract.

   a. $\dfrac{2x}{x^2 + 3x - 10} - \dfrac{x+2}{x^2 + 3x - 10}$

   _____

   b. $\dfrac{7}{x+4} - \dfrac{5}{3x + 12}$

   _____

a. $\dfrac{x}{x^2-7x+10}-\dfrac{2}{x^2-7x+10}$

$$=\dfrac{x-2}{x^2-7x+10}$$

Subtract numerators.
Keep the common denominator.

$$=\dfrac{x-2}{(x-2)(x-5)}\quad\text{Factor the denominator.}$$

$$=\dfrac{1}{x-5}\quad\text{Factor out }x-2.$$

b. $\dfrac{8}{x+8}-\dfrac{4}{x}=\dfrac{8x}{x(x+8)}-\dfrac{4(x+8)}{x(x+8)}$

$$\text{The LCD is }x(x+8).$$

$$=\dfrac{8x}{x(x+8)}-\dfrac{4x+32}{x(x+8)}$$

$$\text{Distributive property}$$

$$=\dfrac{8x-4x-32}{x(x-8)}\quad\text{Subtract.}$$

$$=\dfrac{4x-32}{x(x-8)}\quad\text{Combine like terms.}$$

$$=\dfrac{4(x-8)}{x(x-8)}=\dfrac{4}{x}\quad\begin{array}{l}\text{Factor the}\\\text{numerator.}\\\text{Lowest terms}\end{array}$$

5. Add.

$$\dfrac{6p}{p-4}+\dfrac{p+8}{4-p}$$

The denominators are opposites, so choose one or the other as the common denominator. We choose $p-4$.

$$\dfrac{6p}{p-4}+\dfrac{p+8}{4-p}=\dfrac{6p}{p-4}+\dfrac{(p+8)(-1)}{(4-p)(-1)}$$

$$\text{Multiply }\tfrac{p+8}{4-p}\text{ by }\tfrac{-1}{-1}.$$

$$=\dfrac{6p}{p-4}+\dfrac{-p-8}{p-4}$$

$$=\dfrac{6p+(-p-8)}{p-4}$$

$$\text{Add numerators.}$$

$$=\dfrac{5p-8}{p-4}\quad\text{Combine like terms.}$$

5. Add.

$$\dfrac{12}{x-8}+\dfrac{x+8}{8-x}\quad\underline{\hspace{3cm}}$$

Alternatively, we could use $4 - p$ as the common denominator and rewrite the first expression.

$$\frac{6p}{p-4} + \frac{p+8}{4-p} = \frac{6p(-1)}{(p-4)(-1)} + \frac{p+8}{4-p}$$

$$= \frac{-6p}{4-p} + \frac{p+8}{4-p}$$

$$= \frac{-6p + (p+8)}{4-p}$$

$$= \frac{-5p + 8}{4-p}$$

6. Add and subtract as indicated.

$$\frac{2}{x} - \frac{5}{x-2} + \frac{10}{x^2 - 2x}$$

$$\frac{2}{x} - \frac{5}{x-2} + \frac{10}{x^2 - 2x}$$

$$= \frac{2}{x} - \frac{5}{x-2} + \frac{10}{x(x-2)} \quad \text{Factor the third denominator.}$$

$$= \frac{2(x-2)}{x(x-2)} - \frac{5x}{x(x-2)} + \frac{10}{x(x-2)}$$
The LCD is $x(x-2)$.
Fundamental property

$$= \frac{2(x-2) - 5x + 10}{x(x-2)} \quad \text{Add and subtract the numerators.}$$

$$= \frac{2x - 4 - 5x + 10}{x(x-2)} \quad \text{Distributive property}$$

$$= \frac{-3x + 6}{x(x-2)} \quad \text{Combine like terms.}$$

$$= \frac{-3(x-2)}{x(x-2)} \quad \text{Factor the numerator.}$$

$$= -\frac{3}{x} \quad \text{Lowest terms.}$$

6. Add and subtract as indicated.

$$\frac{12}{x^2 + 3x} - \frac{3}{x} + \frac{4}{x+3}$$

_____

7. Subtract.

$$\frac{m}{m^2-4}-\frac{1-m}{m^2+4m+4}$$

$$\frac{m}{m^2-4}-\frac{1-m}{m^2+4m+4}$$

$$=\frac{m}{(m-2)(m+2)}-\frac{(1-m)}{(m+2)^2}\quad \text{Factor the denominators.}$$

$$=\frac{m(m+2)}{(m-2)(m+2)^2}-\frac{(1-m)(m-2)}{(m-2)(m+2)^2}$$

The LCD is $(m-2)(m+2)^2$.

$$=\frac{m(m+2)-(1-m)(m-2)}{(m-2)(m+2)^2}$$

Subtract numerators.

$$=\frac{m^2+2m-\left(m-2-m^2+2m\right)}{(m-2)(m+2)^2}\quad \text{Multiply.}$$

$$=\frac{m^2+2m-m+2+m^2-2m}{(m-2)(m+2)^2}$$

Distributive property

$$=\frac{2m^2-m+2}{(m-2)(m+2)^2}\quad \text{Combine like terms.}$$

8. Add.

$$\frac{4z}{z^2+6z+8}+\frac{2z-1}{z^2+5z+6}$$

$$\frac{4z}{z^2+6z+8}+\frac{2z-1}{z^2+5z+6}$$

$$=\frac{4z}{(z+4)(z+2)}+\frac{2z-1}{(z+3)(z+2)}$$

Factor the denominators.

$$=\frac{4z(z+3)}{(z+4)(z+2)(z+3)}+\frac{(2z-1)(z+4)}{(z+4)(z+3)(z+2)}$$

The LCD is $(z+4)(z+2)(z+3)$.

$$=\frac{4z^2+12z}{(z+4)(z+2)(z+3)}+\frac{2z^2+7z-4}{(z+4)(z+3)(z+2)}$$

Multiply in the numerators.

$$=\frac{6z^2+19z-4}{(z+4)(z+2)(z+3)}\quad \begin{array}{l}\text{Add the numerators.}\\ \text{Lowest terms}\end{array}$$

7. Subtract.

$$\frac{4z}{z^2+6z+8}-\frac{2z-1}{z^2+5z+6}$$

---

8. Add.

$$\frac{2y+9}{y^2+6y+8}+\frac{y+3}{y^2+2y-8}$$

Objective 1 Add and subtract rational expressions with the same denominator.

For extra help, see Example 1 on page 372 of your text, the Section Lecture video for Section 7.2, and Exercise Solution Clip 9.

Add or subtract. Write the answer in lowest terms.

1. $\dfrac{n}{m+3} - \dfrac{-3n+7}{m+3}$

   1. _____

2. $\dfrac{r}{r^2-s^2} + \dfrac{s}{r^2-s^2}$

   2. _____

3. $\dfrac{k}{k^2+3k-10} - \dfrac{2}{k^2+3k-10}$

   3. _____

Objective 2 Find a least common denominator.

For extra help, see Example 2 on page 373 of your text, the Section Lecture video for Section 7.2, and Exercise Solution Clip 23.

Suppose that the expressions given are denominators of fractions. Find the least common denominator, LCD, for each group.

4. $5a+10, \ a^2+2a$

   4. _____

5. $r^2+5r+4, \ r^2+r$

   5. _____

6. $p-4, \ p^2-16, \ p^2+8p+16$

   6. _____

Objective 3 Add and subtract rational expressions with different denominators.
For extra help, see Examples 3–8 on pages 374–377 of your text, the Section Lecture video
for Section 7.2, and Exercise Solution Clips 43, 57, 61, 67, 81, and 89.

Add or subtract as indicated. Write all answers in lowest terms.

7. $\dfrac{3r+4}{3} + \dfrac{6r+4}{6}$

7. _____

8. $\dfrac{9}{2t} + \dfrac{4}{t^2}$

8. _____

9. $\dfrac{12}{p^2+6p+9} - \dfrac{2}{p+3}$

9. _____

10. $\dfrac{1}{x^2-1} - \dfrac{1}{x^2+3x+2}$

10. _____

11. $\dfrac{1-3x}{4x^2-1} + \dfrac{3x-5}{2x^2+5x+2}$

11. _____

12. $\dfrac{5r}{r+2s} - \dfrac{3r}{-r-2s}$

12. _____

13. $\dfrac{2}{3-m} - \dfrac{2}{m-3} + \dfrac{3}{m^2-9}$

13. _____

14. $\dfrac{4z}{z^2+6z+8}-\dfrac{2z-1}{z^2-z-6}$ **14.** _____

15. $\dfrac{a+3b}{b^2+2ab+a^2}+\dfrac{a-b}{3b^2+4ab+a^2}$ **15.** _____

Chapter 7 RATIONAL EXPRESSIONS AND FUNCTIONS

7.3   Complex Fractions

Learning Objectives
1   Simplify complex fractions by simplifying the numerator and denominator (Method 1).
2   Simplify complex fractions by multiplying by a common denominator (Method 2).
3   Compare the two methods of simplifying complex fractions.
4   Simplify rational expressions with negative exponents.

Key Terms

Use the vocabulary terms listed below to complete each statement in exercises 1–2.

    complex fraction   **least common denominator**

1.   A _____ is an expression with one or more fractions in the numerator, denominator, or both.

2.   When using Method 2 (for simplifying complex fractions), multiply the numerator and denominator of the complex fraction by the _____ of the fractions in the numerator and the fractions in the denominator of the complex fraction. Simplify the resulting fraction if possible.

Guided Examples

Review this example for Objective 1:
1.   Use Method 1 to simplify each complex fraction.

    a. $\dfrac{\dfrac{rs}{3r^2}}{\dfrac{s^2}{3}}$ **b.** $\dfrac{\dfrac{4}{x-4}+3}{5-\dfrac{3}{x-4}}$

    a. Both the numerator and denominator are already simplified, so we divide by multiplying the numerator by the reciprocal of the denominator.

$$\frac{\dfrac{rs}{3r^2}}{\dfrac{s^2}{3}}=\frac{rs}{3r^2}\div\frac{s^2}{3}=\frac{rs}{3r^2}\cdot\frac{3}{s^2}=\frac{1}{rs}$$

Now Try:
1.   For each rational function, find all numbers that are not in the domain. Then give the domain, using set-builder notation.

    a. $\dfrac{\dfrac{x+1}{3x}}{\dfrac{x+1}{6x}}$

    _____

    b. $\dfrac{\dfrac{2}{a+2}-4}{\dfrac{1}{a+2}-3}$

    _____

b. Start by writing the numerator and the denominator as single fractions.

$$\frac{\dfrac{4}{x-4}+3}{5-\dfrac{3}{x-4}} = \frac{\dfrac{4}{x-4}+\dfrac{3(x-4)}{x-4}}{\dfrac{5(x-4)}{x-4}-\dfrac{3}{x-4}}$$

$$= \frac{\dfrac{4+3(x-4)}{x-4}}{\dfrac{5(x-4)-3}{x-4}}$$

$$= \frac{\dfrac{4+3x-12}{x-4}}{\dfrac{5x-20-3}{x-4}} = \frac{\dfrac{3x-8}{x-4}}{\dfrac{5x-23}{x-4}}$$

Now multiply the numerator by the reciprocal of the denominator.

$$\frac{\dfrac{3x-8}{x-4}}{\dfrac{5x-23}{x-4}} = \frac{3x-8}{x-4}\cdot\frac{x-4}{5x-23} = \frac{3x-8}{5x-23}$$

Review this example for Objective 2:

2. Use Method 2 to simplify each complex fraction.

   a. $\dfrac{\dfrac{1}{3x^3}+\dfrac{3}{4x^2}}{\dfrac{1}{2x^2}-\dfrac{1}{x}}$ **b.** $\dfrac{\dfrac{n}{4}-\dfrac{1}{n}}{1+\dfrac{n+4}{n}}$

   a. The LCD for all the denominators is $12x^3$.

   $$\frac{\dfrac{1}{3x^3}+\dfrac{3}{4x^2}}{\dfrac{1}{2x^2}-\dfrac{1}{x}} = \frac{\left(\dfrac{1}{3x^3}+\dfrac{3}{4x^2}\right)12x^3}{\left(\dfrac{1}{2x^2}-\dfrac{1}{x}\right)12x^3}$$

   $$= \frac{4+9x}{6x-12x^2}$$

   $$= \frac{9x+4}{-6x(2x-1)}$$

Now Try:

2. Use Method 2 to simplify each complex fraction.

   a. $\dfrac{3-\dfrac{6}{a}}{\dfrac{1}{a}+\dfrac{3a}{5}}$

   _____

   b. $\dfrac{\dfrac{2}{3x^2}-\dfrac{3}{2x^3}}{\dfrac{5}{2x}+\dfrac{1}{4x^2}}$

   _____

b. $\dfrac{\dfrac{n}{4}-\dfrac{1}{n}}{1+\dfrac{n+4}{n}} = \dfrac{\dfrac{n}{4}-\dfrac{1}{n}}{1+\dfrac{n+4}{n}}\cdot 1$ Identity property of multiplication

$= \dfrac{\left(\dfrac{n}{4}-\dfrac{1}{n}\right)\cdot 4n}{\left(1+\dfrac{n+4}{n}\right)\cdot 4n}$

The LCD of all the fractions is $4n$. Multiply the numerator and denominator by $4n$.

$= \dfrac{n^2-4}{4n+4(n+4)}$ Distributive property

$= \dfrac{n^2-4}{4n+4n+16}$ Distributive property

$= \dfrac{n^2-4}{8n+16}$ Combine like terms.

$= \dfrac{(n+2)(n-2)}{8(n+2)}$ Factor numerator and denominator.

$= \dfrac{n-2}{8}$ Simplify.

Review this example for Objective 3:

3. Simplify each complex fraction by both methods.

a. $\dfrac{2x-y^2}{x+\dfrac{y^2}{x}}$ **b.** $\dfrac{\dfrac{4}{x-4}+3}{5-\dfrac{3}{x-4}}$

a. Method 1:

$\dfrac{2x-y^2}{x+\dfrac{y^2}{x}} = \dfrac{2x-y^2}{\dfrac{x^2+y^2}{x}} = \dfrac{\dfrac{2x-y^2}{1}}{\dfrac{x^2+y^2}{x}}$

$= \dfrac{2x-y^2}{1}\cdot\dfrac{x}{x^2+y^2}$

$= \dfrac{x\left(2x-y^2\right)}{x^2+y^2}$

Now Try:

3. Simplify each complex fraction by both methods.

a. $\dfrac{\dfrac{1}{t+1}-1}{\dfrac{1}{t-1}+1}$

_____

b. $\dfrac{\dfrac{4}{x+4}}{\dfrac{3}{x^2-16}}$ _____

_____

Method 2:

$$\frac{2x - y^2}{x + \dfrac{y^2}{x}} = \frac{\left(2x - y^2\right) \cdot x}{\left(x + \dfrac{y^2}{x}\right) \cdot x}$$

$$= \frac{x\left(2x - y^2\right)}{x^2 + y^2}$$

b. Method 1:

$$\frac{\dfrac{4}{x-4} + 3}{5 - \dfrac{3}{x-4}} = \frac{\dfrac{4(1)}{x-4} + \dfrac{3(x-4)}{x-4}}{\dfrac{5(x-4)}{x-4} - \dfrac{3}{x-4}}$$

$$= \frac{\dfrac{4 + 3x - 12}{x-4}}{\dfrac{5x - 20 - 3}{x-4}}$$

$$= \frac{\dfrac{3x - 8}{x-4}}{\dfrac{5x - 23}{x-4}}$$

$$= \frac{3x-8}{x-4} \cdot \frac{x-4}{5x-23}$$

$$= \frac{3x-8}{5x-23}$$

Method 2:

$$\frac{\dfrac{4}{x-4} + 3}{5 - \dfrac{3}{x-4}} = \frac{\left(\dfrac{4}{x-4} + 3\right) \cdot (x-4)}{\left(5 - \dfrac{3}{x-4}\right) \cdot (x-4)}$$

$$= \frac{4 + 3(x-4)}{5(x-4) - 3}$$

$$= \frac{4 + 3x - 12}{5x - 20 - 3}$$

$$= \frac{3x-8}{5x-23}$$

Review this example for Objective 4:

4. Simplify the expression, using only positive exponents in the answer.

$$\frac{3x^{-2}+5x^{-3}}{10x^{-1}+6}$$

$$\frac{3x^{-2}+5x^{-3}}{10x^{-1}+6} = \frac{\dfrac{3}{x^2}+\dfrac{5}{x^3}}{\dfrac{10}{x}+6} \qquad \text{Write with positive exponents}$$

$$= \frac{\left(\dfrac{3}{x^2}+\dfrac{5}{x^3}\right)\cdot x^3}{\left(\dfrac{10}{x}+6\right)\cdot x^3}$$

Use Method 2 to simplify.
Multiply numerator and
denominator by the LCD, x^3.

$$= \frac{3x+5}{10x^2+6x^3} \qquad \begin{array}{l}\text{Distributive property}\\ \text{Lowest terms}\end{array}$$

Now Try:

4. Simplify the expression, using only positive exponents in the answer.

$$\frac{1+x^{-1}}{(x+y)^{-1}} \qquad \underline{\hspace{2cm}}$$

Objective 1   Simplify complex fractions by simplifying the numerator and denominator (Method 1).

For extra help, see Example 1 on pages 380–381 of your text, the Section Lecture video for Section 7.3, and Exercise Solution Clips 7, 11, and 19.

Use Method 1 to simplify.

1. $\dfrac{\dfrac{3}{k}+1}{\dfrac{3+k}{2}}$

1. _____

2. $\dfrac{\dfrac{1}{t}+\dfrac{1}{z}}{\dfrac{1}{z+t}}$

2. _____

3. $\dfrac{\dfrac{3}{w-4} - \dfrac{3}{w+4}}{\dfrac{1}{w+4} + \dfrac{1}{w^2-16}}$

3. _____

Objective 2 Simplify complex fractions by multiplying by a common denominator (Method 2).

For extra help, see Example 2 on pages 381–382 of your text, the Section Lecture video for Section 7.3, and Exercise Solution Clips 7, 11, and 19.

Use Method 2 to simplify.

4. $\dfrac{\dfrac{9}{x^2} - 1}{\dfrac{3}{x} - 1}$

4. _____

5. $\dfrac{\dfrac{x-2}{x+2}}{\dfrac{x}{x-2}}$

5. _____

6. $\dfrac{\dfrac{r}{r+1} + 1}{\dfrac{2r+1}{r-1}}$

6. _____

Objective 3 Compare the two methods of simplifying complex fractions.
For extra help, see Example 3 on pages 382−383 of your text, the Section Lecture video for
Section 7.3, and Exercise Solution Clips 7, 11, and 19.

Use either method to simplify each complex fraction.

7. $\dfrac{\dfrac{25k^2 - m^2}{4k}}{\dfrac{5k + m}{7k}}$

7. _____

8. $\dfrac{\dfrac{1}{t+1} - 1}{\dfrac{1}{t-1} + 1}$

8. _____

9. $\dfrac{\dfrac{4}{x} - \dfrac{1}{2}}{\dfrac{5}{x} + \dfrac{1}{3}}$

9. _____

10. $\dfrac{\dfrac{1}{m-1}+4}{\dfrac{2}{m-1}-4}$

10. _____

11. $\dfrac{\dfrac{4}{s+3}-\dfrac{2}{s-3}}{\dfrac{5}{s^2-9}}$

11. _____

12. $\dfrac{\dfrac{6}{k+1}-\dfrac{5}{k-3}}{\dfrac{3}{k-3}+\dfrac{2}{k+2}}$

12. _____

Objective 4 Simplify rational expressions with negative exponents.
For extra help, see Example 4 on pages 383–384 of your text, the Section Lecture video for Section 7.3, and Exercise Solution Clip 37.

Simplify each expression, using only positive exponents in the answer.

13. $\dfrac{2x^{-1}+y^2}{z^{-3}}$

13. _____

14. $\left(A^{-2} - B^{-2}\right)^{-1}$

14. _____

15. $\dfrac{4x^{-2}}{2 + 6y^{-3}}$

15. _____

Chapter 7 RATIONAL EXPRESSIONS AND FUNCTIONS

7.4   Equations with Rational Expressions and Graphs

Learning Objectives
1   Determine the domain of the variable in a rational equation.
2   Solve rational equations.
3   Recognize the graph of a rational function.

Key Terms

Use the vocabulary terms listed below to complete each statement in exercises 1–5.

   domain of the variable in a rational equation   **discontinuous**

   reciprocal function   **vertical asymptote**   **horizontal asymptote**

1.   A rational function in simplest form $f(x) = \dfrac{P(x)}{x-a}$ has the line $x = a$ as a

      _____.

2.   The _____ is the intersection (overlap) of the domains
      of the rational expressions in the equation.

3.   A graph of a function is _____ if there are one or more breaks in
      the graph.

4.   A horizontal line that a graph approaches as |x| gets larger without bound is called a

      _____.

5.   The function defined by $f(x) = \dfrac{1}{x}$ is called the _____.

Guided Examples

Review this example for Objective 1:
1.   Find the domain of the variable in each
     equation.

   a. $\dfrac{1}{x} + \dfrac{2}{x+1} = 0$

   b. $\dfrac{x}{2x+2} = \dfrac{-2x}{4x+4} + \dfrac{2x-3}{x+1}$

Now Try:
1.   For each rational function, find
      all numbers that are not in the
      domain. Then give the domain,
      using set-builder notation.

   a. $\dfrac{10}{x-7} + \dfrac{7}{x+8} = \dfrac{1}{2}$

      _____

a. The domains of the three rational expressions in the equation are, in order, $\{x \mid x \neq 0\}, \{x \mid x \neq -1\}$, and $\{x \mid x \text{ is a real number}\}$. The intersection of these three domains is all real numbers except 0 and −1, written using set-builder notation as $\{x \mid x \neq -1, 0\}$.

b. The domains of the three rational expression in the equation are, in order, $\{x \mid x \neq -1\}, \{x \mid x \neq -1\}$, and $\{x \mid x \neq -1\}$. The intersection of these three domains is all real numbers except −1, written using set-builder notation as $\{x \mid x \neq -1\}$.

b. $\dfrac{4}{x^2 - 2x} + \dfrac{x}{x^2 - 4} = \dfrac{1}{x}$

_____

Review these examples for Objective 2:

2. Solve.

$$\frac{4}{5x} + \frac{3}{2x} = \frac{23}{50}$$

Step 1: The domain is $\{x \mid x \neq 0\}$.

Step 2: Multiply each side of the equation by the LCD, $50x$.

$$50x\left(\frac{4}{5x} + \frac{3}{2x}\right) = \left(\frac{23}{50}\right)50x$$

Step 3: Solve the resulting equation.

$$50x\left(\frac{4}{5x}\right) + 50x\left(\frac{3}{2x}\right) = \frac{23}{50} \cdot 50x$$

$$\text{Distributive property}$$
$$40 + 75 = 23x$$
$$\text{Distributive property}$$
$$115 = 23x \quad \text{Add.}$$
$$5 = x \quad \text{Divide by 23.}$$

Now Try:

2. Solve.

$$\frac{1}{2n} + \frac{1}{4n^2} = \frac{1}{4n}$$

_____

Step 4: Check:

$$\frac{4}{5(5)}+\frac{3}{2(5)}\overset{?}{=}\frac{23}{50}$$

$$\frac{4}{25}+\frac{15}{10}\overset{?}{=}\frac{23}{50}$$

$$\frac{8}{50}+\frac{15}{10}\overset{?}{=}\frac{23}{50}$$

$$\frac{23}{50}=\frac{23}{50}\;\checkmark$$

3. Solve.

$$\frac{x-12}{4x}+\frac{3}{x}=1$$

The domain excludes 0. The LCD is $4x$.

$$4x\left(\frac{x-12}{4x}+\frac{3}{x}\right)=1\cdot 4x$$

     Multiply each side by the LCD, $4x$.

$$4x\left(\frac{x-12}{4x}\right)+4x\left(\frac{3}{x}\right)=4x \quad \text{Distributive property}$$

$$x-12+12=4x \quad \text{Multiply.}$$

$$0=3x \quad \text{Add; subtract } x.$$

$$0=x \quad \text{Divide by 3.}$$

Since the proposed solution, 0, is not in the domain, it cannot be a solution of the equation. Therefore the equation has no solution, and the solution set is $\varnothing$.

4. Solve.

$$\frac{r+5}{r^2-16}=\frac{3}{r-4}+\frac{1}{r+4}$$

The LCD for all the denominators is $(r-4)(r+4)$. Note that 4 and -4 cannot be solutions of the equation.

3. Solve.

$$\frac{10}{x^2-2x}+\frac{4}{x}=\frac{5}{x-2}$$

_____

4. Solve.

$$\frac{2}{x+1}-\frac{4}{x-1}=\frac{8}{x^2-1}$$

_____

    265

$$(r-4)(r+4)\frac{r+5}{(r-4)(r+4)}$$

$$= (r-4)(r+4)\left(\frac{3}{r-4}+\frac{1}{r+4}\right)$$

Multiply by the LCD.

$$(r-4)(r+4)\frac{r+5}{(r-4)(r+4)}$$

$$= (r-4)(r+4)\left(\frac{3}{r-4}\right)$$

$$+ (r-4)(r+4)\left(\frac{1}{r+4}\right)$$

Distributive property

$$r+5 = 3(r+4)+1(r-4)$$

Divide out the common factors.

$$r+5 = 3r+12+r-4$$

Distributive property

$r+5 = 4r+8$ Combine like terms.

$-3r = 3$ Subtract $4r$; subtract 5.

$r = -1$ Divide by -3.

Check that $\{-1\}$ is the solution set.

5. Solve.

$$\frac{3}{x^2-1}-\frac{3x}{2x+2}=-\frac{2}{3}$$

Factor the denominators on the left and then determine the LCD.

$$x^2-1 = (x+1)(x-1)$$
$$2x+2 = 2(x+1)$$
$$\text{LCD} = 3\cdot 2(x+1)(x-1) = 6(x+1)(x-1)$$

Multiply both sides by the LCD to clear the fractions.

$$6(x+1)(x-1)\left(\frac{3}{(x+1)(x-1)}-\frac{3x}{2(x+1)}\right)$$

$$= 6(x+1)(x-1)\left(-\frac{2}{3}\right)$$

5. Solve.

$$\frac{9}{n}-\frac{8}{n+1}=10$$

_____

$$6(x+1)(x-1)\frac{3}{(x+1)(x-1)}$$

$$+6(x+1)(x-1)\left(-\frac{3x}{2(x+1)}\right)$$

$$=6(x+1)(x-1)\left(-\frac{2}{3}\right)$$

Distributive property

$$6 \cdot 3 + 3(x-1)(-3x) = -4(x+1)(x-1)$$

Divide out common factors.

$$18 - 9x^2 + 9x = -4x^2 + 4 \quad \text{Multiply.}$$

$$-5x^2 + 9x + 14 = 0$$

Subtract; combine terms.

$$5x^2 - 9x - 14 = 0 \quad \text{Multiply by } -1.$$

$$(5x-14)(x+1) = 0 \quad \text{Factor.}$$

$$5x - 14 = 0 \quad \text{or} \quad x + 1 = 0 \quad \text{Zero-factor}$$

$$x = \frac{14}{5} \qquad\qquad x = -1 \quad \text{property}$$

Since -1 make a denominator equal 0, -1 is *not* a solution. Check that $\left\{\frac{14}{5}\right\}$ is a solution.

Review this example for Objective 3:

6. Graph, and give the equations of the vertical and horizontal asymptotes.

$$f(x) = \frac{-2}{x+3}$$

Some ordered pairs that belong to the function are listed below.

x	$f(x)=\dfrac{-2}{x+3}$	x	$f(x)=\dfrac{-2}{x+3}$
-6	$\frac{2}{3}$	1	$-\frac{1}{2}$
-5	1	2	$-\frac{2}{5}$
-4	2	3	$-\frac{1}{3}$
-2	-2	4	$-\frac{2}{7}$
-1	-1	5	$-\frac{1}{4}$
0	$-\frac{2}{3}$	6	$-\frac{2}{9}$

There is no point on the graph for $x = -3$ because -3 is excluded from the domain of the rational function.

Now Try:

6. Graph, and give the equations of the vertical and horizontal asymptotes.

$$f(x) = -\frac{1}{x}$$

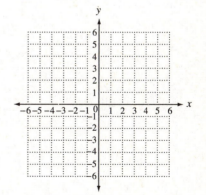

Vertical asymptote: _____

Horizontal asymptote: _____

The dashed line $x = -3$ represents the vertical asymptote and is not part of the graph. The graph gets closer to the vertical asymptote as the x-values get closer to -3. Since the x-values approach 0 as $|x|$ increases, the horizontal asymptote is $y = 0$.

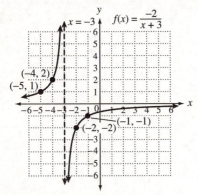

Objective 1 Determine the domain of the variable in a rational equation.

For extra help, see Example 1 on page 386 of your text, the Section Lecture video for Section 7.4, and Exercise Solution Clip 1.

(a) Without actually solving the equations below, list all possible numbers that would have to be rejected if they appeared as potential solutions. (b) Then give the domain using set notation.

1. $\dfrac{x-6}{x+6} + 3 = \dfrac{2}{x^2 + 7x + 6}$

    1. a. _____

       b. _____

2. $1 - \dfrac{2}{x} = \dfrac{-8x}{x^2 + 4x}$

    2. a. _____

       b. _____

3. $\dfrac{1}{x^2 + 3x + 2} - 2 = \dfrac{2}{x^2 + 5x + 6}$

    3. a. _____

       b. _____

Objective 2 Solve rational equations.

For extra help, see Example 2–5 on page 387–389 of your text, the Section Lecture video for Section 7.4, and Exercise Solution Clips 15, 33, 41, and 45.

Solve and check.

4. $\dfrac{3}{k} - \dfrac{2}{k+2} = \dfrac{7}{3}$ **4.** _____

5. $\dfrac{8}{2m+4} + \dfrac{2}{3m+6} = \dfrac{7}{9}$ **5.** _____

6. $\dfrac{2}{z-1} + \dfrac{3}{z+1} - \dfrac{17}{24} = 0$ **6.** _____

7. $\dfrac{5a+1}{2a+2} = \dfrac{5a-5}{5a+5} + \dfrac{3a+1}{a+1}$

7. _____

8. $\dfrac{1}{b+2} - \dfrac{5}{b^2+9b+14} = \dfrac{-3}{b+7}$

8. _____

9. $\dfrac{-16}{n^2-8n+12} = \dfrac{3}{n-2} + \dfrac{n}{n-6}$

9. _____

10. $\dfrac{4}{y+2} - \dfrac{3}{y+3} = \dfrac{8}{y^2+5y+6}$

10. _____

Name: Date:

Instructor: Section:

Objective 3 Recognize the graph of a rational function.

For extra help, see Example 6 on page 390 of your text, the Section Lecture video for Section 7.4, and Exercise Solution Clips 49 and 59.

Graph each rational function. Give the equations of the vertical and horizontal asymptotes.

11. $f(x) = \dfrac{5}{x}$

11.

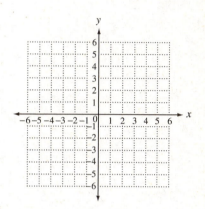

Vertical asymptote: _____

Horizontal asymptote: _____

12. $f(x) = \dfrac{2}{x-1}$

12.

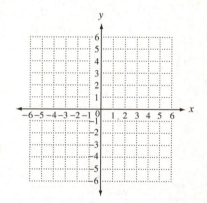

Vertical asymptote: _____

Horizontal asymptote: _____

13. $f(x) = \dfrac{1}{x-3}$

13.

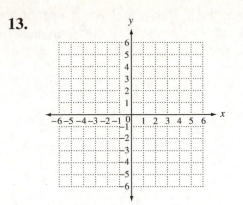

Vertical asymptote: _____

Horizontal asymptote: _____

14. $f(x) = \dfrac{1}{x-4}$

14.

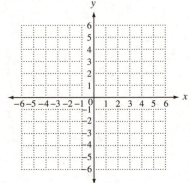

Vertical asymptote: _____

Horizontal asymptote: _____

15. $f(x) = \dfrac{3}{x+2}$

15.

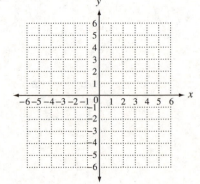

Vertical asymptote: _____

Horizontal asymptote: _____

Chapter 7 RATIONAL EXPRESSIONS AND FUNCTIONS

7.5   Applications of Rational Expressions

Learning Objectives

1   Find the value of an unknown variable in a formula.
2   Solve a formula for a specified variable.
3   Solve applications by using proportions.
4   Solve applications about distance, rate, and time.
5   Solve applications about work rates.

Key Terms

Use the vocabulary terms listed below to complete each statement in exercises 1–3.

   ratio   **proportion**   **rate of work**

1.   A _____ is a comparison of two quantities.

2.   If a job can be accomplished in t units of time, then the _____ is $\dfrac{1}{t}$ job per unit of time.

3.   A _____ is a statement that two ratios are equal.

Guided Examples

Review this example for Objective 1:

1.   The height of a trapezoid is given by the formula $h = \dfrac{2A}{B+b}$. Find B, if $A = 40$, $h = 8$ and $b = 3$.

   Substitute the given values into the equation and solve for B.

$$h = \frac{2A}{B+b}$$

$$8 = \frac{2(40)}{B+3} \quad \text{Let } A = 40,\ b = 3,\ h = 8.$$

$$8(B+3) = \frac{80}{B+3}(B+3)$$

   Multiply by the LCD, $B + 3$.

$$8B + 24 = 80 \quad \text{Distributive property}$$
$$8B = 56 \quad \text{Subtract 24.}$$
$$B = 7 \quad \text{Divide by 8.}$$

Now Try:

1.   The formula for the slope of a line is $m = \dfrac{y_1 - y_2}{x_1 - x_2}$. Find x_2 if $m = 3$, $y_1 = 12$, $y_2 = 8$, and $x_1 = 5$.

   _____

Review these examples for Objective 2:

2. Solve the following formula for R_1.

$$\frac{1}{R} = \frac{1}{R_1} + \frac{1}{R_2}$$

Start by multiplying by the LCD, RR_1R_2.

$$RR_1R_2\left(\frac{1}{R}\right) = RR_1R_2\left(\frac{1}{R_1} + \frac{1}{R_2}\right)$$

$$RR_1R_2\left(\frac{1}{R}\right) = RR_1R_2\left(\frac{1}{R_1}\right) + RR_1R_2\left(\frac{1}{R_2}\right)$$

Distributive property

$$R_1R_2 = RR_2 + RR_1 \quad \text{Simplify.}$$

$$R_1R_2 - RR_1 = RR_2$$

Subtract RR_1 to get both terms with R_1 on same side.

$$R_1(R_2 - R) = RR_2 \quad \text{Factor.}$$

$$R_1 = \frac{RR_2}{R_2 - R} \quad \text{Divide by } R_2 - R.$$

3. Solve the following formula for r.

$$P = \frac{A}{1 + rt}$$

$$P = \frac{A}{1 + rt}$$

$$P(1 + rt) = \frac{A}{1 + rt}(1 + rt)$$

Multiply by $(1 + rt)$.

$$P + Prt = A \quad \text{Distributive property}$$

$$Prt = A - P \quad \text{Subtract } P.$$

$$r = \frac{A - P}{Pt} \quad \text{Divide by } Pt.$$

Now Try:

2. Solve $\dfrac{1}{R} = \dfrac{1}{R_1} + \dfrac{1}{R_2}$ for R.

   _____

3. $A = \dfrac{R_1R_2}{R_1 + R_r}$ for R_r

   _____

Review these examples for Objective 3:

4. In 2009, about 5 of every 100 Americans identified themselves as Asian-American. The population at that time was about 307 million. How many Asian-Americans were there? (Source: U.S. Census)

Step 1: Read the problem.

Step 2: Assign a variable. Let x = the number of Asian-Americans.

Now Try:

4. In 2009, about 13 of every 100 Americans identified themselves as African-American. The population at that time was about 307 million. How many African-Americans were there? (Source: U.S. Census)

   _____

Step 3: Write an equation. Use a proportion.

$$\frac{5}{100} = \frac{x}{307}$$

Step 4: Solve.

$$\frac{5}{100} = \frac{x}{307}$$

$$(100)(307)\left(\frac{5}{100}\right) = \left(\frac{x}{307}\right)(100)(307)$$

Multiply by the LCD, (100)(307).

$$1535 = 100x \quad \text{Simplify.}$$

$$15.35 = x \quad \text{Divide by 100.}$$

Step 5: State the answer. There were 15.35 million Asian-Americans in 2009.

Step 6: Check that the ratio of 15.35 million to 307 million equals $\frac{5}{100}$.

5. Ryan's car uses 5 gallons of gasoline to travel 120 miles. If he has 4 gallons of gas in the car, how much more gasoline will he need to travel 288 miles?

Step 1: **Read** the problem.

Step 2: **Assign** a variable. Let $x =$ the additional number of gallons of gas needed.

Step 3: Write an equation. Use a proportion.

$$\frac{\text{gallons} \rightarrow}{\text{miles} \rightarrow} \frac{5}{120} = \frac{4+x}{288} \frac{\leftarrow \text{gallons}}{\leftarrow \text{miles}}$$

Step 4: **Solve**.

$$\frac{5}{120} = \frac{4+x}{288}$$

$$(288)(5) = (120)(4+x) \quad \text{If } \frac{a}{b} = \frac{c}{d}, \text{ then } ad = bc.$$

$$1440 = 480 + 120x \quad \text{Multiply; distributive property}$$

$$960 = 120x \quad \text{Subtract 480.}$$

$$8 = x \quad \text{Divide by 120.}$$

Step 5: **State the answer**. Ryan will need 8 more gallons of gas.

5. Beth's car uses 12 gallons of gasoline to travel 336 miles. If she has 5 gallons of gas in the car, how much more gasoline will she need to travel 252 miles?

_____

Step 6: **Check**. The original 4 gallons plus 8 gallons equals 12 gallons.

$$\frac{5}{120} \overset{?}{=} \frac{12}{288}$$

$$\frac{1}{24} = \frac{1}{24} \checkmark$$

Review these examples for Objective 4:	**Now Try:**
6. Mark can row 5 miles per hour in still water. It takes him as long to row 4 miles upstream as 16 miles downstream. How fast is the current?	**6.** A boat goes 6 miles per hour in still water. It takes as long to go 40 miles upstream as 80 miles downstream. Find the speed of the current.

Step 1 **Read** the problem carefully. We must find the speed of the current.

Step 2 **Assign a variable.**

Let x = the speed of the current. Traveling upstream (against the current) slows Mark down, so his rate is the difference between his rate in still water and the rate of the current, that is $5 - x$ mph. Traveling downstream (with the current) speeds him up, so his rate is the sum of his rate in still water and the rate of the current, $5 + x$ mph. We can summarize the given information in a table. Use the formula $d = rt$ or $t = \frac{d}{r}$.

	r	d	t
upstream	$5 - x$	4	$\dfrac{4}{5 - x}$
downstream	$5 + x$	16	$\dfrac{16}{5 + x}$

Step 3 **Write an equation.** The times are equal so we have

$$\frac{4}{5 - x} = \frac{16}{5 + x}.$$

Step 4 **Solve.**

$$\frac{4}{5-x} = \frac{16}{5+x}$$

$$(5-x)(5+x)\left(\frac{4}{5-x}\right)$$

$$= (5-x)(5+x)\left(\frac{16}{5+x}\right)$$

Multiply by the LCD, $(5-x)(5+x)$.

$$4(5+x) = (5-x)16 \qquad \text{Divide out common factors.}$$

$$20 + 4x = 80 - 16x \qquad \text{Distributive property}$$

$$20x = 60 \qquad \text{Add } 16x; \text{ subtract } 20.$$

$$x = 3 \qquad \text{Divide by } 20.$$

Step 5 **State the answer.**

The speed of the current is 3 miles per hour.

Step 6 **Check** the solution in the words of the original problem. Mike can row upstream $5 - 3 = 2$ miles per hour. It will take him $\frac{4}{2} = 2$ hours to row four miles upstream. Mike can row downstream $5 + 3 = 8$ miles per hour. It will takes him $\frac{16}{8} = 2$ hours to row 16 miles downstream.

The time upstream equals the time downstream, as required.

7. Bev and Mike drove 140 miles before stopping at a rest area. After leaving the rest area, they traveled an additional 90 miles at an average speed that was 4 miles per hour faster than their speed before stopping. If the entire trip took 4 hours of driving time, what was the average speed during each part of the trip?

Step 1: **Read** the problem carefully. We must find the average speed during each part of the trip.

Step 2: **Assign a variable.**

Let $x =$ the rate during the first part of the trip. Then $x + 4 =$ the rate during the second part of the trip.

7. Marlene ran 5 miles on her treadmill followed by a 1 mile cool-down walk. If she runs twice as fast as she walks and she spent 70 minutes on the treadmill, what was her walking rate in miles per hour?

_____

Express the times in terms of the know distances and the variable rates.

time for the first part of the trip: $t = \dfrac{d}{r} = \dfrac{140}{x}$

time for the second part of the trip:

$t = \dfrac{d}{r} = \dfrac{90}{x+4}$

	d	r	t
First part	140	x	$\dfrac{140}{x}$
Second part	90	$x+4$	$\dfrac{90}{x+4}$

Step 3: **Write an equation.**

$\dfrac{140}{x} + \dfrac{90}{x+4} = 4$

Step 4: **Solve.**

$$\dfrac{140}{x} + \dfrac{90}{x+4} = 4$$

$$x(x+4)\left(\dfrac{140}{x} + \dfrac{90}{x+4}\right) = x(x+4)(4)$$

Multiply by the LCD, $x(x+4)$.

$$x(x+4)\left(\dfrac{140}{x}\right) + x(x+4)\left(\dfrac{90}{x+4}\right)$$
$$= x(x+4)(4)$$

Distributive property

$$140x + 560 + 90x = 4x^2 + 16x$$

$$4x^2 - 214x - 560 = 0 \quad \text{Standard form}$$

$$2x^2 - 107x - 280 = 0 \quad \text{Divide by 2.}$$

$$(x-56)(2x+5) = 0 \quad \text{Factor.}$$

$$x - 56 = 0 \quad \text{or} \quad 2x + 5 = 0$$
$$x = 56 \qquad\qquad 2x = -5$$

Zero-factor property; solve.

$$x = -\dfrac{5}{2}$$

We discard the negative answer since rate (speed) cannot be negative.

Step 5: **State the answer.** The rate for the first part of the trip was 56 miles per hour, and the rate for the second part of the trip was 56 + 4 = 60 miles per hour.

Step 6: **Check** the solution in the words of the original problem.

Review this example for Objective 5:

8. Kelly can clean the house in 6 hours, but it takes Linda 4 hours. How long would it take them to clean the house if they worked together?

   Step 1 **Read** the problem carefully. We must find how long will it take them to clean the house if they work together.

   Step 2 **Assign a variable.**
   Let x = the time working together.
   We can summarize the given information in a table.

	Rate	Time working together	Part of the job
Kelly	$\frac{1}{6}$	x	$\frac{x}{6}$
Linda	$\frac{1}{4}$	x	$\frac{x}{4}$

   Step 3 **Write an equation.**

   $$\underbrace{\frac{x}{6}}_{\substack{\text{Part done} \\ \text{by Kelly}}} + \underbrace{\frac{x}{4}}_{\substack{\text{Part done} \\ \text{by Linda}}} = \underbrace{1}_{\substack{\text{1 whole} \\ \text{job}}}$$

   Step 4 **Solve.**

   $$\frac{x}{6} + \frac{x}{4} = 1$$

   $$24\left(\frac{x}{6} + \frac{x}{4}\right) = 24 \cdot 1$$

   Multiply by the LCD, 24.

   $$24\left(\frac{x}{6}\right) + 24\left(\frac{x}{4}\right) = 24 \cdot 1 \quad \text{Distributive property}$$

   $$4x + 6x = 24 \quad \text{Multiply.}$$
   $$10x = 24 \quad \text{Add.}$$
   $$x = \frac{24}{10} = 2.4 \quad \text{Divide by 10.}$$

   Step 5 **State the answer.**
   It will take Kelly and Linda 2.4 hours to clean the house working together.

   Step 6 **Check** to be sure that answer is correct.

Now Try:

8. Nina can wash the walls in a certain room in 2 hours and Mark can wash these walls in 5 hours. How long would it take them to complete the task if they work together?

   _____

Objective 1 Find the value of an unknown variable in a formula.

For extra help, see Example 1 on page 396 of your text, the Section Lecture video for Section 7.5, and Exercise Solution Clip 5.

Find the value of the variable indicated.

1. If $F = \dfrac{GmM}{d^2}$, $F = 150$, $G = 32$, $M = 50$, and $d = 10$,   1. _____

   find m.

2. If $\dfrac{1}{R} = \dfrac{1}{R_1} + \dfrac{1}{R_2}$, $R = 6$, and $R_1 = 12$, find R_2.   2. _____

3. If $B = \dfrac{3V}{h}$, $B = 12$, and $V = 48$, find h.   3. _____

Objective 2 Solve a formula for a specified variable.

For extra help, see Examples 2 and 3 on page 396–397 of your text, the Section Lecture video for Section 7.5, and Exercise Solution Clips 11 and 19.

Solve the formula for the specified variable.

4. $\dfrac{V_1 P_1}{T_1} = \dfrac{V_2 P_2}{T_2}$ for T_2 4. _____

5. $S_n = \dfrac{n}{2}(a_1 + a_n)$ for a_n 5. _____

6. $I = \dfrac{nE}{R + nr}$ for R

6. _____

Objective 3 Solve applications by using proportions.

For extra help, see Examples 4 and 5 on page 397–398 of your text, the Section Lecture video for Section 7.5, and Exercise Solution Clip 39.

Use a proportion to solve each problem.

7. In a certain midwestern city in a recent year, there were 500 crimes committed per 100,000 population. If the population of that city was 350,000, how many crimes were committed?

7. _____

8. If Colette can address her wedding invitations in 4 ½ hours, what is her rate (in job per hour)?

8. _____

9. George paid $0.93 in sales tax on a purchase of $15.50. How much sales tax would he pay on a purchase of $480 at the same tax rate?

9. _____

Objective 4 Solve applications about distance, rate, and time.

For extra help, see Examples 6 and 7 on page 398–400 of your text, the Section Lecture video for Section 7.5, and Exercise Solution Clips 45 and 51.

Solve each problem.

10. Joan averages 10 miles per hour riding her bike to town. Averaging 30 miles per hour by car takes her 2 hours less time. How far does she travel to town?

10. _____

11. Clara's boat goes 14 miles per hour. Find the speed of the current in the river if she can go 8 miles downstream in the same time as she can go 6 miles upstream.

11. _____

12. Pauline and Pete agree to meet in Columbia. Pauline travels 120 miles, while Pete travels 80 miles. If Pauline's speed is 20 miles per hour greater than Pete's and they both spend the same amount of time traveling, at what speed does each travel?

12. _____

Objective 5 Solve applications about work rates.

For extra help, see Example 8 on page 401 of your text, the Section Lecture video for Section 7.5, and Exercise Solution Clip 53.

Solve each problem.

13. Rose and Max want to paint a chair. Max can do it in 3 hours while Rose can do it in 4 hours. How long will it take them working together?

13. _____

14. Fred can seal an asphalt driveway in $\frac{1}{3}$ the time it takes John. Working together, it takes them $1\frac{1}{2}$ hours. How long would it have taken Fred working alone?

14. _____

15. A vat of chocolate can be filled by an inlet pipe in 12 hours. An outlet pipe can empty the vat in 15 hours. How long will it take to fill the vat if both pipes are left open?

15. _____

Chapter 7 RATIONAL EXPRESSIONS AND FUNCTIONS

7.6   Variation

Learning Objectives
1   Write an equation expressing direct variation.
2   Find the constant of variation, and solve direct variation problems.
3   Solve inverse variation problems.
4   Solve joint variation problems.
5   Solve combined variation problems.

Key Terms

Use the vocabulary terms listed below to complete each statement in exercises 1–4.

varies directly   varies inversely   proportional   constant of variation

1.   In the equations for direct and inverse variation, k is the _____.

2.   If there exists a real number k such that $y = \dfrac{k}{x}$, then y _____ as x.

3.   The equation $y = kxz$ is used to represent a situation where _____ as x and z.

4.   If y varies directly as x and there exists a real number k such that $y = kx$, then y is said to be _____ to x.

Guided Examples

Review these examples for Objective 2:

1.   If 12 gallons of gasoline cost $34.68, how much does 1 gallon of gasoline cost?

Let g represent the number of gallons of gasoline and let C represent the total cost of the gasoline. Then the variation equation is
$C = kg$ C varies directly as g.

$$C = kg$$
$$34.68 = 12k$$
$$2.89 = k$$

The cost per gallon is $2.89.

Now Try:

1.   One week a manufacturer sold 1200 items for a total profit of $30,000. What was the profit for one item.

_____

2. A person's weight on the moon varies directly with the person's weight on earth. A 120-pound person would weigh about 20 pounds on the moon. How much would a 150-pound person weigh on the moon?

If m represents the person's weight on the moon and w represents the person's weight on earth. Then, the variation equation is $m = kw$.

$$m = kw$$
$$20 = k \cdot 120 \quad \text{Let } m = 20 \text{ and } w = 120.$$
$$\frac{20}{120} = \frac{1}{6} = k \quad \text{Solve for } k; \text{ lowest terms}$$

Now, substitute $\frac{1}{6}$ for k and 150 for w in the variation equation.

$$m = kw$$
$$m = \frac{1}{6} \cdot 150 = 25$$

A 150-pound person will weigh 25 pounds on the moon.

3. The surface area of a sphere varies directly as the square of its radius. If the surface area of a sphere with a radius of 12 inches is 576π square inches, find the surface area of a sphere with a radius of 3 inches.

$$A = kr^2 \quad \text{Equation for direct variation}$$
$$576\pi = k \cdot 12^2 \quad \text{Substitute given values.}$$
$$576\pi = 144k$$
$$\frac{576\pi}{144} = 4\pi = k \quad \text{Constant of variation}$$

Since $A = kr^2$ and $k = \dfrac{576\pi}{144} = 4\pi$ we have

$A = 4\pi r^2$.

Now find A when r is 3.

$$A = 4\pi \cdot 3^2 = 4\pi \cdot 9 = 36\pi$$

The surface area of a sphere with a radius of 3 inches is 36π square inches.

2. The pressure exerted by a certain liquid at a given point varies directly as the depth of the point beneath the surface of the liquid. The pressure at 10 feet is 50 pounds per square inch (psi). What is the pressure at 25 feet?

_____

3. The area of a circle varies directly as the square of the radius. A circle with a radius of 5 centimeters has an area of 78.5 square centimeters. Find the area if the radius changes to 7 centimeters.

_____

Review these examples for Objective 3:

4. The current in a simple electrical circuit varies inversely as the resistance. If the current is 50 amps (an *ampere* is a unit for measuring current) when the resistance is 10 ohms (an *ohm* is a unit for measuring resistance), find the current if the resistance is 5 ohms.

    Let c = the current in amps and let r = the resistance in ohms.

    $c = \dfrac{k}{r}$ Equation for indirect variation

    $50 = \dfrac{k}{10}$ Substitute given values.

    $500 = k \leftarrow$ Constant of variation

    Now use $c = \dfrac{500}{r}$.

    $c = \dfrac{500}{5} = 100$

    The current is 100 amps when the resistance is 5 ohms.

5. With constant power, the resistance used in a simple electric circuit varies inversely as the square of the current. If the resistance is 120 ohms when the current is 12 amps, find the resistance if the current is reduced to 9 amps.

    Let R represent resistance (in ohms) and

    I = current (in amps). Then $R = \dfrac{k}{I^2}$.

    First, we solve for the constant of variation by substituting 120 for R and 12 for I.

    $120 = \dfrac{k}{12^2}$

    $120 \cdot 12^2 = k$

    Now use the value for k and 9 for I to find R.

    $R = \dfrac{120 \cdot 12^2}{9^2} \approx 213.3$

    The resistance is about 213.3 ohms when the current is 9 amps.

Now Try:

4. The speed of a pulley varies inversely as its diameter. One kind of pulley, with a diameter of 3 inches, turns at 150 revolutions per minute. Find the speed of a similar pulley with a diameter of 5 inches.

    _____

5. If y varies inversely as x^3, and $y = 9$ when $x = 2$, find y when $x = 4$.

    _____

Review this example for Objective 4:

6. For a fixed interest rate, interest varies jointly as the principal and the time in years. If $5000 invested for 4 years earns $900, how much interest will $6000 invested for 3 years earn at the same interest rate?

    Let I = the interest, p = the principal, and t = the time in years. Then, $I = kpt$.

    $$I = kpt$$
    $$900 = k \cdot 5000 \cdot 4 \quad \text{Substitute given values.}$$
    $$\frac{900}{20,000} = \frac{9}{200} = k$$

    Now use $k = \frac{9}{200}$.

    $$I = \frac{9}{200} \cdot 6000 \cdot 3 = 810$$

    $6000 invested for three years will earn $810 in interest.

Now Try:

6. The strength of a rectangular beam varies jointly as its width and the square of its depth. If the strength of a beam 2 inches wide by 10 inches deep is 1000 pounds per square inch, what is the strength of a beam 4 inches wide and 8 inches deep?

    _____

Review this example for Objective 5:

7. The number of hours h that it takes w workers to assemble x machines varies directly as the number of machines and inversely as the number of workers. If four workers can assemble 12 machines in four hours, how many workers are needed to assemble 36 machines in eight hours?

    The variation equation is $h = \dfrac{kx}{w}$.

    To find k, let $h = 4$, $x = 12$, and $w = 4$.

    $$4 = \frac{k \cdot 12}{4}$$
    $$k = \frac{4 \cdot 4}{12} = \frac{4}{3}$$

    Now find w when $x = 36$ and $h = 8$.

    $$8 = \frac{\frac{4}{3} \cdot 36}{w}$$
    $$8w = 48 \quad \text{Multiply by } w; \text{ multiply.}$$
    $$w = 6 \quad \text{Divide by 8.}$$

    Eight workers are needed to assemble 36 machines in eight hours.

Now Try:

7. The volume of a gas varies directly as its temperature and inversely as its pressure. The volume of a gas at 85° C at a pressure of 12 kg/cm^2 is 300 cm^3. What is the volume when the pressure is 20 kg/cm^2 and the temperature is 30° C?

    _____

Objective 1   Write an equation expressing direct variation.
Objective 2   Find the constant of variation, and solve direct variation problems.
For extra help, see Examples 1–3 on pages 407–409 of your text, the Section Lecture video for Section 7.6, and Exercise Solution Clip 37, 39, and 45.

Solve each problem.

1.   If c varies directly as the square of d, and $c = 100$ when $d = 5$, find c when $d = 3$.

1. _____

2.   A car travels 342 mile on 12 gallons of gasoline. How many gallons of gasoline are needed to travel 228 miles?

2. _____

3.   The force required to compress a spring varies directly as the change in length of the spring. If a force of 20 newtons is required to compress a spring 2 centimeters in length, how much force is required to compress a spring of length 10 centimeters?

3. _____

Objective 3 Solve inverse variation problems.
For extra help, see Examples 4 and 5 on pages 410–411 of your text, the Section Lecture video for Section 7.6, and Exercise Solution Clip 49.

Solve each problem.

4.   If d varies inversely as c, and $d = 18$ when $c = \frac{1}{3}$, find d when $c = \frac{2}{5}$.

4. _____

5.   The illumination produced by a light source varies inversely as the square of the distance from the source. If the illumination produced 4 feet from a light source is 75 foot-candles, find the illumination produced 9 feet from the same source.

5. _____

6. The weight of an object varies inversely as the square of   6. _____
    its distance from the center of the earth. If an object 8000
    miles from the center of the earth weighs 90 pounds, find
    its weight when it is 12,000 miles from the center of the
    earth.

7. The speed of a pulley varies inversely as its diameter.   7. _____
    One kind of pulley, with a diameter of 3 inches, turns at
    150 revolutions per minute. Find the speed of a similar
    pulley with diameter of 5 inches.

Objective 4 Solve joint variation problems.
For extra help, see Example 6 on page 411 of your text, the Section Lecture video for Section
7.6, and Exercise Solution Clip 51.

Solve each problem.

8. Suppose d varies jointly as f^2 and g^2, and $d = 384$ 8. _____
    when $f = 3$ and $g = 8$. Find d when $f = 6$ and $g = 2$.

9. Kinetic energy varies jointly as the mass and the square   9. _____
    of the velocity. A mass of 8 grams and a velocity of 5
    centimeters per second has a kinetic energy of 100 ergs.
    Find the kinetic energy for a mass of 6 grams and a
    velocity of 9 centimeters per second.

10. The work w (in joules) done when lifting an object is   10. _____
    jointly proportional to the product of the mass m (in kg)
    of the object and the height h (in meter) the object is
    lifted. If the work done when a 120 kg object is lifted
    1.8 meters above the ground is 2116.8 joules, how much
    work is done when lifting a 100kg object 1.5 meters
    above the ground?

11. The absolute temperature of an ideal gas varies jointly   11. _____
    as its pressure and its volume. If the absolute
    temperature is 250° when the pressure is 25 pounds per
    square centimeter and the volume is 50 cubic
    centimeters, find the absolute temperature when the
    pressure is 50 pounds per square centimeter and the
    volume is 75 cubic centimeters.

Objective 5 Solve combined variation problems.
For extra help, see Example 7 on page 412 of your text, the Section Lecture video for Section
7.6, and Exercise Solution Clip 59.

Solve each problem.

12. p varies jointly as P, V, and t and inversely as v and T.   12. _____
    Suppose $p = 65.625$ when $P = 50$, $V = 9$, $t = 350$, $v = 8$,
    and $T = 300$. Find p when $P = 60$, $V = 8$,
    $t = 300$, $v = 6$, and $T = 200$.

13. The volume of a gas varies inversely as the pressure and   13. _____
    directly as the temperature. If a certain gas occupies a
    volume of 1.3 liters at 300 K and a pressure of 18
    kilograms per square centimeter, find the volume at 340
    K and a pressure of 24 kilograms per square centimeter.

14. The time required to lay a sidewalk varies directly as its   14. _____
    length and inversely as the number of people who are
    working on the job. If three people can lay a sidewalk
    100 feet long in 15 hours, how long would it take two
    people to lay a sidewalk 40 feet long?

15. When an object is moving in a circular path, the centripetal force varies directly as the square of the velocity and inversely as the radius of the circle. A stone that is whirled at the end of a string 50 centimeters long at 900 centimeters per second has a centripetal force of 3,240,000 dynes. Find the centripetal force if the stone is whirled at the end of a string 75 centimeters long at 1500 centimeters per second.

15. _____

Chapter 8 ROOTS, RADICALS, AND ROOT FUNCTIONS

8.1   Radical Expressions and Graphs

Learning Objectives
1   Find roots of numbers.
2   Find principal roots.
3   Graph functions defined by radical expressions.
4   Find nth roots of nth powers.
5   Use a calculator to find roots.

Key Terms

Use the vocabulary terms listed below to complete each statement in exercises 1–7.

    radicand   **index (order)**   **radical**   **principal root**

    radical expression   **square root function**

    cube root function

1.   A _____ is a radical sign and the number or expression that appears under it.

2.   The _____ is the number or expression under the radical sign.

3. $f(x) = \sqrt[3]{x}$ is called the _____.

4.   The domain and range of $f(x) = \sqrt{x}$, _____, are both $[0, \infty)$.

5.   In the expression $\sqrt[5]{8}$, the _____ is 5.

6.   A _____ is an algebraic expression containing a radical.

7.   For even indexes, the symbols $\sqrt{}, \sqrt[4]{}, \sqrt[6]{}, \ldots, \sqrt[n]{}$, are used for nonnegative roots, which are called _____.

Guided Examples

Review this example for Objective 1:

1. Simplify.

   a. $\sqrt[3]{216}$ **b.** $\sqrt[4]{81}$

   c. $\sqrt[4]{\dfrac{16}{81}}$ **d.** $\sqrt[5]{0.00243}$

   a. $\sqrt[3]{216} = 6$ **b.** $\sqrt[4]{81} = 3$

   c. $\sqrt[4]{\dfrac{16}{81}} = \dfrac{2}{3}$ **d.** $\sqrt[5]{0.00243} = 0.3$

Now Try:

1. Simplify.

   a. $\sqrt[3]{125}$ _____

   b. $\sqrt[4]{10{,}000}$ _____

   c. $\sqrt[3]{\dfrac{512}{27}}$ _____

   d. $\sqrt[5]{0.00032}$ _____

Review this example for Objective 2:

2. Find each root.

   a. $\sqrt{900}$ **b.** $-\sqrt{144}$

   c. $\sqrt[3]{-216}$ **d.** $\sqrt[4]{-625}$

   a. The radical $\sqrt{900}$ represents the positive or principal square root of 900. Since $30^2 = 900$, $\sqrt{900} = 30$.

   b. $-\sqrt{144}$ represents the negative square root of 144. Therefore, $-\sqrt{144} = -12$.

   c. $\sqrt[3]{-216} = -6$

   d. $\sqrt[4]{-625}$ is not a real number.

Now Try:

2. Find each root.

   a. $\sqrt{484}$ _____

   b. $-\sqrt{324}$ _____

   c. $\sqrt[5]{-1024}$ _____

   d. $\sqrt[6]{-64}$ _____

Review this example for Objective 3:

3. Graph each function. Give the domain and range.

   a. $f(x) = \sqrt{x-1}$ **b.** $f(x) = \sqrt[3]{x} + 1$

   a. Create a table of values.

x	$f(x) = \sqrt{x-1}$
1	$\sqrt{1-1} = 0$
5	$\sqrt{5-1} = 2$
10	$\sqrt{10-1} = 3$

   For the radicand to be nonnegative, we must have $x-1 \geq 0$ or $x \geq 1$. Therefore, the domain is $[1, \infty)$. Function values are nonnegative, so the range is $[0, \infty)$.

   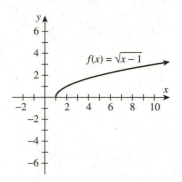

   b. Create a table of values.

x	$f(x) = \sqrt[3]{x} + 1$
-8	$\sqrt[3]{-8} + 1 = -1$
-1	$\sqrt[3]{-1} + 1 = 0$
0	$\sqrt[3]{0} + 1 = 1$
1	$\sqrt[3]{1} + 1 = 2$
8	$\sqrt[3]{8} + 1 = 3$

Now Try:

3. Graph each function. Give the domain and range.

   a. $f(x) = \sqrt{x} - 1$

   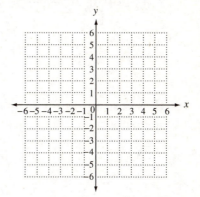

   Domain: _____

   Range: _____

   b. $f(x) = \sqrt[3]{x} + 1$

   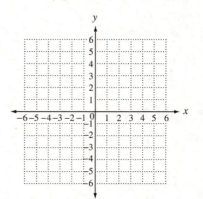

   Domain: _____

   Range: _____

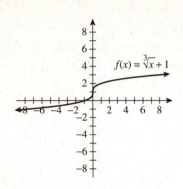

$f(x) = \sqrt[3]{x} + 1$

Review these examples for Objective 4:

4. Find each square root.

   a. $\sqrt{33^2}$ **b.** $\sqrt{(-33)^2}$

   c. $\sqrt{r^2}$ **d.** $\sqrt{(-r)^2}$

   a. $\sqrt{33^2} = |33| = 33$

   b. $\sqrt{(-33)^2} = |-33| = 33$

   c. $\sqrt{r^2} = |r|$

   d. $\sqrt{(-r)^2} = |-r| = |r|$

5. Simplify each root.

   a. $\sqrt[5]{(-3)^5}$ **b.** $\sqrt[4]{(-5)^4}$

   c. $-\sqrt[6]{(-8)^6}$ **d.** $-\sqrt{r^{12}}$

   e. $\sqrt[4]{x^{20}}$ **f.** $\sqrt[5]{s^{20}}$

   a. $\sqrt[5]{(-3)^5} = -3$ n is odd.

   b. $\sqrt[4]{(-5)^4} = |-5| = 5$ n is even.
   Use absolute value.

   c. $-\sqrt[6]{(-8)^6} = -|-8| = -8$ n is even.
   Use absolute value.

   d. $-\sqrt{r^{12}} = -\left|r^6\right| = -r^6$ For all r, $\left|r^6\right| = r^6$.

Now Try:

4. Find each square root.

   a. $\sqrt{73^2}$ _____

   b. $\sqrt{(-37)^2}$ _____

   c. $\sqrt{a^2}$ _____

   d. $\sqrt{(-a)^2}$ _____

5. Simplify each root.

   a. $\sqrt[7]{(-5)^7}$ _____

   b. $\sqrt[6]{(-4)^6}$ _____

   c. $-\sqrt[4]{(-2)^4}$ _____

   d. $-\sqrt{x^8}$ _____

   e. $\sqrt[6]{x^{30}}$ _____

   f. $\sqrt[3]{w^{30}}$ _____

e. $\sqrt[4]{x^{20}} = \left| x^5 \right|$

Use absolute value to guarantee that the result is not negative.

f. $\sqrt[5]{s^{20}} = s^4$, because $s^{20} = \left(s^4 \right)^5$.

Review these examples for Objective 5:

6. Use a calculator to approximate each radical to three decimal places.

  a. $-\sqrt{420}$ **b.** $\sqrt[3]{21}$

  c. $\sqrt[4]{84}$

```
-√(420)
         -20.49390153
3√(21)
         2.758924176
4*√84
         3.027400104
```

  a. $-\sqrt{420} \approx -20.494$

  b. $\sqrt[3]{21} \approx 2.759$

  c. $\sqrt[4]{84} \approx 3.027$

7. The minimum speed at which a driver is traveling when a car skids to a stop can be estimated by the formula $s = \sqrt{30fd}$, where s is the speed of the car in miles per hour when the brakes are applied, f is a constant drag factor, and d is the length of the skid marks in feet. Suppose that a car skids to a top leaving a skid of 90 feet. Assume that the drag factor is 0.7. How fast was the driver going? Give your answer to the nearest tenth.

We must find s when $f = 0.7$ and $d = 90$.

$s = \sqrt{30fd}$

$s = \sqrt{30 \cdot 0.7 \cdot 90} = \sqrt{1890} \approx 43.5$

The driver was traveling at about 43.5 miles per hour at the start of the skid.

Now Try:

6. Use a calculator to approximate each radical to three decimal places.

  a. $-\sqrt{138}$ _____

  b. $\sqrt[5]{725}$ _____

  c. $\sqrt[4]{725}$ _____

7. Use the formula at the left to find out how fast a driver was traveling if the skid mark is 20 feet and the drag factor is 0.9. Give your answer to the nearest tenth.

_____

Name: 　　　　　　　　　　　　　Date:

Instructor: 　　　　　　　　　　Section:

Objective 1 Find roots of numbers.

Objective 2 Find principal roots

For extra help, see Examples 1 and 2 on pages 428–429 of your text, the Section Lecture video for Section 8.1, and Exercise Solution Clips 5 and 7.

Find all real numbers that are not in the domain of the function.

1. $-\sqrt{\dfrac{625}{484}}$ 　　　　　　　　　　　　1. _____

2. $\sqrt[3]{-125}$ 　　　　　　　　　　　　　　2. _____

3. $\sqrt[4]{-81}$ 　　　　　　　　　　　　　　3. _____

4. $-\sqrt[3]{-64}$ 　　　　　　　　　　　　　4. _____

Objective 3 Graph functions defined by radical expressions.

For extra help, see Example 3 on page 430 of your text, the Section Lecture video for Section 8.1, and Exercise Solution Clip 41.

Graph each function by creating a table of values. Give the domain and the range.

5. $f(x) = \sqrt{x+6}$ 　　　　　　　　　　5.

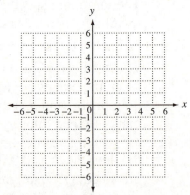

Domain: _____

Range: _____

6. $f(x) = \sqrt[3]{x} - 4$

6.

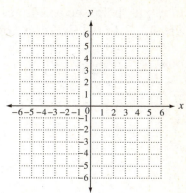

Domain: _____

Range: _____

7. $f(x) = 3 - \sqrt[3]{x}$

7.

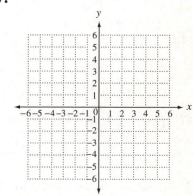

Domain: _____

Range: _____

8. $f(x) = \sqrt{x} + 2$

8.

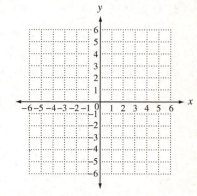

Domain: _____

Range: _____

Objective 4 Find *n*th roots of *n*th powers.
For extra help, see Examples 4 and 5 on page 431 of your text, the Section Lecture video for Section 8.1, and Exercise Solution Clip 49 and 53.

Simplify each root.

9. $\sqrt{(-9)^2}$

9. _____

10. $-\sqrt{y^6}$

10. _____

11. $\sqrt[6]{a^{18}}$

11. _____

12. $-\sqrt[4]{x^{28}}$

12. _____

Objective 5 Use a calculator to find roots.
For extra help, see Examples 6 and 7 on page 432 of your text, the Section Lecture video for Section 8.1, and Exercise Solution Clips 71 and 85.

Use a calculator to find a decimal approximation. Give the answer to the nearest thousandth.

13. $\sqrt[3]{701}$

13. _____

14. $-\sqrt{990}$

14. _____

15. The time t in seconds for one complete swing of a simple pendulum, where L is the length of the pendulum in feet is $t = 2\pi\sqrt{\dfrac{L}{32}}$. Find the time of a complete swing of a 4-ft pendulum to the nearest tenth of a second.

15. _____

Chapter 8 ROOTS, RADICALS, AND ROOT FUNCTIONS

8.2   Rational Exponents

Learning Objectives
1   Use exponential notation for *n*th roots.
2   Define and use expressions of the form $a^{m/n}$.
3   Convert between radicals and rational exponents.
4   Use the rules for exponents with rational exponents.

Key Terms

Use the vocabulary terms listed below to complete each statement in exercises 1–4.

product rule for exponents   **quotient rule for exponents**

power rule for exponents   **radical form of** $a^{m/n}$

1.   The _____ is written as $\sqrt[n]{a^m} = \left(\sqrt[n]{a}\right)^m$

2. $\left(x^2 y^3\right)^4 = x^8 y^{12}$ is an example of the _____.

3. $w^5 w^3 = w^8$ is an example of the _____.

4. $\dfrac{z^6}{z^4} = z^2$ is an example of the _____.

Guided Examples

Review this example for Objective 1:
1.   Evaluate each exponential.

    a. $(-8)^{1/3}$ **b.** $25^{1/2}$

    c. $-81^{1/4}$ **d.** $(-81)^{1/4}$

    e. $-8^{1/3}$ **f.** $\left(\frac{1}{125}\right)^{1/3}$

The denominator of the exponent is the index.

    a. $(-8)^{1/3} = \sqrt[3]{-8} = -2$

    b. $25^{1/2} = \sqrt{25} = 5$

Now Try:
1.   Evaluate each exponential.

    a. $216^{1/3}$ _____

    b. $(-216)^{1/3}$ _____

    c. $1024^{1/10}$ _____

    d. $(-1024)^{1/10}$ _____

    e. $-1024^{1/10}$ _____

c. $-81^{1/4} = -\sqrt[4]{81} = -3$

d. $(-81)^{1/4} = \sqrt[4]{-81}$, which is not a real number.

e. $-8^{1/3} = -\sqrt[3]{8} = -2$

f. $\left(\frac{1}{125}\right)^{1/3} = \sqrt[3]{\frac{1}{125}} = \frac{1}{5}$

f. $\left(\frac{1}{16}\right)^{1/4}$ _____

Review these examples for Objective 2:

2. Evaluate each exponential.

    a. $100^{3/2}$ **b.** $125^{2/3}$

    c. $-729^{5/6}$ **d.** $(-729)^{5/6}$

    e. $(-125)^{2/3}$

    a. $100^{3/2} = \left(100^{1/2}\right)^3 = 10^3 = 1000$

    b. $125^{2/3} = \left(125^{1/3}\right)^2 = 5^2 = 25$

    c. $-729^{5/6} = -\left(729^{1/6}\right)^5 = -3^5 = -243$

    d. $(-729)^{5/6}$ is not a real number because $(-729)^{1/6}$ is not a real number.

    e. $(-125)^{2/3} = \left((-125)^{1/3}\right)^2 = (-5)^2 = 25$

Now Try:

2. Evaluate each exponential.

    a. $27^{2/3}$ _____

    b. $16^{3/4}$ _____

    c. $-36^{3/2}$ _____

    d. $(-36)^{3/2}$ _____

    e. $(-27)^{2/3}$ _____

3. Evaluate each exponential.

    a. $100^{-3/2}$ **b.** $216^{-2/3}$

    c. $-729^{-5/6}$

    a. $100^{-3/2} = \dfrac{1}{100^{3/2}} = \dfrac{1}{\left(100^{1/2}\right)^3}$

              $= \dfrac{1}{10^3} = \dfrac{1}{1000}$

3. Evaluate.

    a. $32^{-2/5}$ _____

    b. $-125^{-4/3}$ _____

    c. $\left(\frac{27}{64}\right)^{-2/3}$ _____

b. $216^{-2/3} = \dfrac{1}{216^{2/3}} = \dfrac{1}{\left(216^{1/3}\right)^2}$

$\quad = \dfrac{1}{6^2} = \dfrac{1}{36}$

c. $-729^{-5/6} = -\dfrac{1}{729^{5/6}} = -\dfrac{1}{\left(729^{1/6}\right)^5}$

$\quad = -\dfrac{1}{3^5} = -\dfrac{1}{243}$

Review this example for Objective 3:

4. Write each exponential as a radical. Assume that all variables represent positive real numbers. In parts g and h, write each radical as an exponential.

  a. $(-9)^{1/3}$ **b.** $27^{3/2}$

  c. $8x^{3/4}$ **d.** $2x^{2/5} - (4x)^{5/6}$

  e. $x^{-4/5}$ **f.** $\left(x^3 + y^2\right)^{1/5}$

  g. $\sqrt[4]{3^2}$ **h.** $\sqrt[3]{x^{12}}$

  a. $(-9)^{1/3} = \sqrt[3]{-9}$

  b. $27^{3/2} = \left(\sqrt{27}\right)^3$

  c. $8x^{3/4} = 8\left(\sqrt[4]{x}\right)^3$

  d. $2x^{2/5} - (4x)^{5/6} = 2\left(\sqrt[5]{x}\right)^2 - \left(\sqrt[6]{4x}\right)^5$

  e. $x^{-4/5} = \dfrac{1}{x^{4/5}} = \dfrac{1}{\left(\sqrt[5]{x}\right)^4}$

  f. $\left(x^3 + y^2\right)^{1/5} = \sqrt[5]{x^3 + y^2}$

  g. $\sqrt[4]{3^2} = \left(3^2\right)^{1/4} = 3^{1/2}$

  h. $\sqrt[3]{x^{12}} = \left(x^{12}\right)^{1/3} = x^4$

Now Try:

4. Write each exponential as a radical. Assume that all variables represent positive real numbers. In parts g and h, write each radical as an exponential.

  a. $216^{1/4}$ _____

  b. $18^{2/3}$ _____

  c. $5x^{5/4}$ _____

  d. $(2x)^{4/3} - 3x^{2/5}$ _____

  e. $x^{-3/2}$ _____

  f. $\left(x^2 - y^2\right)^{1/4}$ _____

  g. $\sqrt[4]{10^2}$ _____

  h. $\sqrt[20]{x^5}$ _____

Review these examples for Objective 4:

5. Write with only positive exponents. Assume that all variables represent positive real numbers.

    a. $13^{4/5} \cdot 13^{1/2}$ **b.** $\dfrac{8^{-3/4}}{8^{-1/4}}$

    c. $\left(\dfrac{w^{7/4} w^{-1/2}}{w^{3/4}} \right)^2$ **d.** $\dfrac{\left(c^6 x^2 \right)^{2/3}}{x^3}$

    e. $a^{3/4} \left(a^{2/3} - a^{1/2} \right)$

    a. $13^{4/5} \cdot 13^{1/2} = 13^{4/5 + 1/2}$ Product rule

    $= 13^{13/10}$ Add exponents.

    b. $\dfrac{8^{-3/4}}{8^{-1/4}} = 8^{-3/4 - (-1/4)}$ Quotient rule

    $= 8^{-1/2}$ Subtract exponents.

    $= \dfrac{1}{8^{1/2}}$ $a^{-1} = \dfrac{1}{a}$

    c. $\left(\dfrac{w^{7/4} w^{-1/2}}{w^{3/4}} \right)^2 = \left(\dfrac{w^{(7/4 - 1/2)}}{w^{3/4}} \right)^2$

    Product rule

    $= \left(\dfrac{w^{5/4}}{w^{3/4}} \right)^2$ Product rule

    $= \left(w^{5/4 - 3/4} \right)^2$ Quotient rule

    $= \left(w^{1/2} \right)^2$ Quotient rule

    $= w$ Power rule

Now Try:

5. Write with only positive exponents. Assume that all variables represent positive real numbers.

    a. $5^{3/4} \cdot 5^{9/4}$ _____

    b. $\dfrac{a^{4/5}}{a^{2/3}}$ _____

    c. $\left(\dfrac{x^{-1} y^{2/3}}{x^{1/3} y^{1/2}} \right)^{-6}$ _____

    d. $\dfrac{a^{2/3} \cdot a^{-1/3}}{\left(a^{-1/6} \right)^3}$ _____

    e. $r^{1/2} \left(r^{2/3} + r^{8/3} \right)$ _____

d. $\dfrac{\left(c^6 x^2\right)^{2/3}}{x^3} = \dfrac{\left(c^6 x^2\right)^{2/3}}{x^3}$ Power rule

$= \dfrac{c^{6(2/3)} x^{2(2/3)}}{x^3}$ Power rule

$= \dfrac{c^4 x^{4/3}}{x^3}$ Multiply exponents.

$= c^4 x^{4/3-3}$ Quotient rule

$= c^4 x^{-5/3}$ Subtract exponents.

$= \dfrac{c^4}{x^{5/3}}$ $a^{-1} = \dfrac{1}{a}$

e. $a^{3/4}\left(a^{2/3} - a^{1/2}\right)$

$= a^{3/4}\left(a^{2/3}\right) - a^{3/4}\left(a^{1/2}\right)$

   Distributive property

$= a^{3/4+2/3} - a^{3/4+1/2}$

   Product property

$= a^{17/12} - a^{5/4}$ Add exponents.

6. Write all radicals as exponentials, and then apply the rules for rational exponents. Leave answers in exponential form. Assume that all variables represent positive real numbers.

a. $\sqrt[4]{x^3} \cdot \sqrt[5]{2x}$ **b.** $\dfrac{\sqrt[3]{y^5}}{\sqrt{y^3}}$

c. $\sqrt{\sqrt[4]{y^3}}$

a. $\sqrt[4]{x^3} \cdot \sqrt[5]{2x} = x^{3/4} \cdot (2x)^{1/5}$

   Convert to rational exponents.

$= x^{3/4} \cdot 2^{1/5} \cdot x^{1/5}$

   Power rule

$= 2^{1/5} x^{3/4+1/5}$

   Commutative property
   Product rule

$= 2^{1/5} x^{19/20}$

   Add exponents.

6. Write all radicals as exponentials, and then apply the rules for rational exponents. Leave answers in exponential form. Assume that all variables represent positive real numbers.

a. $\sqrt[6]{x^3} \cdot \sqrt[3]{x^2}$ _____

b. $\dfrac{\sqrt[4]{y^5}}{\sqrt{y^4}}$ _____

c. $\sqrt[3]{\sqrt[4]{x^3}}$ _____

b. $\dfrac{\sqrt[3]{y^5}}{\sqrt{y^3}} = \dfrac{y^{5/3}}{y^{1/3}}$ Convert to rational exponents.

$= y^{5/3-1/3}$ Quotient rule

$= y^{1/6}$ Subtract exponents.

c. $\sqrt{\sqrt[4]{y^3}} = \sqrt{y^{4/3}}$ Convert inside radical to rational exponents.

$= \left(y^{4/3}\right)^{1/2}$ Convert to rational exponents.

$= y^{2/3}$ Power rule

Objective 1 Use exponential notation for nth roots.

For extra help, see Example 1 on page 436 of your text, the Section Lecture video for Section 8.2, and Exercise Solution Clip 11.

Evaluate each exponential.

1. $(-32)^{1/5}$ **1.** _____

2. $\left(\dfrac{1}{256}\right)^{1/4}$ **2.** _____

3. $-144^{1/2}$ **3.** _____

Objective 2 Define and use expressions of the form $a^{m/n}$.

For extra help, see Examples 2 and 3 on pages 436–437 of your text, the Section Lecture video for Section 8.2, and Exercise Solution Clips 23 and 31.

Evaluate each exponential.

4. $-81^{5/4}$ **4.** _____

5. $216^{-2/3}$ **5.** _____

6. $\left(\dfrac{125}{64}\right)^{-2/3}$ **6.** _____

7. $243^{3/5}$

7. _____

Objective 3 Convert between radicals and rational exponents.

For extra help, see Example 4 on page 438–439 of your text, the Section Lecture video for Section 8.2, and Exercise Solution Clip 39.

Write with radicals. Assume that all variable represent positive real numbers.

8. $(4y)^{2/5} + (5x)^{1/5}$

8. _____

9. $\left(2x^4 - 3y^2\right)^{-4/3}$

9. _____

Simplify by first converting to rational exponents. Assume that all variables represent positive real numbers.

10. $\sqrt[9]{125^3}$

10. _____

11. $\sqrt[8]{a^2}$

11. _____

Objective 4 Use the rules for exponents with rational exponents.

For extra help, see Example 5 and 6 on pages 439–441 of your text, the Section Lecture video for Section 8.2, and Exercise Solution Clips 61 and 91.

Simplify each expression. Write all answers with positive exponents. Assume that all variables represent positive real numbers.

12. $\dfrac{\left(x^{-3}y^2\right)^{2/3}}{\left(x^2y^{-5}\right)^{2/5}}$

12. _____

13. $d^{-8/7}\left(d^{5/4} + 4d^3\right)$ **13.** _____

Write with rational exponents, and then apply the properties of exponents. Assume that all radicands represent positive real numbers. Give answers in exponential form.

14. $\sqrt[4]{y^5} \cdot \sqrt[5]{y^6}$ **14.** _____

15. $\sqrt[4]{\sqrt[4]{x^5}}$ **15.** _____

Chapter 8 ROOTS, RADICALS, AND ROOT FUNCTIONS

8.3   Simplifying Radical Expressions

Learning Objectives
1   Use the product rule for radicals.
2   Use the quotient rule for radicals.
3   Simplify radicals.
4   Simplify products and quotients of radicals with different indexes.
5   Use the Pythagorean theorem.
6   Use the distance formula.

Key Terms

Use the vocabulary terms listed below to complete each statement in exercises 1–5.

   simplified radical   product rule for radicals   quotient rule for radicals

   distance formula   Pythagorean theorem

1.   The _____ states that, in a right triangle, the sum of the squares of
      the legs equals the square of the hypotenuse.

2.   The rule $\sqrt[n]{a} \cdot \sqrt[n]{b} = \sqrt[n]{ab}$ is called the _____.

3.   The rule $\sqrt[n]{\dfrac{a}{b}} = \dfrac{\sqrt[n]{a}}{\sqrt[n]{b}}$ is called the _____.

4.   The formula $d = \sqrt{(x_2 - x_1)^2 + (y_2 - y_1)^2}$ is called the _____.

5.   A _____ has the following properties: the radicand has no
      factor raised to a power greater than or equal to the index; the radicand has no
      fractions; no denominator contains a radical; and exponents in the radicand and the
      index of the radical have greatest common factor 1.

Guided Examples

Review these examples for Objective 1:	**Now Try:**
1. Multiply. Assume that all variables represent positive real numbers.	1. Multiply. Assume that all variables represent positive real numbers.
a. $\sqrt{13} \cdot \sqrt{5}$ **b.** $\sqrt{5} \cdot \sqrt{2r}$	**a.** $\sqrt{2} \cdot \sqrt{7}$ _____
a. $\sqrt{13} \cdot \sqrt{5} = \sqrt{13 \cdot 5} = \sqrt{65}$	

b. $\sqrt{5} \cdot \sqrt{2r} = \sqrt{5 \cdot 2r} = \sqrt{10r}$

2. Multiply. Assume that all variables represent positive real numbers.

   a. $\sqrt[4]{2} \cdot \sqrt[4]{2x}$ **b.** $\sqrt[3]{8x} \cdot \sqrt[3]{2y^2}$

   c. $\sqrt[5]{6r^2} \cdot \sqrt[5]{4r^4}$ **d.** $\sqrt[5]{3} \cdot \sqrt[4]{6}$

   a. $\sqrt[4]{2} \cdot \sqrt[4]{2x} = \sqrt[4]{2 \cdot 2x} = \sqrt[4]{4x}$

   b. $\sqrt[3]{8x} \cdot \sqrt[3]{2y^2} = \sqrt[3]{8x \cdot 2y^2} = \sqrt[3]{16xy^2}$

   c. $\sqrt[5]{6r^2} \cdot \sqrt[5]{4r^4} = \sqrt[5]{6r^2 \cdot 4r^4} = \sqrt[5]{24r^6}$

   d. $\sqrt[5]{3} \cdot \sqrt[4]{6}$ cannot be simplified using the product rule for radicals because the indexes (5 and 4) are different.

b. $\sqrt{3x} \cdot \sqrt{7}$ _____

2. Multiply. Assume that all variables represent positive real numbers.

   a. $\sqrt[3]{3} \cdot \sqrt[3]{7}$ _____

   b. $\sqrt[3]{7x} \cdot \sqrt[3]{5y}$ _____

   c. $\sqrt[5]{4w} \cdot \sqrt[5]{2w^3}$ _____

   d. $\sqrt{3} \cdot \sqrt[5]{64}$ _____

Review this example for Objective 2:

3. Simplify. Assume that all variables represent positive real numbers.

   a. $\sqrt{\dfrac{16}{9}}$ **b.** $\sqrt{\dfrac{5}{64}}$

   c. $\sqrt[3]{-\dfrac{27}{8}}$ **d.** $\sqrt[3]{\dfrac{\ell^2}{27}}$

   e. $\sqrt{\dfrac{z^4}{36}}$

   a. $\sqrt{\dfrac{16}{9}} = \dfrac{\sqrt{16}}{\sqrt{9}} = \dfrac{4}{3}$

   b. $\sqrt{\dfrac{5}{64}} = \dfrac{\sqrt{5}}{\sqrt{64}} = \dfrac{\sqrt{5}}{8}$

   c. $\sqrt[3]{-\dfrac{27}{8}} = \dfrac{\sqrt[3]{-27}}{\sqrt[3]{8}} = \dfrac{-3}{2} = -\dfrac{3}{2}$

   d. $\sqrt[3]{\dfrac{\ell^2}{27}} = \dfrac{\sqrt[3]{\ell^2}}{\sqrt[3]{27}} = \dfrac{\sqrt[3]{\ell^2}}{3}$

   e. $\sqrt{\dfrac{z^4}{36}} = \dfrac{\sqrt{z^4}}{\sqrt{36}} = \dfrac{z^2}{6}$

Now Try:

3. Simplify. Assume that all variables represent positive real numbers.

   a. $\sqrt{\dfrac{36}{49}}$ _____

   b. $\sqrt{\dfrac{13}{81}}$ _____

   c. $\sqrt[3]{-\dfrac{343}{125}}$ _____

   d. $\sqrt[5]{\dfrac{7x}{32}}$ _____

   e. $\sqrt[3]{-\dfrac{a^6}{125}}$ _____

Review these examples for Objective 3:

4. Simplify.

   a. $\sqrt{24}$ **b.** $\sqrt{288}$

   c. $\sqrt{35}$ **d.** $-\sqrt[3]{81}$

   e. $\sqrt[4]{512}$

   a. $\sqrt{24} = \sqrt{4 \cdot 6}$ Factor; 4 is a perfect square.

   $= \sqrt{4} \cdot \sqrt{6}$ Product rule

   $= 2\sqrt{6}$ $\sqrt{4} = 2$

   b. $\sqrt{288} = \sqrt{144 \cdot 2}$ Factor; 144 is a perfect square.

   $= \sqrt{144} \cdot \sqrt{2}$ Product rule

   $= 12\sqrt{2}$ $\sqrt{144} = 12$

   c. The number 35 has no perfect square factors, so $\sqrt{35}$ cannot be simplified further.

   d. $-\sqrt[3]{81} = -\sqrt[3]{27 \cdot 3} = -\sqrt[3]{27} \cdot \sqrt[3]{3} = -3\sqrt[3]{3}$

   e. $\sqrt[4]{512} = \sqrt[4]{2^9} = \sqrt[4]{2^8 \cdot 2}$

   $= \sqrt[4]{2^8} \sqrt[4]{2} = 2^2 \sqrt[4]{2} = 4\sqrt[4]{2}$

5. Simplify. Assume that all variables represent positive real numbers.

   a. $\sqrt{81x^7}$ **b.** $\sqrt{56x^7 y^6}$

   c. $\sqrt[3]{-270b^4 c^8}$ **d.** $-\sqrt[6]{448a^7 b^7}$

   Note that it is not necessary to place variables in absolute value bars because all variables represent positive real numbers.

   a. $\sqrt{81x^7} = \sqrt{9^2 \cdot x^6 \cdot x} = 9x^3 \sqrt{x}$

   b. $\sqrt{56x^7 y^6} = \sqrt{4 \cdot 14 \cdot x^6 \cdot x \cdot y^6}$

   $= \sqrt{4} \cdot \sqrt{14} \cdot \sqrt{x^6 \cdot y^6} \cdot \sqrt{x}$

   $= 2x^3 y^3 \sqrt{14x}$

   c. $\sqrt[3]{-270b^4 c^8} = \sqrt[3]{\left(-27b^3 c^6\right)\left(10bc^2\right)}$

   $= \sqrt[3]{-27b^3 c^6} \cdot \sqrt[3]{10bc^2}$

   $= -3bc^2 \sqrt[3]{10bc^2}$

Now Try:

4. Simplify.

   a. $\sqrt{84}$ _____

   b. $\sqrt{162}$ _____

   c. $\sqrt{95}$ _____

   d. $-\sqrt[3]{256}$ _____

   e. $\sqrt[5]{512}$ _____

5. Simplify. Assume that all variables represent positive real numbers.

   a. $\sqrt{100y^{12}}$ _____

   b. $\sqrt{48m^5 r^9}$ _____

   c. $\sqrt[3]{-32n^7 t^5}$ _____

   d. $-\sqrt[4]{405x^3 y^9}$ _____

d. $-\sqrt[6]{448a^7b^7} = -\sqrt[6]{\left(64a^6b^6\right)(7ab)}$

$= -\sqrt[6]{64a^6b^6} \cdot \sqrt[6]{7ab}$

$= -2ab\sqrt[6]{7ab}$

6. Simplify. Assume that all variables represent positive real numbers.

a. $\sqrt[24]{5^4}$ **b.** $\sqrt[12]{x^8}$

a. $\sqrt[24]{5^4} = \left(5^4\right)^{1/24} = 5^{4/24} = 5^{1/6} = \sqrt[6]{5}$

b. $\sqrt[12]{x^8} = \left(x^8\right)^{1/12} = x^{8/12} = x^{2/3} = \sqrt[3]{x^2}$

6. Simplify. Assume that all variables represent positive real numbers.

a. $\sqrt[12]{9^9}$ _____

b. $\sqrt[30]{z^{24}}$ _____

Review this example for Objective 4:

7. Simplify $\sqrt{3} \cdot \sqrt[5]{6}$.

Because the different indexes, 2 and 5, have a least common multiple of 10, use rational exponents to write each radical as a tenth root.

$\sqrt{3} = 3^{1/2} = 3^{5/10} = \sqrt[10]{3^5} = \sqrt[10]{243}$

$\sqrt[5]{6} = 6^{1/5} = 6^{2/10} = \sqrt[10]{6^2} = \sqrt[10]{36}$

$\sqrt{3} \cdot \sqrt[5]{6} = \sqrt[10]{243} \cdot \sqrt[10]{36}$

$= \sqrt[10]{243 \cdot 36}$

$= \sqrt[10]{8748}$

Now Try:

7. Simplify.

$\sqrt[3]{3} \cdot \sqrt[6]{7}$ _____

Review this example for Objective 5:

8. Find the length of the unknown side in each triangle.

a.

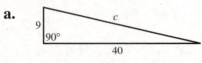

b.

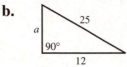

Now Try:

8. Find the length of the unknown side in each triangle.

a.

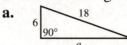

_____

b.

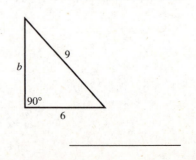

_____

a. $a^2 + b^2 = c^2$ Pythagorean theorem

$40^2 + 9^2 = c^2$ $a = 40$, $b = 9$

$1600 + 81 = c^2$ Square.

$1681 = c^2$ Add.

$41 = c$ $\sqrt{1681} = 41$

b. $a^2 + b^2 = c^2$ Pythagorean theorem

$12^2 + b^2 = 25^2$ $a = 12$, $c = 25$

$144 + b^2 = 625$ Square.

$b^2 = 481$ Subtract 144.

$b = \sqrt{481} \approx 21.9$

Review this example for Objective 6:

9. Find the distance between the points $(2, -2)$ and $(-6, 1)$.

Use the distance formula.
Let $(x_1, y_1) = (2, -2)$ and
$(x_2, y_2) = (-6, 1)$.

$d = \sqrt{(x_2 - x_1)^2 + (y_2 - y_1)^2}$

$= \sqrt{(-6 - 2)^2 + (1 - (-2))^2}$

$= \sqrt{(-8)^2 + 3^2}$

$= \sqrt{64 + 9} = \sqrt{73}$

Now Try:

9. Find the distance between the points $(-1, -2)$ and $(-4, 3)$.

_____

Objective 1 Use the product rule for radicals.
For extra help, see Examples 1 and 2 on page 444 of your text, the Section Lecture video for Section 8.3, and Exercise Solution Clips 5 and 13.

Multiply if possible, using the product rule. Assume that all variables represent positive real numbers.

1. $\sqrt{7x} \cdot \sqrt{6t}$

1. _____

2. $\sqrt[6]{9t} \cdot \sqrt[6]{3t^2}$

2. _____

3. $\sqrt{3} \cdot \sqrt[3]{7}$

3. _____

Objective 2   Use the quotient rule for radicals.
For extra help, see Example 3 on pages 444–445 of your text, the Section Lecture video for Section 8.3, and Exercise Solution Clip 21.

Simplify each radical. Assume that all variables represent positive real numbers.

4. $\sqrt{\dfrac{25}{16}}$

4. _____

5. $\sqrt[3]{\dfrac{a^6}{125}}$

5. _____

6. $-\sqrt[4]{\dfrac{p}{16}}$

6. _____

Objective 3   Simplify radicals.
For extra help, see Examples 4–6 on pages 445–447 of your text, the Section Lecture video for Section 8.3, and Exercise Solution Clips 37, 77, and 95.

Express each radical in simplified form. Assume that all variables represent positive real numbers.

7. $\sqrt[3]{54x^{11}}$

7. _____

8. $\sqrt[3]{80x^9c^7}$

8. _____

9. $\sqrt[4]{16x^{12}y^{10}}$

9. _____

Objective 4 Simplify products and quotients of radicals with different indexes.
For extra help, see Example 7 on page 447 of your text, the Section Lecture video for Section 8.3, and Exercise Solution Clip 101.

Simplify by first writing the radicals as radicals with the same index. Then multiply. Assume that all variables represent positive real numbers.

10. $\sqrt{6} \cdot \sqrt[3]{5}$

10. _____

11. $\sqrt{3} \cdot \sqrt[5]{64}$

11. _____

Objective 5 Use the Pythagorean theorem.
For extra help, see Example 8 on page 448 of your text, the Section Lecture video for Section 8.3, and Exercise Solution Clip 107.

Find the unknown length in each right triangle. Simplify the answer if possible.

12.

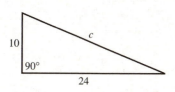

12. _____

13.

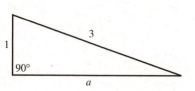

13. _____

Objective 6 Use the distance formula.

For extra help, see Example 9 on page 450 of your text, the Section Lecture video for Section 8.3, and Exercise Solution Clip 115.

Find the distance between each pair of points.

14. $(2, -3)$ and $(-5, 4)$

14. _____

15. $\left(4\sqrt{3}, 2\sqrt{5}\right)$ and $\left(3\sqrt{3}, -\sqrt{5}\right)$

15. _____

Chapter 8 · ROOTS, RADICALS, AND ROOT FUNCTIONS

8.4   Adding and Subtracting Radical Expressions

Learning Objectives
1   Simplify radical expressions involving addition and subtraction

Key Terms

Use the vocabulary terms listed below to complete each statement in exercises 1–2.

like radicals   unlike radicals

1.   The expressions $2\sqrt{2}$ and $6\sqrt[3]{2}$ are _____.

2.   The expressions $2\sqrt{2}$ and $7\sqrt{2}$ are _____.

Guided Examples

Review these examples for Objective 1:

1.   Add or subtract to simplify each radical expression.

a. $3\sqrt{13} + 5\sqrt{52}$

b. $\sqrt{48x} - \sqrt{12x}$, $x \geq 0$

c. $7\sqrt{3} - 6\sqrt{21}$

a. $3\sqrt{13} + 5\sqrt{52} = 3\sqrt{13} + 5\sqrt{4}\sqrt{13}$

$$\text{Product rule}$$
$$= 3\sqrt{13} + 5 \cdot 2\sqrt{13}$$
$$\sqrt{4} = 2$$
$$= 3\sqrt{13} + 10\sqrt{13}$$
$$\text{Multiply.}$$
$$= 13\sqrt{13} \quad \text{Add.}$$

b. $\sqrt{48x} - \sqrt{12x} = \sqrt{16}\sqrt{3x} - \sqrt{4}\sqrt{3x}$

$$\text{Product rule}$$
$$= 4\sqrt{3x} - 2\sqrt{3x}$$
$$\sqrt{16} = 4; \sqrt{4} = 2$$
$$= 2\sqrt{3x} \quad \text{Subtract.}$$

c. $7\sqrt{3} - 6\sqrt{21}$

The radicands differ and are already simplified, so the expression cannot be simplified further.

Now Try:

1.   Add or subtract to simplify each radical expression.

a. $3\sqrt{54} - 5\sqrt{24}$

_____

b. $3\sqrt{18z} + 2\sqrt{8z}$, $z \geq 0$

_____

c. $3\sqrt{7} + 2\sqrt{6}$

_____

2. Add or subtract to simplify each radical expression. Assume that all variables represent positive real numbers.

    a. $7\sqrt[4]{32} - 9\sqrt[4]{2}$

    b. $6\sqrt[3]{27x^5 r} + 2x\sqrt[3]{x^2 r}$

    c. $2x\sqrt[4]{64x} + 5x\sqrt{4x}$

    a. $7\sqrt[4]{32} - 9\sqrt[4]{2} = 7\sqrt[4]{16 \cdot 2} - 9\sqrt[4]{2}$

                           Factor.

$$= 7\sqrt[4]{16} \cdot \sqrt[4]{2} - 9\sqrt[4]{2}$$

                           Product rule

$$= 7 \cdot 2\sqrt[4]{2} - 9\sqrt[4]{2}$$

                           $\sqrt[4]{16} = 2$

$$= 14\sqrt[4]{2} - 9\sqrt[4]{2}$$

                           Multiply.

$$= 5\sqrt[4]{2} \quad \text{Subtract.}$$

    b. $6\sqrt[3]{27x^5 r} + 2x\sqrt[3]{x^2 r}$

$$= 6\sqrt[3]{3^3 x^3 x^2 r} + 2x\sqrt[3]{x^2 r} \quad \text{Factor}$$

$$= 6 \cdot 3x\sqrt[3]{x^2 r} + 2x\sqrt[3]{x^2 r}$$

        Product rule; find the cube roots.

$$= 18x\sqrt[3]{x^2 r} + 2x\sqrt[3]{x^2 r} \quad \text{Multiply.}$$

$$= 20x\sqrt[3]{x^2 r} \quad \text{Add.}$$

    c. $2x\sqrt[4]{64x} + 5x\sqrt{4x}$

$$= 2x\sqrt[4]{16 \cdot 4x} + 5x\sqrt{4x} \quad \text{Factor.}$$

$$= 2 \cdot 2x\sqrt[4]{4x} + 5x\sqrt{4x} \quad \begin{array}{l}\text{Product rule;}\\\text{find the roots.}\end{array}$$

$$= 4x\sqrt[4]{4x} + 5x\sqrt{4x} \quad \text{Multiply.}$$

The radicands are the same, but since the indexes are different, this expression cannot be simplified further.

3. Perform the indicated operations. Assume that all variables represent positive real numbers.

    a. $\dfrac{\sqrt{32}}{3} + \dfrac{\sqrt{8}}{\sqrt{18}}$

---

2. Add or subtract to simplify each radical expression. Assume that all variables represent positive real numbers.

    a. $7\sqrt[3]{54} - 6\sqrt[3]{128}$ _____

    b. $\sqrt[4]{32y^2 z^5} + 3z\sqrt[4]{2y^2 z}$

                          _____

    c. $2\sqrt{45x} + \sqrt[3]{40x}$

                          _____

3. Perform the indicated operations. Assume that all variables represent positive real numbers.

    a. $\sqrt{\dfrac{10}{18}} + \dfrac{\sqrt{15}}{\sqrt{27}}$ _____

b. $\sqrt[3]{\dfrac{81}{y^6}} + 5\sqrt[3]{\dfrac{27}{y^3}}$

b. $\sqrt[3]{\dfrac{216}{w^6}} + \sqrt{\dfrac{121}{w^4}}$ _____

a. $\dfrac{\sqrt{32}}{3} + \dfrac{\sqrt{8}}{\sqrt{18}} = \dfrac{\sqrt{16\cdot 2}}{3} + \dfrac{\sqrt{4\cdot 2}}{\sqrt{9\cdot 2}}$

                      Factor.

$= \dfrac{4\sqrt{2}}{3} + \dfrac{2\sqrt{2}}{3\sqrt{2}}$ Product rule; find roots.

$= \dfrac{4\sqrt{2}}{3} + \dfrac{2}{3}$ $\dfrac{\sqrt{2}}{\sqrt{2}} = 1$

$= \dfrac{4\sqrt{2} + 2}{3}$ Add.

b. $\sqrt[3]{\dfrac{81}{y^6}} + 5\sqrt[3]{\dfrac{27}{y^3}} = \dfrac{\sqrt[3]{81}}{\sqrt[3]{y^6}} + 5\left(\dfrac{\sqrt[3]{27}}{\sqrt[3]{y^3}}\right)$

                      Quotient rule

$= \dfrac{3\sqrt[3]{3}}{y^2} + 5\left(\dfrac{3}{y}\right)$ Simplify.

$= \dfrac{3\sqrt[3]{3}}{y^2} + \dfrac{15}{y}$ Multiply.

$= \dfrac{3\sqrt[3]{3}}{y^2} + \dfrac{15y}{y^2}$ Write with a common denominator.

$= \dfrac{3\sqrt[3]{3} + 15y}{y^2}$ Add.

Objective 1 Simplify radical expressions involving addition and subtraction.

For extra help, see Examples 1–3 on page 454–455 of your text, the Section Lecture video for Section 8.4, and Exercise Solution Clips 3, 21, and 47.

Simplify. Assume that all variables represent positive real numbers.

1. $3\sqrt{48} + 5\sqrt{27}$

1. _____

2. $2\sqrt{18} - 5\sqrt{32} + 7\sqrt{162}$

2. _____

3. $7\sqrt{3} + 2\sqrt{15}$

3. _____

4. $6\sqrt[3]{135} + 3\sqrt[3]{40}$

4. _____

5. $-3\sqrt[4]{243} - 2\sqrt[4]{48}$

5. _____

6. $y\sqrt[4]{48y} - 6\sqrt[4]{3y^5}$

6. _____

7. $5\sqrt[3]{24r^2z} + 4\sqrt[3]{81r^5z^4}$

7. _____

8. $\sqrt{72x} - 2\sqrt[3]{64x}$

8. _____

9. $2\sqrt[3]{16r} + \sqrt[3]{54r} - \sqrt[3]{16r}$

9. _____

10. $3\sqrt[3]{625x^4} - \sqrt{80x} + \sqrt{45x}$

10. _____

11. $\sqrt{\dfrac{12}{25}} - \sqrt{\dfrac{27}{49}}$

11. _____

12. $\dfrac{3\sqrt{2}}{4} - \sqrt{\dfrac{54}{96}} + \dfrac{\sqrt{8}}{\sqrt{16}}$

12. _____

13. $2\sqrt{\dfrac{k^3}{36}} - 3\sqrt{\dfrac{2k^3}{72}}$

13. _____

14. $\sqrt[3]{\dfrac{y^7}{125}} + y^2 \sqrt[3]{\dfrac{y}{27}}$

14. _____

15. $\sqrt{\dfrac{12y}{t^2}} - 2\sqrt[3]{\dfrac{125y}{t^6}}$

15. _____

Chapter 8 ROOTS, RADICALS, AND ROOT FUNCTIONS

8.5   Multiplying and Dividing Radical Expressions

Learning Objectives
1   Multiply radical expressions.
2   Rationalize denominators with one radical term.
3   Rationalize denominators with binomials involving radicals.
4   Write radical quotients in lowest terms.

Key Terms

Use the vocabulary terms listed below to complete each statement in exercises 1–2.

rationalizing the denominator   **conjugate**

1.   The _____ of $a + b$ is $a - b$.

2.   The process of removing radicals from the denominator so that the denominator contains only rational quantities is called _____.

Guided Examples

Review this example for Objective 1:
1.   Multiply, using the FOIL method.

    a. $\left(-2\sqrt{3} - 3\sqrt{5}\right)\left(5\sqrt{5} - \sqrt{3}\right)$

    b. $\left(\sqrt{6} + \sqrt{5}\right)\left(\sqrt{6} - \sqrt{5}\right)$

    c. $\left(\sqrt{7} + 6\right)^2$

    d. $\left(2 + \sqrt[3]{5}\right)\left(2 - \sqrt[3]{5}\right)$

    a. $\left(-2\sqrt{3} - 3\sqrt{5}\right)\left(5\sqrt{5} - \sqrt{3}\right)$
$$= -2\sqrt{3}\left(5\sqrt{5}\right) - 2\sqrt{3}\left(-\sqrt{3}\right)$$
$$- 3\sqrt{5}\left(5\sqrt{5}\right) - 3\sqrt{5}\left(-\sqrt{3}\right)$$

        Use the FOIL method to multiply.
$$= -10\sqrt{15} + 2 \cdot 3 - 15 \cdot 5 + 3\sqrt{15}$$

        Product rule
$$= -10\sqrt{15} + 6 - 75 + 3\sqrt{15} \quad \text{Multiply.}$$
$$= -7\sqrt{15} - 69 \quad \text{Combine like terms.}$$

Now Try:
1.   Multiply, using the FOIL method.

    a. $\left(4\sqrt{2} + 3\sqrt{3}\right)\left(\sqrt{2} - 7\sqrt{3}\right)$

_____

    b. $\left(2\sqrt{3} - 5\sqrt{2}\right)\left(2\sqrt{3} + 5\sqrt{2}\right)$

_____

    c. $\left(3 - 2\sqrt{x}\right)^2, \; x \geq 0$

_____

    d. $\left(4 + \sqrt[3]{2}\right)\left(4 - \sqrt[3]{2}\right)$

_____

b. $\left(\sqrt{6}+\sqrt{5}\right)\left(\sqrt{6}-\sqrt{5}\right)$

$= \left(\sqrt{6}\right)^2 - \left(\sqrt{5}\right)^2$

$\qquad\qquad (x+y)(x-y) = x^2 - y^2$

$= 6 - 5 = 1$

c. $\left(\sqrt{7}+6\right)^2 = \left(\sqrt{7}\right)^2 + 2 \cdot \sqrt{7} \cdot 6 + 6^2$

$\qquad\qquad (x+y)^2 = x^2 + 2xy + y^2$

$= 7 + 12\sqrt{7} + 36$ Multiply.

$= 43 + 12\sqrt{7}$ Combine like terms.

d. This is the difference of squares.

$\left(2 + \sqrt[3]{5}\right)\left(2 - \sqrt[3]{5}\right) = (2)^2 - \left(\sqrt[3]{5}\right)^2$

$= 4 - \sqrt[3]{5^2}$

$= 4 - \sqrt[3]{25}$

Review these examples for Objective 2:

2. Rationalize each denominator.

a. $\dfrac{12}{\sqrt{3}}$ 　　　　　**b.** $\dfrac{\sqrt{3}}{8\sqrt{5}}$

c. $-\dfrac{15}{\sqrt{27}}$

a. $\dfrac{12}{\sqrt{3}} = \dfrac{12}{\sqrt{3}} \cdot \dfrac{\sqrt{3}}{\sqrt{3}}$ Multiply by $\frac{\sqrt{3}}{\sqrt{3}} = 1$.

$= \dfrac{12\sqrt{3}}{3}$ $\sqrt{3} \cdot \sqrt{3} = 3$

$= 4\sqrt{3}$ Lowest terms

b. $\dfrac{\sqrt{3}}{8\sqrt{5}} = \dfrac{\sqrt{3}}{8\sqrt{5}} \cdot \dfrac{\sqrt{5}}{\sqrt{5}}$ Multiply by $\frac{\sqrt{5}}{\sqrt{5}} = 1$.

$= \dfrac{\sqrt{3} \cdot \sqrt{5}}{8 \cdot 5}$ $\sqrt{5} \cdot \sqrt{5} = 5$

$= \dfrac{\sqrt{15}}{40}$ Product rule; multiply

Now Try:

2. Rationalize each denominator.

a. $\dfrac{10}{\sqrt{15}}$ _____

b. $-\dfrac{6}{\sqrt{28}}$ _____

c. $\dfrac{3\sqrt{15}}{\sqrt{75}}$ _____

c. $-\dfrac{15}{\sqrt{27}} = -\dfrac{15}{\sqrt{9 \cdot 3}} = -\dfrac{15}{3\sqrt{3}} = -\dfrac{5}{\sqrt{3}}$

Simplify denominator; lowest terms

$= -\dfrac{5}{\sqrt{3}} \cdot \dfrac{\sqrt{3}}{\sqrt{3}}$ Multiply by $\dfrac{\sqrt{3}}{\sqrt{3}} = 1$.

$= -\dfrac{5\sqrt{3}}{3}$ $\sqrt{3} \cdot \sqrt{3} = 3$

3. Simplify each radical.

a. $-\sqrt{\dfrac{27}{98}}$

b. $\sqrt{\dfrac{20b^4}{a^5}}, \; a \geq 0, \, b \geq 0$

a. $-\sqrt{\dfrac{27}{98}} = -\dfrac{\sqrt{27}}{\sqrt{98}}$ Quotient rule

$= -\dfrac{\sqrt{9 \cdot 3}}{\sqrt{49 \cdot 2}}$ Factor.

$= -\dfrac{3\sqrt{3}}{7\sqrt{2}}$ Product rule

$= -\dfrac{3\sqrt{3}}{7\sqrt{2}} \cdot \dfrac{\sqrt{2}}{\sqrt{2}}$ Multiply by $\dfrac{\sqrt{2}}{\sqrt{2}}$.

$= -\dfrac{3\sqrt{6}}{7 \cdot 2}$ Product rule

$= -\dfrac{3\sqrt{6}}{14}$ Multiply.

b. $\sqrt{\dfrac{20b^4}{a^5}} = \dfrac{\sqrt{20b^4}}{\sqrt{a^5}}$

$= \dfrac{2b^2\sqrt{5}}{a^2\sqrt{a}}$ Product rule

$= \dfrac{2b^2\sqrt{5}}{a^2\sqrt{a}} \cdot \dfrac{\sqrt{a}}{\sqrt{a}}$ Multiply by $\dfrac{\sqrt{a}}{\sqrt{a}}$.

$= \dfrac{2b^2\sqrt{5a}}{a^2 \cdot a}$ Product rule

$= \dfrac{2b^2\sqrt{5a}}{a^3}$ Multiply.

3. Simplify each radical.

a. $-\sqrt{\dfrac{45}{32}}$ _____

b. $\sqrt{\dfrac{162x^3}{t^5}}, \; x \geq 0, \, t \geq 0$

_____

4. Simplify.

   a. $\sqrt[3]{\dfrac{16}{9}}$

   b. $\sqrt[4]{\dfrac{128s^{10}}{r^3}}$, $r \ge 0$, $s \ge 0$

   a. $\sqrt[3]{\dfrac{16}{9}} = \dfrac{\sqrt[3]{8 \cdot 2}}{\sqrt[3]{9}} = \dfrac{2\sqrt[3]{2}}{\sqrt[3]{9}}$ Quotient rule; product rule

        $= \dfrac{2\sqrt[3]{2}}{\sqrt[3]{9}} \cdot \dfrac{\sqrt[3]{81}}{\sqrt[3]{81}}$ Multiply by $\dfrac{\sqrt[3]{9^2}}{\sqrt[3]{9^2}} = \dfrac{\sqrt[3]{81}}{\sqrt[3]{81}}$.

        $= \dfrac{2\sqrt[3]{162}}{\sqrt[3]{729}}$ Multiply.

        $= \dfrac{2 \cdot 3\sqrt[3]{6}}{9}$ $\begin{aligned}\sqrt[3]{729} &= 9\\ \sqrt[3]{162} &= \sqrt[3]{27 \cdot 6}\\ &= 3\sqrt[3]{6}.\end{aligned}$

        $= \dfrac{2\sqrt[3]{6}}{3}$ Simplify.

   b. $\sqrt[4]{\dfrac{128s^{10}}{r^3}} = \dfrac{\sqrt[4]{128s^{10}}}{\sqrt[4]{r^3}}$ Quotient rule

        $= \dfrac{\sqrt[4]{2^4 \cdot 2^3 \cdot s^8 \cdot s^2}}{\sqrt[4]{r^3}}$ Factor.

        $= \dfrac{2s^2\sqrt[4]{2^3 s^2}}{\sqrt[4]{r^3}}$ Product rule

        $= \dfrac{2s^2\sqrt[4]{2^3 s^2}}{\sqrt[4]{r^3}} \cdot \dfrac{\sqrt[4]{r}}{\sqrt[4]{r}}$ Multiply by $\dfrac{\sqrt[4]{r}}{\sqrt[4]{r}}$.

        $= \dfrac{2s^2\sqrt[4]{2^3 s^2 r}}{\sqrt[4]{r^4}}$ Product rule

        $= \dfrac{2s^2\sqrt[4]{2^3 s^2 r}}{r}$ $\sqrt[4]{r^4} = r$

4. Simplify.

   a. $\sqrt[3]{\dfrac{8}{100}}$ _____

   b. $\sqrt[4]{\dfrac{64t^6}{x^7}}$, $t \ge 0$, $x \ge 0$

                    _____

Review this example for Objective 3:

5. Rationalize each denominator.

a. $\dfrac{5}{5+\sqrt{2}}$

b. $-\dfrac{4}{\sqrt{7}-\sqrt{5}}$

c. $\dfrac{2-\sqrt{3}}{2-\sqrt{5}}$

d. $\dfrac{2}{3-\sqrt{x}}, \; x \neq 0$

a. $\dfrac{5}{5+\sqrt{2}} = \dfrac{5\left(5-\sqrt{2}\right)}{\left(5+\sqrt{2}\right)\left(5-\sqrt{2}\right)}$

Multiply the numerator and denominator by the conjugate of the denominator.

$= \dfrac{5\left(5-\sqrt{2}\right)}{5^2 - \left(\sqrt{2}\right)^2}$

$(x+y)(x-y) = x^2 - y^2$

$= \dfrac{25-5\sqrt{2}}{25-2} = \dfrac{25-5\sqrt{2}}{23}$

b. $-\dfrac{4}{\sqrt{7}-\sqrt{5}} = -\dfrac{4\left(\sqrt{7}+\sqrt{5}\right)}{\left(\sqrt{7}-\sqrt{5}\right)\left(\sqrt{7}+\sqrt{5}\right)}$

Multiply the numerator and denominator by the conjugate of the denominator.

$= -\dfrac{4\left(\sqrt{7}+\sqrt{5}\right)}{\left(\sqrt{7}\right)^2 - \left(\sqrt{5}\right)^2}$

$(x+y)(x-y) = x^2 - y^2$

$= -\dfrac{4\sqrt{7}+4\sqrt{5}}{7-5}$

$= -\dfrac{4\sqrt{7}+4\sqrt{5}}{2}$

$= -2\sqrt{7} - 2\sqrt{5}$

Now Try:

5. Rationalize each denominator.

a. $\dfrac{4}{\sqrt{3}+2}$

_____

b. $-\dfrac{5}{\sqrt{5}-\sqrt{3}}$

_____

c. $\dfrac{\sqrt{3}+\sqrt{2}}{\sqrt{3}-\sqrt{2}}$

_____

d. $\dfrac{3}{\sqrt{x}-\sqrt{3y}}$

_____

c. $\dfrac{2-\sqrt{3}}{2-\sqrt{5}} = \dfrac{(2-\sqrt{3})(2+\sqrt{5})}{(2-\sqrt{5})(2+\sqrt{5})}$

Multiply the numerator and denominator by the conjugate of the denominator.

$= \dfrac{4+2\sqrt{5}-2\sqrt{3}-\sqrt{15}}{2^2-(\sqrt{5})^2}$

$(x+y)(x-y) = x^2 - y^2$

$= \dfrac{4+2\sqrt{5}-2\sqrt{3}-\sqrt{15}}{4-5}$

$= \dfrac{4+2\sqrt{5}-2\sqrt{3}-\sqrt{15}}{-1}$

$= -4-2\sqrt{5}+2\sqrt{3}+\sqrt{15}$

d. $\dfrac{2}{3-\sqrt{2x}}, x \neq 0$

$= \dfrac{2(3+\sqrt{2x})}{(3-\sqrt{2x})(3+\sqrt{2x})}$

Multiply the numerator and denominator by the conjugate of the denominator.

$= \dfrac{6+2\sqrt{2x}}{3^2-(\sqrt{2x})^2} = \dfrac{6+2\sqrt{2x}}{9-2x}$

Review this example for Objective 4:

6. Write each quotient in lowest terms.

  a. $\dfrac{72\sqrt{2}-16\sqrt{7}}{24}$

  b. $\dfrac{7y-\sqrt{98y^5}}{14y}, y \geq 0$

  a. $\dfrac{72\sqrt{2}-16\sqrt{7}}{24}$

  $= \dfrac{8(9\sqrt{2}-2\sqrt{7})}{24}$ Factor.

  $= \dfrac{9\sqrt{2}-2\sqrt{7}}{3}$

  Divide numerator and denominator by 8.

Now Try:

6. Write each quotient in lowest terms.

  a. $\dfrac{9+6\sqrt{15}}{12}$ _____

  b. $\dfrac{2x-\sqrt{8x^3}}{4x}, x \geq 0$

  _____

b. $\dfrac{7y - \sqrt{98y^5}}{14y} = \dfrac{7y - 7y^2\sqrt{2y}}{14y}$

$$\sqrt{98y^5} = \sqrt{2 \cdot 49 \cdot y^4 \cdot y}$$
$$= 7y^2\sqrt{2y}$$

$$= \dfrac{7y\left(1 - y\sqrt{2y}\right)}{14y} \quad \text{Factor.}$$

$$= \dfrac{1 - y\sqrt{2y}}{2}$$

Divide out the common factor.

Objective 1 Multiply radical expressions

For extra help, see Example 1 on pages 458–459 of your text, the Section Lecture video for Section 8.5, and Exercise Solution Clip 13.

Multiply and then simplify each product. Assume that all variables represent positive real numbers.

1. $\left(3 + \sqrt{2}\right)\left(2 + \sqrt{7}\right)$

1. _____

2. $\left(\sqrt{10} + \sqrt{3}\right)\left(\sqrt{6} - \sqrt{11}\right)$

2. _____

3. $\left(3 + \sqrt[3]{5}\right)\left(3 - \sqrt[3]{5}\right)$

3. _____

4. $\left(2\sqrt{x} - 3\right)\left(3\sqrt{x} - 2\right)$

4. _____

Objective 2 Rationalize denominators with one radical term.
For extra help, see Examples 2–4 on pages 459–461 of your text, the Section Lecture video for Section 8.5, and Exercise Solution Clips 43, 55, and 71.

Simplify. Assume that all variables represent positive real numbers.

5. $\sqrt{\dfrac{19}{32}}$

5. _____

6. $\sqrt[3]{\dfrac{a^4}{b}}$

6. _____

7. $\sqrt{\dfrac{75y^5}{8x^2}}$

7. _____

8. $\sqrt[3]{\dfrac{7}{36}}$

8. _____

Objective 3 Rationalize denominators with binomials involving radicals.
For extra help, see Example 5 on page 462 of your text, the Section Lecture video for Section 8.5, and Exercise Solution Clip 83.

Rationalize the denominator in each expression. Assume that all variables represent positive real numbers.

9. $\dfrac{-6}{\sqrt{7}+3}$

9. _____

10. $\dfrac{26}{\sqrt{11} + \sqrt{2}}$

10. _____

11. $\dfrac{3 - \sqrt{7}}{\sqrt{3} - 7}$

11. _____

12. $\dfrac{3\sqrt{x}}{2\sqrt{t} + \sqrt{x}}$

12. _____

Objective 4 Write radical quotients in lowest terms.
For extra help, see Example 6 on page 463 of your text, the Section Lecture video for Section 8.5, and Exercise Solution Clip 103.

Write each expression in lowest terms. Assume that all variables represent positive real numbers.

13. $\dfrac{5 + 2\sqrt{75}}{25}$

13. _____

14. $\dfrac{16 - 12\sqrt{72}}{24}$

14. _____

15. $\dfrac{3x^2 + \sqrt{18x^9}}{6x^3}$

15. _____

Chapter 8 ROOTS, RADICALS, AND ROOT FUNCTIONS

8.6   **Solving Equations with Radicals**

Learning Objectives
1   Solve radical equations by using the power rule.
2   Solve radical equations that require additional steps.
3   Solve radical equations with indexes greater than 2.
4   Use the power rule to solve a formula for a specified variable.

Key Terms

Use the vocabulary terms listed below to complete each statement in exercises 1–3.

     radical equation   **power rule**   **extraneous solution**

1.   The _____ guarantees that, if each side of a given equation is raised to the same power, then all the solutions of the original equation are among the solutions of the resulting equation.

2.   An equation with a variable in the radicand is a _____.

3.   A number is called an _____ if, when checked in the original equation, a false statement results.

Guided Examples

Review these examples for Objective 1:

1. Solve $\sqrt{12 - x} = 8$.

$$\sqrt{12 - x} = 8$$

$$\left(\sqrt{12 - x}\right)^2 = 8^2 \quad \text{Squaring property}$$

$$12 - x = 64 \quad \left(\sqrt{a}\right)^2 = a$$

$$-x = 52 \quad \text{Subtract 12.}$$

$$x = -52 \quad \text{Multiply by } -1.$$

Check:

$$\sqrt{12 - x} = 8$$

$$\sqrt{12 - (-52)} \overset{?}{=} 8$$

$$\sqrt{64} \overset{?}{=} 8$$

$$8 = 8 \checkmark$$

Now Try:

1. Solve $\sqrt{7x - 6} = 8$.

   _____

2. Solve $\sqrt{12p+1}+7=0$.

$$\sqrt{12p+1}+7=0$$

$\sqrt{12p+1}=-7$ Subtract 7.

$\left(\sqrt{12p+1}\right)^2=(-7)^2$ Square each side.

$12p+1=49$ Apply the exponents.

$12p=48$ Subtract 1.

$p=4$ Divide by 12.

Check:

$$\sqrt{12p+1}+7=0$$

$$\sqrt{12(4)+1}+7\overset{?}{=}0$$

$$\sqrt{49}+7\overset{?}{=}0$$

$$7+7\overset{?}{=}0$$

$$14\neq0$$

The false result shows that the proposed solution of 4 is not a solution of the original equation. It is extraneous. The solution set is $\varnothing$.

2. Solve $\sqrt{4x-19}+5=0$

_____

Review these examples for Objective 2:

3. Solve $\sqrt{x+3}=x-3$.

$$\sqrt{x+3}=x-3$$

$\left(\sqrt{x+3}\right)^2=(x-3)^2$ Square each side.

$$x+3=x^2-6x+9$$

$0=x^2-7x+6$ Standard form

$0=(x-1)(x-6)$ Factor.

$x-1=0$ or $x-6=0$ Zero-factor property

$x=1$ $x=6$

Check:

$$\sqrt{x+3}=x-3$$

$\sqrt{1+3}\overset{?}{=}1-3$ $\sqrt{6+3}\overset{?}{=}6-3$

$\sqrt{4}\overset{?}{=}-2$ $\sqrt{9}\overset{?}{=}3$

$2=-2$ False $3=3$ ✓

Only 6 is a valid solution. (1 is an extraneous solution.) The solution set is $\{6\}$.

Now Try:

3. Solve $\sqrt{x+11}=x-1$.

_____

Name: Date:

Instructor: Section:

4. Solve $\sqrt{x^2 + 7x - 14} = x + 2$.

$$\sqrt{x^2 + 7x - 14} = x + 2.$$

$$\left(\sqrt{x^2 + 7x - 14}\right)^2 = (x + 2)^2$$

                         Square each side.

$$x^2 + 7x - 14 = x^2 + 4x + 4$$

$$7x - 14 = 4x + 4 \quad \text{Subtract } x^2.$$

$$3x = 18 \quad \text{Subtract } 4x; \text{ add } 14.$$

$$x = 6 \quad \text{Divide by } 3.$$

Check:

$$\sqrt{x^2 + 7x - 14} = x + 2$$

$$\sqrt{(6)^2 + 7(6) - 14} \overset{?}{=} 6 + 2$$

$$\sqrt{36 + 42 - 14} \overset{?}{=} 8$$

$$\sqrt{64} \overset{?}{=} 8$$

$$8 = 8 \checkmark$$

The solution set is $\{6\}$.

5. Solve $\sqrt{3x} - 4 = \sqrt{x - 2}$.

$$\sqrt{3x} - 4 = \sqrt{x - 2}$$

$$\left(\sqrt{3x} - 4\right)^2 = \left(\sqrt{x - 2}\right)^2$$

                     Square both sides.

$$3x - 8\sqrt{3x} + 16 = x - 2$$

$$-8\sqrt{3x} = -2x - 18$$

                     Isolate the radical.

$$\left(-8\sqrt{3x}\right)^2 = (-2x - 18)^2$$

                     Square both sides.

$$192x = 4x^2 + 72x + 324$$

$$0 = 4x^2 - 120x + 324$$

                     Standard form

$$0 = 4(x - 27)(x - 3)$$

                     Factor.

$$0 = (x - 27)(x - 3)$$

                     Divide by 4.

$$x - 27 = 0 \quad \text{or} \quad x - 3 = 0 \quad \text{Zero-factor property}$$

$$x = 27 \qquad\qquad x = 3$$

4. Solve $\sqrt{x^2 + 8x - 17} = x + 1$.

_____

5. Solve $\sqrt{3x + 4} = \sqrt{9x} - 2$.

_____

Check:

$$\sqrt{3x} - 4 = \sqrt{x} - 2$$

$$\sqrt{3(27)} - 4 \overset{?}{=} \sqrt{27} - 2 \quad \Bigg| \quad \sqrt{3(3)} - 4 \overset{?}{=} \sqrt{3} - 2$$

$$\sqrt{81} - 4 \overset{?}{=} \sqrt{25} \quad \Bigg| \quad \sqrt{9} - 4 \overset{?}{=} \sqrt{1}$$

$$9 - 4 \overset{?}{=} 5 \quad \Bigg| \quad 3 - 4 \overset{?}{=} 1$$

$$5 = 5 \checkmark \quad \Bigg| \quad -1 = 1 \quad \text{False}$$

Only 27 is a valid solution. (3 is an extraneous solution.) The solution set is {27}.

Review this example for Objective 3:

6. Solve $\sqrt[3]{5r - 6} = \sqrt[3]{3r + 4}$.

$$\sqrt[3]{5r - 6} = \sqrt[3]{3r + 4}$$

$$\left(\sqrt[3]{5r - 6}\right)^3 = \left(\sqrt[3]{3r + 4}\right)^3 \quad \text{Cube both sides.}$$

$$5r - 6 = 3r + 4$$

$$2r = 10 \quad \text{Subtract } 2r; \text{ add } 6.$$

$$r = 5 \quad \text{Divide by } 2.$$

Check:

$$\sqrt[3]{5r - 6} = \sqrt[3]{3r + 4}$$

$$\sqrt[3]{5(5) - 6} \overset{?}{=} \sqrt[3]{3(5) + 4}$$

$$\sqrt[3]{19} = \sqrt[3]{19} \checkmark$$

The solution set is {5}.

Now Try:

6. Solve $\sqrt[4]{8x + 5} = \sqrt[4]{7x + 7}$

_____

Review this example for Objective 4:

7. Solve the formula $d = \sqrt{\dfrac{H}{1.6n}}$ for n.

$$d = \sqrt{\frac{H}{1.6n}}$$

$$d^2 = \left(\sqrt{\frac{H}{1.6n}}\right)^2 \quad \text{Square both sides.}$$

$$d^2 = \frac{H}{1.6n}$$

$$1.6nd^2 = H \quad \text{Multiply by } 1.6n.$$

$$n = \frac{H}{1.6d^2} \quad \text{Divide by } 16d^2.$$

Now Try:

7. Solve the formula $r = \sqrt{\dfrac{3v}{\pi h}}$ for h.

_____

Name: Date:
Instructor: Section:

Objective 1 Solve radical equations by using the power rule.

For extra help, see Examples 1 and 2 on pages 468–469 of your text, the Section Lecture video for Section 8.6, and Exercise Solution Clips 9 and 11.

Solve each equation.

1. $\sqrt{3n-8}=5$ 1. _____

2. $\sqrt{3w+4}=7$ 2. _____

3. $\sqrt{a+4}+6=0$ 3. _____

4. $\sqrt{5r-4}-9=0$ 4. _____

Objective 2 Solve radical equations that require additional steps.

For extra help, see Examples 3–5 on pages 469–471 of your text, the Section Lecture video for Section 8.6, and Exercise Solution Clips 27, 29, and 51.

Solve each equation.

5. $\sqrt{3q-8}=q-2$ 5. _____

6. $\sqrt{9a^2+6a-23}=3a+5$ 6. _____

7. $2w + 1 - \sqrt{4w^2 + 3w + 2} = 0$ 7. _____

8. $\sqrt{3k + 7} + \sqrt{k + 1} = 2$ 8. _____

9. $\sqrt{10d + 6} - \sqrt{4d + 4} = 2$ 9. _____

Objective 3 Solve radical equations with indexes greater than 2.

For extra help, see Example 6 on page 471 of your text, the Section Lecture video for Section 8.6, and Exercise Solution Clip 37.

Solve each equation.

10. $\sqrt[3]{7x - 5} = \sqrt[3]{3x + 7}$ 10. _____

11. $\sqrt[5]{2t + 1} - 1 = 0$ 11. _____

12. $\sqrt[4]{2m + 1} = \sqrt[4]{m + 22}$ 12. _____

Objective 4 Use the power rule to solve a formula for a specified variable.
For extra help, see Example 7 on page 472 of your text, the Section Lecture video for
Section 8.6, and Exercise Solution Clip 65.

Solve each formula for the indicated variable.

13. $Z = \sqrt{\dfrac{L}{C}}$, for L 13. _____

14. $f = \dfrac{1}{2\pi\sqrt{LC}}$, for C 14. _____

15. $N = \dfrac{1}{2\pi}\sqrt{\dfrac{a}{r}}$, for r 15. _____

Chapter 8 ROOTS, RADICALS, AND ROOT FUNCTIONS

8.7   Complex Numbers

Learning Objectives

1   Simplify numbers of the form $\sqrt{-b}$, where $b > 0$.
2   Recognize subsets of the complex numbers.
3   Add and subtract complex numbers.
4   Multiply complex numbers.
5   Divide complex numbers.
6   Find powers of i.

Key Terms

Use the vocabulary terms listed below to complete each statement in exercises 1–6.

complex number   **real part**   **imaginary part**

pure imaginary number   **standard form (of a complex number)**

complex conjugate

1.   A _____ is a number that can be written in the form $a + bi$, where a and b are real numbers.

2.   The _____ of $a + bi$ is $a - bi$.

3.   The _____ of $a + bi$ is bi.

4.   The _____ of $a + bi$ is a.

5.   A complex number is in _____ if it is written in the form $a + bi$.

6.   A complex number $a + bi$ with $a = 0$ and $b \neq 0$ is called a
     _____.

Guided Examples

Review these examples for Objective 1:

1.   Write each number as a product of a real number and i.

   a. $\sqrt{-36}$ **b.** $-\sqrt{-121}$

   c. $\sqrt{-3}$ **d.** $\sqrt{-75}$

   a. $\sqrt{-36} = i\sqrt{36} = 6i$

   b. $-\sqrt{-121} = -i\sqrt{121} = -11i$

Now Try:

1.   Write each number as a product of a real number and i.

   a. $\sqrt{-16}$ _____

   b. $-\sqrt{-144}$ _____

   c. $\sqrt{-11}$ _____

c. $\sqrt{-3} = i\sqrt{3}$

d. $\sqrt{-75} = i\sqrt{75} = i\sqrt{25 \cdot 3} = 5i\sqrt{3}$

d. $\sqrt{-128}$ _____

2. Multiply.

  a. $\sqrt{-4} \cdot \sqrt{-36}$ **b.** $\sqrt{-6} \cdot \sqrt{-7}$

  c. $\sqrt{-21} \cdot \sqrt{-7}$ **d.** $\sqrt{-14} \cdot \sqrt{3}$

  a. $\sqrt{-4} \cdot \sqrt{-36} = i\sqrt{4} \cdot i\sqrt{36}$ $\sqrt{-b} = i\sqrt{b}$

           $= i \cdot 2 \cdot i \cdot 6$ Take square roots.

           $= 12i^2$ Multiply.

           $= -12$ $i^2 = -1$

  b. $\sqrt{-6} \cdot \sqrt{-7} = i\sqrt{6} \cdot i\sqrt{7}$ $\sqrt{-b} = i\sqrt{b}$

           $= i^2\sqrt{6 \cdot 7}$ Multiply.
                          Product rule

           $= -\sqrt{42}$ $i^2 = -1$

  c. $\sqrt{-21} \cdot \sqrt{-7} = i\sqrt{21} \cdot i\sqrt{7}$ $\sqrt{-b} = i\sqrt{b}$

           $= i^2\sqrt{21 \cdot 7}$ Product rule

           $= -\sqrt{7^2 \cdot 3}$ $i^2 = -1$; factor.

           $= -7\sqrt{3}$

  d. $\sqrt{-14} \cdot \sqrt{3} = i\sqrt{14} \cdot \sqrt{3}$

           $= i\sqrt{42}$

2. Multiply.

  a. $\sqrt{-9} \cdot \sqrt{-121}$ _____

  b. $\sqrt{-3} \cdot \sqrt{-7}$ _____

  c. $\sqrt{-24} \cdot \sqrt{-6}$ _____

  d. $\sqrt{-3} \cdot \sqrt{7}$ _____

3. Divide.

  a. $\dfrac{\sqrt{-125}}{\sqrt{-5}}$ **b.** $\dfrac{\sqrt{-28}}{\sqrt{7}}$

  a. $\dfrac{\sqrt{-125}}{\sqrt{-5}} = \dfrac{i\sqrt{125}}{i\sqrt{5}}$

            $= \sqrt{\dfrac{125}{5}}$ Quotient rule

            $= \sqrt{25}$ Divide.

            $= 5$

  b. $\dfrac{\sqrt{-28}}{\sqrt{7}} = \dfrac{i\sqrt{28}}{\sqrt{7}} = i\sqrt{\dfrac{28}{7}} = i\sqrt{4} = 2i$

3. Divide.

  a. $\dfrac{\sqrt{-200}}{\sqrt{-8}}$ _____

  b. $\dfrac{\sqrt{-80}}{\sqrt{5}}$ _____

Review these examples for Objective 3:

4. Add.

    a. $(-2+9i)+(10-3i)$

    b. $(-1-5i)+(2+5i)+(-8+2i)$

    a. $(-2+9i)+(10-3i)$

        $= (-2+10)+(9i-3i)$

        $= 8+6i$

    b. $(-1-5i)+(2+5i)+(-8+2i)$

        $= \left[-1+2+(-8)\right]+(-5i+5i+2i)$

        $= -7+2i$

5. Subtract.

    a. $(-1-5i)-(2+7i)$

    b. $(7-9i)-(-5-6i)$

    c. $(8+3i)-(5+3i)$

    a. $(-1-5i)-(2+7i)$

        $= (-1-2)+(-5i-7i)$

        $= -3-12i$

    b. $(7-9i)-(-5-6i)$

        $= \left[7-(-5)\right]+\left[-9i-(-6i)\right]$

        $= 12+(-3i)$

        $= 12-3i$

    c. $(8+3i)-(5+3i)$

        $= (8-5)+(3i-3i)$

        $= 3$

Review this example for Objective 4:

6. Multiply.

    a. $6i(2-7i)$

    b. $(3+2i)(5-i)$

    c. $(2-5i)(2+5i)$

    a. $6i(2-7i)=6i(2)+6i(-7i)$ Distributive property

        $= 12i-42i^2$ Multiply.

        $= 12i-42(-1)$ $i^2 = -1$

        $= 42+12i$ Multiply. Standard form

Now Try:

4. Add.

    a. $(-4-7i)+(6-2i)$

    _____

    b. $(6+3i)+(-7+i)+(-9-8i)$

    _____

5. Subtract.

    a. $(3+3i)-(-8+4i)$

    _____

    b. $7-(2-3i)$

    _____

    c. $(12+2i)-(12-2i)$

    _____

Now Try:

6. Multiply.

    a. $-2i(-4-7i)$

    _____

    b. $(12+2i)(-1+i)$

    _____

    c. $(1+3i)^2$

    _____

b. $(3+2i)(5-i)$

$= 3(5) + 3(-i) + 2i(5) + 2i(-i)$ FOIL

$= 15 + 3i + 10i - 2i^2$ Multiply.

$= 15 + 7i - 2(-1)$ Combine like terms.
$i^2 = -1$

$= 15 + 7i + 2$ Multiply.

$= 17 + 7i$ Combine like terms.

c. $(2-5i)(2+5i) = 2^2 - (5i)^2$

$(a+b)(a-b) = a^2 - b^2$

$= 4 - 25i^2$ Multiply.

$= 4 - 25(-1)$ $i^2 = -1$

$= 4 + 25$ Multiply.

$= 29$ Add.

Review this example for Objective 5:

7. Find each quotient.

a. $\dfrac{6-i}{2-3i}$ **b.** $\dfrac{3-i}{i}$

To divide complex numbers, multiply both the numerator and denominator by the conjugate of the denominator.

a. $\dfrac{6-i}{2-3i} = \dfrac{(6-i)(2+3i)}{(2-3i)(2+3i)}$

$= \dfrac{12 + 18i - 2i - 3i^2}{2^2 + 3^2}$

Use FOIL in the numerator.
In the denominator,
$(a-bi)(a+bi) = a^2 + b^2$.

$= \dfrac{15 + 16i}{13}$ $-3i^2 = -3(-1) = 3$;
Combine like terms.

b. $\dfrac{3-i}{i} = \dfrac{(3-i)(-i)}{i(-i)}$ The complex conjugate of i is $-i$.

$= \dfrac{-3i + i^2}{-i^2}$ Multiply.

$= \dfrac{-3i - 1}{-(-1)}$ $i^2 = -1$

$= -1 - 3i$ Lowest terms

Now Try:

7. Find each quotient.

a. $\dfrac{4+i}{5-2i}$

_____

b. $\dfrac{2-i}{4i}$

_____

Review this example for Objective 6:

8. Find each power of i.

    a. i^{100} **b.** i^{27}

    c. i^{22} **d.** i^{49}

    a. $i^{100} = \left(i^4\right)^{25} = 1^{25} = 1$

    b. $i^{27} = i^{24} \cdot i^3 = \left(i^4\right)^6 \cdot (-i) = 1^6 \cdot (-i) = -i$

    c. $i^{22} = i^{20} \cdot i^2 = \left(i^4\right)^5 \cdot (-1) = 1^5 \cdot (-1) = -1$

    d. $i^{49} = i^{48} \cdot i = \left(i^4\right)^{12} \cdot i = 1^{12} \cdot i = i$

Now Try:

8. Find each power of i.

    a. i^{-48} _____

    b. i^{77} _____

    c. i^{66} _____

    d. i^{55} _____

Objective 1 Simplify numbers of the form $\sqrt{-b}$, where $b > 0$.

For extra help, see Examples 1–3 on pages 475–476 of your text, the Section Lecture video for Section 8.7, and Exercise Solution Clips 7, 15 and 21.

Simplify.

1. $\sqrt{-288}$

1. _____

2. $-\sqrt{-72}$

2. _____

Multiply or divide as indicated.

3. $\sqrt{-6} \cdot \sqrt{-3} \cdot \sqrt{2}$

3. _____

4. $\dfrac{\sqrt{-56} \cdot \sqrt{-6}}{\sqrt{16}}$

4. _____

Objective 2 Recognize subsets of the complex numbers.

For extra help, see page 476 of your text and the Section Lecture video for Section 8.7.

The real numbers are a subset of the complex numbers. Classify the complex number as real, nonreal complex, *or* pure imaginary.

5. $7 - 3i$ 5. _____

6. $\sqrt{5}$ 6. _____

Objective 3 Add and subtract complex numbers.

For extra help, see Examples 4 and 5 on page 477 of your text, the Section Lecture video for Section 8.7, and Exercise Solution Clip 29 and 33.

Add or subtract as indicated. Write your answers in standard form.

7. $(-7 - 2i) - (-3 - 3i)$ 7. _____

8. $\left[(8 + 4i) - (5 - 3i) \right] + (4 - 2i)$ 8. _____

Objective 4 Multiply complex numbers.

For extra help, see Example 6 on pages 477–478 of your text, the Section Lecture video for Section 8.7, and Exercise Solution Clip 47.

Multiply.

9. $(2 + 5i)(3 - i)$ 9. _____

10. $4i(-1 + 2i)^2$ 10. _____

Objective 5 Divide complex numbers.
For extra help, see Example 7 on page 478–479 of your text, the Section Lecture video for Section 8.7, and Exercise Solution Clip 67.

Find each quotient.

11. $\dfrac{1+i}{2-i}$

11. _____

12. $\dfrac{5+2i}{9-4i}$

12. _____

13. $\dfrac{7}{5+2i}$

13. _____

Objective 6 Find powers of *i*.
For extra help, see Example 8 on page 479 of your text, the Section Lecture video for Section 8.7, and Exercise Solution Clip 75.

Find powers of i.

14. i^{-62}

14. _____

15. i^{11}

15. _____

Chapter 9 QUADRATIC EQUATIONS, INEQUALITIES, AND FUNCTIONS

9.1 The Square Root Property and Completing the Square

Learning Objectives
1   Review the zero-factor property.
2   Learn the square root property.
3   Solve quadratic equations of the form $(ax + b)^2 = c$ by extending the square root property.
4   Solve quadratic equations by completing the square.
5   Solve quadratic equations with solutions that are not real numbers.

Key Terms

Use the vocabulary terms listed below to complete each statement in exercises 1–4.

quadratic equation   **standard form**   **zero-factor property**

square root property

1.   A quadratic equation written in the form $ax^2 + bx + c = 0$, where a, b, and c are real numbers, with $a \neq 0$, is written in _____.

2.   A _____ is an equation that can be written in the form
     $ax^2 + bx + c = 0$,   where a, b, and c are real numbers, with $a \neq 0$.

3.   The _____ states that if $x^2 = k$, then $x = \sqrt{k}$ or $x = -\sqrt{k}$.

4.   The _____ states that if two numbers have a product of 0, then at least one of the numbers must be 0.

Guided Examples

Review this example for Objective 1:
1.   Use the zero-factor property to solve
     $2x^2 + 5x - 3 = 0$.

$$2x^2 + 5x - 3 = 0$$
$$(2x - 1)(x + 3) = 0 \quad \text{Factor.}$$
$$2x - 1 = 0 \ \text{ or } \ x + 3 = 0 \quad \text{Zero-factor property}$$
$$2x = 1 \qquad\quad x = -3 \ \text{Solve each equation.}$$
$$x = \tfrac{1}{2}$$

Now Try:
1.   Use the zero-factor property to solve $3x^2 = 5x + 28$.

     _____

Review these examples for Objective 2:

2. Solve each equation.

    a. $x^2 = 20$ **b.** $3x^2 + 11 = 35$

    a. $x^2 = 20$

    $\quad x = \pm\sqrt{20}$ Square root property

    $\quad x = \pm 2\sqrt{5}$ Simplify.

    b. $3x^2 + 11 = 35$

    $\quad\quad 3x^2 = 24$ Subtract 11.

    $\quad\quad x^2 = 8$ Divide by 3.

    $\quad\quad x = \pm\sqrt{8}$ Square root property

    $\quad\quad x = \pm 2\sqrt{2}$ Simplify

3. Use Galileo's formula to determine how long it will take a penny dropped from the 86th floor Observatory deck of the Empire State Building to reach the ground. The deck is 1050 feet above the ground. Round your answer to the nearest tenth.

    Galileo's formula is $d = 16t^2$, where d is the distance in feet that an object falls and t is the time in seconds.

    $$d = 16t^2$$
    $$1050 = 16t^2$$
    $$65.625 = t^2 \quad \text{Divide by 16.}$$
    $$t = \sqrt{65.625} \text{ or } t = -\sqrt{65.625}$$
    $$\text{Square root property}$$

    Time cannot be negative, so we discard the negative solution.

    $$t = \sqrt{65.625} \approx 8.1$$

    The penny will take about 8.1 seconds to reach the ground.

Now Try:

2. Solve each equation.

    a. $x^2 = 84$ _____

    b. $12 - 4y^2 = 9$ _____

3. A child dropped a ball from a hotel balcony that is 113 ft above the ground. Use Galileo's formula to determine how long it takes for the ball to reach the ground. Round your answer to the nearest tenth.

    _____

Review these examples for Objective 3:

4. Solve $(x-2)^2 = 25$.

$$(x-2)^2 = 25$$
$$x-2 = \sqrt{25} \quad \text{or} \quad x-2 = -\sqrt{25}$$

                    Square root property

$$x-2 = 5 \qquad x-2 = -5 \quad \sqrt{25} = \pm 5$$
$$x = 7 \qquad\qquad x = -3 \quad \text{Add 2.}$$

The solution set is $\{-3, 7\}$.

5. Solve $(3x+4)^2 = 40$.

$$(3x+4)^2 = 40$$
$$3x+4 = \sqrt{40} \quad \text{or} \quad 3x+4 = -\sqrt{40}$$

                    Square root property

$$3x+4 = 2\sqrt{10} \qquad 3x+4 = -2\sqrt{10}$$

                  $\sqrt{40} = \sqrt{4 \cdot 10} = 2\sqrt{10}$

$$3x = 2\sqrt{10} - 4 \quad \text{or} \quad 3x = -2\sqrt{10} - 4 \quad \text{Subtract 4.}$$

$$x = \frac{2\sqrt{10} - 4}{3} \qquad x = \frac{-2\sqrt{10} - 4}{3} \quad \begin{array}{l}\text{Divide}\\ \text{by 3.}\end{array}$$

The solution set is $\left\{ \dfrac{2\sqrt{10} - 4}{3}, \dfrac{-2\sqrt{10} - 4}{3} \right\}$.

Review these examples for Objective 4:

6. Solve $x^2 - 12x + 24 = 0$.

$$x^2 - 12x + 24 = 0$$
$$x^2 - 12x = -24 \quad \text{Subtract 24.}$$
$$x^2 - 12x + 36 = -24 + 36$$

                  $2kx = -12$, so $k = -6$

                  and $k^2 = 36$.

$$(x-6)^2 = 12 \qquad \text{Factor.}$$

Now use the square root property to solve for x.

$$x-6 = -\sqrt{12} \qquad \text{or} \qquad x-6 = \sqrt{12}$$
$$x-6 = -2\sqrt{3} \qquad\qquad x-6 = 2\sqrt{3}$$
$$x = 6 - 2\sqrt{3} \qquad\qquad x = 6 + 2\sqrt{3}$$

The solution set is $\left\{ 6 \pm 2\sqrt{3} \right\}$.

Now Try:

4. Solve $(x-1)^2 = 16$.

_____

5. Solve $(2x+5)^2 = 32$.

_____

6. Solve $x^2 - 6x + 1 = 0$.

_____

7. Solve $x^2 + 5x + 2 = 0$ by completing the square.

$$x^2 + 5x + 2 = 0$$

$$x^2 + 5x = -2 \quad \text{Subtract 2.}$$

$$x^2 + 5x + \frac{25}{4} = -2 + \frac{25}{4}$$

$$2kx = 5, \text{ so } k = \frac{5}{2}$$

$$\text{and } k^2 = \frac{25}{4}.$$

$$\left(x + \frac{5}{2}\right)^2 = \frac{17}{4} \quad \text{Factor; add.}$$

Now use the square root property to solve for x.

$$x + \frac{5}{2} = -\sqrt{\frac{17}{4}} \qquad \text{or} \qquad x + \frac{5}{2} = \sqrt{\frac{17}{4}}$$

$$x + \frac{5}{2} = -\frac{\sqrt{17}}{2} \qquad\qquad x + \frac{5}{2} = \frac{\sqrt{17}}{2}$$

$$x = -\frac{5}{2} - \frac{\sqrt{17}}{2} \qquad\qquad x = -\frac{5}{2} + \frac{\sqrt{17}}{2}$$

The solution set is $\left\{ -\frac{5}{2} \pm \frac{\sqrt{17}}{2} \right\}$.

8. Solve $4n^2 + 4n - 15 = 0$

In order to complete the square, the coefficient of n^2 must be 1, so start by dividing each side of the equation by 4.

$$4n^2 + 4n - 15 = 0$$

$$n^2 + n - \frac{15}{4} = 0$$

$$n^2 + n = \frac{15}{4} \quad \text{Add } \frac{15}{4}.$$

Now complete the square on the left by taking half the coefficient of n and squaring it, then adding the result to both sides of the equation. The coefficient of n is 1, so add $\left(\frac{1}{2}\right)^2 = \frac{1}{4}$ to each side.

$$n^2 + n + \frac{1}{4} = \frac{15}{4} + \frac{1}{4}$$

$$\left(n + \frac{1}{2}\right)^2 = \frac{16}{4} = 4$$

7. Solve $x^2 - 11x + 8 = 0$ by completing the square.

_____

8. Solve $9x^2 + 6x - 8 = 0$.

_____

Name:
Instructor:

Date:
Section:

Solve the equation by using the square root property.

$$n + \frac{1}{2} = -\sqrt{4} \quad \text{or} \quad n + \frac{1}{2} = \sqrt{4}$$

$$n + \frac{1}{2} = -2 \qquad\qquad n + \frac{1}{2} = 2$$

$$n = -\frac{5}{2} \qquad\qquad\qquad n = \frac{3}{2}$$

Check by substituting each value of n into the original equation. The solution set is $\left\{ -\frac{5}{2}, \frac{3}{2} \right\}$.

Review this example for Objective 5:

9. Solve each equation.

    a. $x^2 = -48$ **b.** $(x-3)^2 = -25$

    c. $100x^2 - 100x + 34 = 0$

    a. $x^2 = -48$

$$x = \sqrt{-48} \quad \text{or} \quad x = -\sqrt{-48} \quad \text{Square root}$$
$$= 4i\sqrt{3} \qquad\qquad = -4i\sqrt{3} \quad \begin{array}{l}\text{property;}\\ \text{simplify.}\end{array}$$

The solution set is $\left\{ \pm 4i\sqrt{3} \right\}$.

    b. $(x-3)^2 = -25$

$$x - 3 = \sqrt{-25} \quad \text{or} \quad x - 3 = -\sqrt{-25}$$
$$\text{Square root property}$$
$$x - 3 = 5i \qquad \text{or} \qquad x - 3 = -5i$$
$$\text{Simplify the radical.}$$
$$x = 3 + 5i \quad \text{or} \qquad x = 3 - 5i$$
$$\text{Add 3.}$$

The solution set is $\left\{ 3 \pm 5i \right\}$

    c. $100x^2 - 100x + 34 = 0$

$$100x^2 - 100x = -34 \quad \text{Subtract 34.}$$
$$x^2 - x = -\frac{34}{100} = -\frac{17}{50}$$
$$\text{Divide by 100.}$$
$$x^2 - x + \frac{1}{4} = -\frac{17}{50} + \frac{1}{4}$$
$$\text{Complete the square.}$$

Now Try:

9. Solve each equation.

    a. $y^2 = -32$ _____

    b. $(x+2)^2 = -49$ _____

    c. $4x^2 - 4x + 17 = 0$

    _____

$$\left(x - \frac{1}{2}\right)^2 = -\frac{9}{100}$$ Factor on the left.
Add on the right.

$$x - \frac{1}{2} = \sqrt{-\frac{9}{100}}$$ or $$x - \frac{1}{2} = -\sqrt{-\frac{9}{100}}$$

Square root property

$$x - \frac{1}{2} = \frac{3}{10}i$$ or $$x - \frac{1}{2} = -\frac{3}{10}i$$

Simplify the radical.

$$x = \frac{1}{2} + \frac{3}{10}i$$ or $$x = \frac{1}{2} - \frac{3}{10}i$$

Add $\frac{1}{2}$.

The solution set is $\left\{\frac{1}{2} \pm \frac{3}{10}i\right\}$.

Objective 1 Review the zero-factor property.

For extra help, see Example 1 on page 496 of your text and the Section Lecture video for Section 9.1.

Use the zero-factor property to solve each equation.

1. $6z^2 + 19z + 10 = 0$ 1. _____

2. $8x^2 + 2x - 15 = 0$ 2. _____

Objective 2 Learn the square root property.

For extra help, see Examples 2 and 3 on pages 497–498 of your text, the Section Lecture video for Section 9.1, and Exercise Solution Clips 13 and 35.

Use the square root property to solve each equation.

3. $12 - 4y^2 = 8$ 3. _____

4. $121x^2 - 24 = 0$ 4. _____

5. The deck of the Golden Gate bridge is about 245 feet above the water. How long will it take a rock dropped from the deck to hit the water? Round your answer to the nearest tenth.

5. _____

Objective 3 Solve quadratic equations of the form $(ax + b)^2 = c$ by extending the square root property.

For extra help, see Examples 4 and 5 on pages 498–499 of your text, the Section Lecture video for Section 9.1, and Exercise Solution Clip 31.

Use the square root property to solve each equation.

6. $(3t + 4)^2 = 49$

6. _____

7. $(6p + 9)^2 = 54$

7. _____

8. $(4t + 3)^2 = 5$

8. _____

Objective 4 Solve quadratic equations by completing the square.

For extra help, see Examples 6–8 on pages 499–501 of your text, the Section Lecture video for Section 9.1, and Exercise Solution Clips 57, 63, and 67.

Solve each equation by completing the square.

9. $r^2 + 8r + 4 = 0$

9. _____

10. $2x^2 - 4x = 1$

10. _____

11. $3x^2 = 4x + 2$

11. _____

Objective 5 Solve quadratic equations with solutions that are not real numbers.
For extra help, see Example 9 on page 502 of your text, the Section Lecture video for
Section 9.1, and Exercise Solution Clip 75.

Solve each equation.

12. $49 + 16w^2 = 0$

12. _____

13. $(7c - 1)^2 + 5 = 0$

13. _____

14. $x^2 - 2x + 3 = 0$

14. _____

15. $6x^2 - 2x = -1$

15. _____

Chapter 9 QUADRATIC EQUATIONS, INEQUALITIES, AND FUNCTIONS

9.2   The Quadratic Formula

Learning Objectives
1   Derive the quadratic formula.
2   Solve quadratic equations by using the quadratic formula.
3   Use the discriminant to determine the number and type of solutions.

Key Terms

Use the vocabulary terms listed below to complete each statement in exercises 1–2.

quadratic formula   **discriminant**

1.   The _____ can be used to solve any quadratic equation.

2.   The value of the _____ is used to determine the number of solutions to a quadratic equation.

Guided Examples

Review these examples for Objective 2:

1.   Solve $5t^2 - 13t + 6 = 0$.

$a = 5, b = -13, c = 6$

$$x = \frac{-b \pm \sqrt{b^2 - 4ac}}{2a}$$

$$x = \frac{-(-13) \pm \sqrt{(-13)^2 - 4(5)(6)}}{2(5)}$$

$$= \frac{13 \pm \sqrt{49}}{10} = \frac{13 \pm 7}{10}$$

$$x = \frac{20}{10} = 2 \text{ or } x = \frac{6}{10} = \frac{3}{5}$$

Be sure to check each solution in the original equation. The solution set is $\left\{\frac{3}{5}, 2\right\}$.

Now Try:

1.   Solve $6x^2 - 17x + 12 = 0$.

_____

2. Solve $4x(x+1)=1$.

Write the equation in standard form.
$$4x(x+1)=1$$
$$4x^2+4x=1 \quad \text{Distributive property}$$
$$4x^2+4x-1=0$$
$$a=4,\ b=4,\ c=-1$$

$$x=\frac{-b\pm\sqrt{b^2-4ac}}{2a}$$

$$x=\frac{-4\pm\sqrt{4^2-4(4)(-1)}}{2(4)}=\frac{-4\pm\sqrt{32}}{8}$$

$$x=\frac{-4\pm4\sqrt{2}}{8}=\frac{-1\pm\sqrt{2}}{2}$$

The solution set is $\left\{\frac{-1\pm\sqrt{2}}{2}\right\}$.

3. Solve $7x^2+5x+3=0$.

$$a=7,\ b=5,\ c=3$$

$$x=\frac{-b\pm\sqrt{b^2-4ac}}{2a}$$

$$x=\frac{-5\pm\sqrt{5^2-4(7)(3)}}{2(7)}$$

$$=\frac{-5\pm\sqrt{-59}}{14}$$

$$=\frac{-5\pm i\sqrt{59}}{14}$$

The solution set is $\left\{\frac{-5\pm i\sqrt{59}}{14}\right\}$.

2. Solve $2x^2=2x+3$.

_____

3. Solve $2x^2=4x-3$.

_____

Review these examples for Objective 3:

4. Find the discriminant. Use it to predict the number and type of solutions for each equation. Tell whether the equation can be solved by factoring or whether the quadratic formula should be used.

a. $16x^2-40x+25=0$

b. $6x^2-3x-10=0$

c. $5y^2-5y+2=0$

d. $3x^2+x=2$

Now Try:

4. Find the discriminant. Use it to predict the number and type of solutions for each equation. Tell whether the equation can be solved by factoring or whether the quadratic formula should be used.

a. $10x^2+21x+9=0$

_____

a. $b^2 - 4ac = (-40)^2 - 4(16)(25)$

$= 1600 - 1600$

$= 0$

Since the discriminant is zero, the equation will have one rational solution. The equation can be solve by factoring.

b. $b^2 - 4ac = (-3)^2 - 4(6)(-10)$

$= 9 + 240$

$= 249$

Since the discriminant is positive, but not the square of an integer, and a, b, and c are integers, the equation will have two irrational solutions. The equation should be solved using the quadratic formula.

c. $b^2 - 4ac = (-5)^2 - 4(5)(2)$

$= 25 - 40$

$= -15$

Since the discriminant is negative, the equation will have two nonreal complex solutions. The quadratic formula should be used to solve it.

d. Write in standard form: $3x^2 + x - 2 = 0$.

$b^2 - 4ac = (1)^2 - 4(3)(-2)$

$= 1 + 24$

$= 25$

Since the discriminant is the square of an integer, and a, b, and c are integers, there are two rational solutions. The equation can be solved by factoring.

5. Find k so that $9x^2 - kx + 49 = 0$ will have exactly one rational solution.

The equation will have only one rational solution if the discriminant is 0.

$b^2 - 4ac = (-k)^2 - 4(9)(49)$

$= k^2 - 1764$

Set the discriminant equal to 0 and solve for k.

$k^2 - 1764 = 0$

$k^2 = 1764$

$k = \pm\sqrt{1764} = \pm 42$

b. $2y^2 + 4y + 8 = 0$

_____

c. $25x^2 - 30x + 9 = 0$

_____

d. $2x^2 - 4x = 8$

_____

5. Find k so that $4x^2 + kx + 49 = 0$ will have exactly one rational solution.

_____

The equation will have only one rational
solution if $k = -42$ or $k = 42$.

Objective 2 Solve quadratic equations by using the quadratic formula.
For extra help, see Examples 1–3 on pages 506–508 of your text, the Section Lecture video
for Section 9.2, and Exercise Solution Clips 5, 9, and 37.

*Use the quadratic formula to solve each equation. (All solutions for these equations are real
numbers.)*

1. $5t^2 - 13t + 6 = 0$

1. _____

2. $6m^2 - 17m + 12 = 0$

2. _____

3. $x(x - 2) = 1$

3. _____

4. $5m^2 + 6m - 1 = 0$

4. _____

5. $7 - 6t - 5t^2 = 0$

5. _____

Use the quadratic formula to solve each equation. (All solutions for these equations are nonreal complex numbers.)

6. $x^2 - 6x + 10 = 0$ **6.** _____

7. $4x^2 + 4x + 5 = 0$ **7.** _____

8. $2x^2 - 5x + 4 = 0$ **8.** _____

Objective 3 Use the discriminant to determine the number and type of solutions.
For extra help, see Examples 4 and 5 on pages 509–510 of your text, the Section Lecture video for Section 9.2, and Exercise Solution Clips 39 and 53.

Use the discriminant to determine whether the solutions for each equation are

     A.   *two rational numbers*   B.   *one rational number*
     C.   *two irrational numbers*   D.   *two nonreal complex numbers.*

Tell whether the equation can be solved by factoring or whether the quadratic formula should be used. Do not actually solve.

9. $4t^2 + 12t + 9 = 0$ **9.** _____

10. $5a^2 - 4a + 1 = 0$ **10.** _____

11. $3r^2 + 5r + 1 = 0$ **11.** _____

12. $5t^2 + 3t = 2$

13. $16y^2 + 9 = -12y$

Find the value of a, b, or c so that each equation will have exactly one rational solution.

14. $ax^2 - 40x + 16 = 0$

15. $49x^2 + bx + 16 = 0$

Chapter 9 QUADRATIC EQUATIONS, INEQUALITIES, AND FUNCTIONS

9.3   Equations Quadratic in Form

Learning Objectives

1 Solve an equation with fractions by writing it in quadratic form.
2 Use quadratic equations to solve applied problems.
3 Solve an equation with radicals by writing it in quadratic form.
4 Solve an equation that is quadratic in form by substitution.

Key Terms

Use the vocabulary terms listed below to complete each statement in exercises 1–2.

quadratic in form   standard form

1.   A nonquadratic equation that is written in the form $au^2 + bu + c = 0$, for $a \neq 0$ and an algebraic expression u, is called _____.

2.   A quadratic equation written in the form $ax^2 + bx + c = 0$, $a \neq 0$, is written in

_____.

Guided Examples

Review this example for Objective 1:

1. Solve $5 + \dfrac{6}{m+1} = \dfrac{14}{m}$.

Clear fractions by multiplying each term by the least common denominator, $m(m + 1)$. Note that the domain must be restricted to $m \neq -1$, $m \neq 0$.

$$m(m+1)\left(5 + \frac{6}{m+1}\right) = m(m+1)\left(\frac{14}{m}\right)$$

$$5m(m+1) + \left(\frac{6}{m+1}\right)m(m+1)$$

$$= m(m+1)\left(\frac{14}{m}\right)$$

Distributive property

$$5m^2 + 5m + 6m = 14m + 14$$

Distributive property

Now Try:

1. Solve $9 - \dfrac{12}{x} = -\dfrac{4}{x^2}$.

_____

$$5m^2 + 11m = 14m + 14$$

Combine like terms.

$$5m^2 - 3m - 14 = 0 \qquad \text{Standard form}$$
$$(5m + 7)(m - 2) = 0 \qquad \text{Factor.}$$

$$5m + 7 = 0 \qquad \text{or} \qquad m - 2 = 0 \qquad \text{Zero-factor}$$
$$5m = -7 \qquad\qquad m = 2 \qquad \text{property}$$
$$m = -\frac{7}{5}$$

The solution set is $\left\{ -\frac{7}{5}, 2 \right\}$.

Review these examples for Objective 2:

2. Amy rows her boat 6 miles upstream and then returns in $2\frac{6}{7}$ hours. The speed of the current is 2 miles per hour. How fast can she row?

   Step 1: Read the problem carefully.

   Step 2: Assign a variable. Let x = the rate that Amy rows in still water. The current slows Amy when she is going upstream, so the Amy's rate going upstream is her rate in still water less the rate of the current, or $x - 2$. Similarly, the current makes Amy row faster when she is going downstream, so her downstream rate is $x + 2$.

   Organize the information using a table. Recall that $d = rt$.

	d	r	t
upstream	6	$x - 2$	$\dfrac{6}{x-2}$
downstream	6	$x + 2$	$\dfrac{6}{x+2}$

   Step 3: Write an equation. The time upstream plus the time downstream equals the total time, $2\frac{6}{7}$ hours.

   $$\frac{6}{x-2} + \frac{6}{x+2} = \frac{20}{7}$$

Now Try:

2. Mike can row 3 miles per hour in still water. It takes him 3 hours and 36 minutes to row 3 miles round trip, upstream and back. Find the speed of the current.

_____

Step 4: Solve the equation. The LCD is $7(x-2)(x+2)$.

$$7(x-2)(x+2)\left(\frac{6}{x-2}+\frac{6}{x+2}\right)$$
$$= 7(x-2)(x+2)\left(\frac{20}{7}\right)$$

Multiply by the LCD.

$$7(x-2)(x+2)\left(\frac{6}{x-2}\right)+$$
$$7(x-2)(x+2)\left(\frac{6}{x+2}\right)$$
$$= 7(x-2)(x+2)\left(\frac{20}{7}\right)$$

Distributive property

$$42x+84+42x-84=20x^2-80$$

Distributive property

$$84x=20x^2-80$$

Combine like terms.

$$20x^2-84x-80=0$$

Standard form

$$4(5x+4)(x-5)=0$$

Factor.

$$5x+4=0 \quad \text{or} \quad x-5=0 \quad \text{Zero-factor}$$
$$5x=-4 \qquad\qquad x=5 \quad \text{property}$$
$$x=-\frac{4}{5}$$

Step 5: State the answer. Amy's rowing speed cannot be $-\frac{4}{5}$ miles per hour. Amy rows at 5 miles per hour.

Step 6: Check that this value satisfies the original equation.

3. It takes two painters working together $4\frac{4}{5}$ hours to paint a house. If each worked alone, one of them could do the job in 4 hours less than the other. How long would it take each painter to complete the job alone?

Step 1: Read the problem carefully. There will be two answers.

3. Working together, Tom and Huck painted a fence in 8 hours. If each worked alone, Tom could paint the fence 12 hours faster than Huck could. How long would it take each to paint the fence alone?

_____

Step 2: Let x = the number of hours for the faster painter to complete the job alone. Then the slower painter could do the entire job in $x - 4$ hours. The slower painter's rate is $\frac{1}{x-4}$ and the faster painter's rate is $\frac{1}{x}$.

Together, they do the job in $4\frac{4}{5}$ hours.

Organize the information using a table. Recall that $d = rt$.

	Rate	Time Working Together	Fractional Part of the Job Done
Slower painter	$\frac{1}{x}$	$\frac{24}{5}$	$\frac{1}{x}\left(\frac{24}{5}\right)$
Faster painter	$\frac{1}{x-4}$	$\frac{24}{5}$	$\frac{1}{x-4}\left(\frac{24}{5}\right)$

Step 3: Write an equation. The sum of the two fractional parts is 1.

$$\frac{1}{x}\left(\frac{24}{5}\right) + \frac{1}{x-4}\left(\frac{24}{5}\right) = 1$$

Step 4: Solve the equation. The LCD is $5x(x-4)$.

$$5x(x-4)\frac{1}{x}\left(\frac{24}{5}\right) + 5x(x-4)\frac{1}{x-4}\left(\frac{24}{5}\right)$$
$$= 5x(x-4)(1)$$

Multiply by the LCD.

$$24x - 96 + 24x = 5x^2 - 20x$$

Distributive property

$$48x - 96 = 5x^2 - 20x$$

Combine like terms.

$5x^2 - 68x + 96 = 0$ Standard form

$(5x-8)(x-12) = 0$ Standard form

$5x - 8 = 0$ or $x - 12 = 0$ Zero-factor

$5x = 8$ $x = 12$ property

$x = \frac{8}{5}$

Step 5: State the answer. If the slower

painter can do the job in $\frac{8}{5}$ hr (or 1.6 hr),

then the faster painter's time to do the job is
$1.6 - 4 = -2.4$ hr, which cannot represent the
time for the slower painter. If the slower
painter's time is 12 hours, then the faster
painter's time is $12 - 4 = 8$ hours.

Step 6: Check that these results satisfy the
original equation.

Review this example for Objective 3:

4. Solve each equation.

    a. $y = \sqrt{y + 42}$

    b. $\sqrt{3}y = \sqrt{28y - 49}$

    a.
    $$y = \sqrt{y + 42}$$
    $$y^2 = y + 42 \quad \text{Square both sides.}$$
    $$y^2 - y - 42 = 0 \qquad \text{Standard form}$$
    $$(y - 7)(y + 6) = 0 \qquad \text{Factor.}$$
    $$y - 7 = 0 \quad \text{or} \quad y + 6 = 0 \quad \text{Zero-factor}$$
    $$y = 7 \qquad\qquad y = -6 \quad \text{property}$$

    We must check each proposed solution
    in the original equation because squaring
    each side of an equation can introduce
    extraneous solutions.

    Check
    $$-6 \overset{?}{=} \sqrt{-6 + 42} \qquad 7 \overset{?}{=} \sqrt{7 + 42}$$
    $$-6 \overset{?}{=} \sqrt{36} \qquad\quad 7 \overset{?}{=} \sqrt{49}$$
    $$-6 = 6 \quad \text{False} \qquad 7 = 7 \quad \text{True}$$

    The solution set is $\{7\}$.

    b.
    $$\sqrt{3}y = \sqrt{28y - 49}$$
    $$3y^2 = 28y - 49$$
    $$\qquad\qquad\quad \text{Square both sides.}$$
    $$3y^2 - 28y + 49 = 0 \quad \text{Standard form}$$
    $$(3y - 7)(y - 7) = 0 \quad \text{Factor.}$$
    $$3y - 7 = 0 \quad \text{or} \quad y - 7 = 0 \quad \text{Zero-factor}$$
    $$3y = 7 \qquad\qquad y = 7 \quad \text{property}$$
    $$y = \frac{7}{3}$$

Now Try:

4. Solve each equation.

    a. $x = \sqrt{x + 2}$ _____

    b. $p = \sqrt{\dfrac{7p - 1}{12}}$ _____

Check

$$\sqrt{3}\left(\frac{7}{3}\right)\overset{?}{=}\sqrt{28\left(\frac{7}{3}\right)-49}$$

$$\frac{7\sqrt{3}}{3}\overset{?}{=}\sqrt{\frac{196}{3}-49}$$

$$\frac{7\sqrt{3}}{3}\overset{?}{=}\sqrt{\frac{49}{3}}$$

$$\frac{7\sqrt{3}}{3}=\frac{7\sqrt{3}}{3} \quad \text{True}$$

$$\sqrt{3}(7)\overset{?}{=}\sqrt{28(7)-49}$$

$$7\sqrt{3}\overset{?}{=}\sqrt{147}$$

$$7\sqrt{3}=7\sqrt{3} \quad \text{True}$$

The solution set is $\left\{\frac{7}{3},7\right\}$.

Review these examples for Objective 4:

5. Define a variable u, and write each equation in the form $au^2+bu+c=0$. Then solve each equation.

   a. $c^4-13c^2+36=0$

   b. $(m+5)^2+6(m+5)+8=0$

   c. $x^{4/3}-20x^{2/3}+36=0$

   a. Let $u=c^2$. Then, the equation becomes $u^2-13u+36=0$.

   $$u^2-13u+36=0$$
   $$(u-4)(u-9)=0 \quad \text{Factor.}$$

$u-4=0$	or	$u-9=0$	Zero-factor
$u=4$		$u=9$	property
$c^2=4$		$c^2=9$	Substitute.
$c=\pm2$		$c=\pm3$	Square root property

   Be sure to check each solution in the original equation. The solution set is $\{\pm2,\pm3\}$

Now Try:

5. Define a variable u, and write each equation in the form $au^2+bu+c=0$. Then solve each equation.

   a. $x^4-5x^2+4=0$

   _____

   b. $(x-5)^2+2(x-5)-35=0$

   _____

   c. $x^{2/3}-2x^{1/3}=3$

   _____

b. Let $u = m + 5$. Then, the equation becomes $u^2 + 6u + 8 = 0$.

$$u^2 + 6u + 8 = 0$$
$$(u + 4)(u + 2) = 0 \quad \text{Factor.}$$

$u + 4 = 0 \quad$ or $\quad u + 2 = 0 \quad$ Zero-factor
$\quad u = -4 \qquad\qquad u = -2 \quad$ property
$m + 5 = -4 \qquad m + 5 = -2 \quad$ Substitute.
$\quad m = -9 \qquad\qquad m = -7 \quad$ Solve.

Be sure to check each solution in the original equation. The solution set is $\{-9, -7\}$.

c. Since $x^{4/3} = \left(x^{2/3}\right)^2$, we let $u = x^{2/3}$.

The equation becomes

$$u^2 - 20u + 36 = 0.$$

$$u^2 - 20u + 36 = 0$$
$$(u - 18)(u - 2) = 0 \quad \text{Factor.}$$

$u - 18 = 0 \quad$ or $\quad u - 2 = 0 \qquad$ Zero-factor
$\quad u = 18 \qquad\qquad u = 2 \qquad$ property
$x^{2/3} = 18 \qquad x^{2/3} = 2 \qquad$ Substitute.
$\quad x = 18^{3/2} \qquad x = 2^{3/2} \quad \left(x^{2/3}\right)^{3/2} = x$
$\quad x = 54\sqrt{2} \qquad x = 2\sqrt{2} \quad$ Simplify.

Be sure to check each solution in the original equation. The solution set is $\left\{2\sqrt{2}, 54\sqrt{2}\right\}$.

6. Solve $x^4 - 4x^2 + 1$.

Let $u = x^2$. Then the equation becomes $u^2 - 4u + 1 = 0$. This equation cannot be solved by factoring, so we use the quadratic formula.

$$u = \frac{-(-4) \pm \sqrt{(-4)^2 - 4(1)(1)}}{2(1)}$$
$$= \frac{4 \pm \sqrt{12}}{2} = \frac{4 \pm 2\sqrt{3}}{2} = 2 \pm \sqrt{3}$$

6. Solve $x^4 - 4x^2 - 8 = 0$.

_____

Now substitute and solve for x:

$$x^2 = 2 - \sqrt{3} \quad \text{or} \quad x^2 = 2 + \sqrt{3}$$

$$x = \pm\sqrt{2 - \sqrt{3}} \qquad x = \pm\sqrt{2 + \sqrt{3}}$$

The solution set is $\left\{ \pm\sqrt{2 - \sqrt{3}}, \pm\sqrt{2 + \sqrt{3}} \right\}$.

Objective 1 Solve an equation with fractions by writing it in quadratic form.

For extra help, see Example 1 on page 512 of your text, the Section Lecture video for Section 9.3, and Exercise Solution Clip 13.

Solve each equation.

1. $1 + \dfrac{1}{x} = \dfrac{6}{x^2}$

1. _____

2. $4 - \dfrac{8}{x - 1} = -\dfrac{35}{x}$

2. _____

3. $\dfrac{5x}{x + 1} + \dfrac{6}{x + 2} = \dfrac{3}{(x + 1)(x + 2)}$

3. _____

Objective 2   Use quadratic equations to solve applied problems.
For extra help, see Examples 2 and 3 on pages 513–515 of your text, the Section Lecture video for Section 9.3, and Exercise Solution Clips 27 and 33.

Solve each problem.

4. A jet plane traveling at a constant speed goes 1200 miles   4. _____
    with the wind, then turns around and travels for 1000
    miles against the wind. If the speed of the wind is 50
    miles per hour and the total flight takes 4 hours, find the
    speed of the plane.

5. Marlene rode a bicycle for 12 miles and then hiked an   5. _____
    additional 8 miles. The total time for the trip was 5
    hours. If her rate when she was riding the bicycle was
    10 miles per hour faster than her rate walking, what was
    each rate?

6. Two pipes together can fill a large tank in 10 hours. One   6. _____
    of the pipes, used alone, takes 15 hours longer than the
    other to fill the tank. How long would each pipe used
    alone take to fill the tank?

7. Working together, Dale and Roger complete a job in 6   7. _____
    hours. It would take Dale 9 hours longer than Roger to
    do the job alone. How long would it take Roger alone?

Objective 3 Solve an equation with radicals by writing it in quadratic form.
For extra help, see Example 4 on page 515 of your text, the Section Lecture video for Section 9.3, and Exercise Solution Clip 37.

Solve each equation.

8. $\sqrt{2}y = \sqrt{6-y}$

8. _____

9. $k - \sqrt{8k-15} = 0$

9. _____

10. $\sqrt{x} = x - 6$

10. _____

11. $y = \sqrt{\dfrac{1-2y}{8}}$

11. _____

Objective 4 Solve an equation that is quadratic in form by substitution.
For extra help, see Examples 5–7 on pages 516–518 of your text, the Section Lecture video for Section 9.3, and Exercise Solution Clips 49 and 57.

Solve each equation.

12. $16m^4 = 25m^2 - 9$

12. _____

13. $\dfrac{1}{(x+6)^2} - \dfrac{7}{2(x+6)} = -\dfrac{3}{2}$

13. _____

14. $m^{-2} + m^{-1} - 20 = 0$

14. _____

15. $x^4 = -x^2 + 20$

15. _____

Chapter 9 QUADRATIC EQUATIONS, INEQUALITIES, AND FUNCTIONS

9.4   Formulas and Further Applications

Learning Objectives
1   Solve formulas for variables involving squares and square roots.
2   Solve applied problems using the Pythagorean theorem.
3   Solve applied problems using area formulas.
4   Solve applied problems using quadratic functions as models.

Key Terms

Use the vocabulary terms listed below to complete each statement in exercises 1–2.

Pythagorean theorem   **quadratic function**

1.   A function defined by an equation of the form $f(x) = ax^2 + bx + c$, for real numbers, a, b, and c, with $a \neq 0$, is a _____.

2.   The _____ states that the sum of the squares of the lengths of the legs of a right triangle equals the square of the length of the hypotenuse.

Guided Examples

Review these examples for Objective 1:
1.   Solve each formula for the given variable.

   a. $D = \sqrt{kh}$ for k

   b. $y = \dfrac{1}{2}gt^2$ for t

   a. The goal is to isolate k on one side.
   $$D = \sqrt{kh}$$
   $$D^2 = kh \qquad \text{Square both sides.}$$
   $$\frac{D^2}{h} = k \qquad \text{Divide by } h.$$

Now Try:
1.   Solve each formula for the given variable.

   a. $p = \dfrac{yz}{\sqrt{6}}$ for z

   _____

   b. $F = \dfrac{mx}{t^2}$ for t

   _____

b. The goal is to isolate t on one side.

$$y = \frac{1}{2}gt^2$$

$$\frac{2y}{g} = t^2 \qquad \text{Multiply by 2; divide by } g.$$

$$\pm\sqrt{\frac{2y}{g}} = t \qquad \text{Square root property}$$

$$t = \pm\frac{\sqrt{2y}}{\sqrt{g}} \cdot \frac{\sqrt{g}}{\sqrt{g}} = \pm\frac{\sqrt{2yg}}{g}$$

Rationalize the deonominator.

2. Solve $rk^2 - 2k = -s$ for k.

Write the equation in standard form and then use the quadratic formula to solve for k.

$$rk^2 - 2k = -s$$

$$rk^2 - 2k + s = 0 \quad \text{Add } s. \text{ Standard form}$$

To use the quadratic formula, let $a = r$, $b = -2$, and $c = s$.

$$k = \frac{-(-2) \pm \sqrt{(-2)^2 - 4rs}}{2r}$$

$$= \frac{2 \pm \sqrt{4 - 4rs}}{2r}$$

$$= \frac{2 \pm \sqrt{4(1-rs)}}{2r} \quad \text{Factor under square root.}$$

$$= \frac{2 \pm 2\sqrt{1-rs}}{2r} \quad \text{Simplify.}$$

$$k = \frac{1 \pm \sqrt{1-rs}}{r} \quad \text{Factor out 2.}$$

2. Solve $p^2q^2 + pkq = k^2$ for q.

_____

Review this example for Objective 2:

3. A 13-foot ladder is leaning against a building. The distance from the bottom of the ladder to the building is 2 feet more than twice the distance from the top of the ladder to the ground. How far is the bottom of the ladder from the building?

Step 1: Read the problem carefully.

Step 2: Assign a variable. Let x = the distance from the top of the ladder to the ground. Then $2x + 2$ = the distance from the bottom of the ladder to the building. Draw a picture to represent the problem.

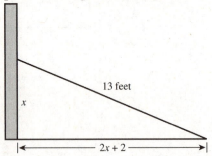

Step 3: Write an equation. Use the Pythagorean theorem: $x^2 + (2x + 2)^2 = 13^2$

Step 4: Solve.

$$x^2 + (2x + 2)^2 = 13^2$$

$x^2 + 4x^2 + 8x + 4 = 169$ Square the binomial.

$5x^2 + 8x - 165 = 0$ Combine like terms. Standard form

$(5x + 33)(x - 5) = 0$ Factor.

$5x + 33 = 0$ or $x - 5 = 0$ Zero-factor property;

$5x = -33$ $x = 5$ Solve for x.

$x = -\dfrac{33}{5}$

Step 5: State the answer. Since length cannot be negative, we disregard the negative solution. The distance from the top of the ladder to the ground is 5 feet. However, we are asked to find the distance from the bottom of the ladder to the building. This distance is 2(5) + 2 = 12 feet.

Step 6: Check. Since $5^2 + 12^2 = 13^2$, the answer is correct.

Now Try:

3. Two cars left an intersection at the same time, one heading south, the other heading east. Some time later the car traveling south had gone 18 miles farther than the car headed east. At that time they were 90 miles apart. How far had each car traveled?

_____

Name: _____   Date: _____

Instructor: _____   Section: _____

Review this example for Objective 3:

4. A fish pond is 3 feet by 4 feet. How wide a strip of concrete can be poured around the pool if there is enough concrete for 44 square feet?

Step 1: Read the problem carefully.

Step 2: Assign a variable. Let x = the width of the strip of concrete. Then $2x + 3$ = the width of the fish pond with the two strips of concrete and $2x + 4$ = the length of the fish pond with the two strips of concrete.

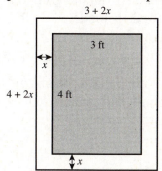

Step 3: Write an equation. The area of the strip is 44 sq ft and the area of the fish pond is 3(4) = 12 sq ft, so the total area of the outer rectangle is 44 + 12 = 56 sq ft.

$$(3 + 2x)(4 + 2x) = 56$$

Step 4: Solve.

$$(3 + 2x)(4 + 2x) = 56$$

$12 + 14x + 4x^2 = 56$ Multiply.

$4x^2 + 14x - 44 = 0$ Subtract 56.
Standard form

$2(2x + 11)(x - 2) = 0$ Factor.

$(2x + 11)(x - 2) = 0$ Divide by 2.

$2x + 11 = 0$ or $x - 2 = 0$ Zero-factor
$2x = -11$ $x = 2$ property;
$x = -\dfrac{11}{2}$ Solve for x.

Step 5: State the answer. Since length cannot be negative, we disregard the negative solution. The concrete strip should be 2 ft wide.

Now Try:

4. A picture 9 inches by 12 inches is to be mounted on a piece of mat board so that there is an even width of mat all around the picture. How wide will the matted border be if the area of the mounted picture is 238 square inches?

_____

Step 6: Check. If $x = 2$, then the area of the large rectangle is

$(3 + 2 \cdot 2)(4 + 2 \cdot 2) = 7 \cdot 8 = 56$ sq ft. The area of the fish pond is $3(4) = 12$ sq ft, so the area of the concrete strip is $56 - 12 = 44$ sq ft.

Review these examples for Objective 4:

5. A certain projectile is located at a distance of $d(t) = 3t^2 - 6t + 1$ feet from its starting point after t seconds. How many seconds will it take the projectile to travel 10 feet?

   Let $d = 10$ in the formula and solve for t.

   $$d = 3t^2 - 6t + 1$$
   $$10 = 3t^2 - 6t + 1$$
   $$9 = 3t^2 - 6t \qquad \text{Subtract 1.}$$
   $$3 = t^2 - 2t \qquad \text{Divide by 3.}$$
   $$3 + 1 = t^2 - 2t + 1 \qquad \text{Complete the square.}$$
   $$4 = (t - 1)^2 \qquad \text{Factor.}$$

   Solve the equation by using the square root property.

   $$t - 1 = -\sqrt{4} \quad \text{or} \quad t - 1 = \sqrt{4}$$
   $$t - 1 = -2 \qquad\qquad t - 1 = 2$$
   $$t = -1 \qquad\qquad\quad t = 3$$

   Since t represents time, we reject the negative solution. It will take 3 seconds for the projectile to travel 10 feet.

6. The profit from the sale of x items is given by the function $P(x) = 2x^2 - 10x - 100$.
   a. What is the profit if 50 items are sold?
   b. How many items must be sold for the profit to equal $1000?

   a. Since 50 were sold, we must find $P50)$.
   $$P(50) = 2(50)^2 - 10(50) - 100$$
   $$= 4400$$
   The profit if 50 items are sold is $4400.

Now Try:

5. A baseball is thrown upward from a building 20 m high with a velocity of 15 m/sec. Its distance from the ground after t seconds is modeled by the function

   $f(t) = -4.9t^2 + 15t + 20$. When will the ball hit the ground? Round your answer to the nearest tenth.

   _____

6. A manufacturer finds that the number of items sold each month, n, is related to the price per item, p, by the equation $p = -2n + 200$. The monthly revenue is given by $R(n) = pn$.
   a. What was the revenue if 30 items were sold?

   _____

b. We must find the value of x that makes $P(x) = 1000$.

$$1000 = 2x^2 - 10x - 100$$
$$0 = 2x^2 - 10x - 1100 \quad \text{Standard form}$$
$$0 = x^2 - 5x - 550 \quad \text{Divide by 2.}$$

Now solve using the quadratic formula and $a = 1$, $b = -5$, $c = -550$.

$$x = \frac{-(-5) \pm \sqrt{(-5)^2 - 4(1)(-550)}}{2(1)}$$
$$= \frac{5 \pm \sqrt{2225}}{2}$$
$$\approx -21 \text{ or } 26$$

There cannot be a negative number of items sold, so disregard the negative solution. The profit will be $1000 is 26 items are sold.

b. How many items were sold if the revenue was $4800?

_____

Objective 1 Solve formulas for variables involving squares and square roots.
For extra help, see Examples 1 and 2 on pages 523–524 of your text, the Section Lecture video for Section 9.4, and Exercise Solution Clips 9 and 15.

Solve each equation for the indicated variable. Leave ± in your answer, where appropriate.

1. $a = \sqrt{bc} + 1$ for c

1. _____

2. $f = \dfrac{m - x}{t^2}$ for t

2. _____

3. $2m = m^2 + x^2 + x^2$ for x

3. _____

4. $m^2x^2 + 2rx = s^2$ for x 4.

Objective 2 Solve applied problems using the Pythagorean theorem.
For extra help, see Example 3 on page 524 of your text, the Section Lecture video for
Section 9.4, and Exercise Solution Clip 31.

Solve each problem. When appropriate, round answers to the nearest tenth.

5.   A child flying a kite has let out 45 feet of string to the   5.
kite. The distance from the kite to the ground is 9 feet
more than the distance from the child to a point directly
below the kite. How high up is the kite?

6.   Two cars left an intersection at the same time, one   6. _____
heading north, the other heading west. Later they were
exactly 95 miles apart. The car headed west had gone
38 miles less than twice as far as the car headed north.
How far had each car traveled?

7.   The longest side of a right triangle is 2 feet more than   7. _____
the middle side and the middle side is 1 foot less than
twice the shortest side. Find the length of the shortest
side.

Objective 3 Solve applied problems using area formulas.

For extra help, see Example 4 on page 525 of your text, the Section Lecture video for Section 9.4, and Exercise Solution Clip 37.

Solve each problem.

8. A rectangle has a length 1 meter less than twice its width. If 1 meter is cut from the length and added to the width, the figure becomes a square with an area of 16 square meters. Find the dimensions of the original rectangle.

8. _____

9. The area of a square is 81 square centimeters. If the same amount is added to one dimension and removed from the other, the resulting rectangle has an area 9 centimeters less than the area of the square. How much is added and subtracted?

9. _____

10. An open box is to be made from a rectangular piece of tin by cutting 2-inch squares out of the corners and folding up the sides. The length of the finished box is to be twice the width. The volume of the box will be 100 cubic inches. Find the dimensions of the rectangular piece of tin.

10. _____

11. A rectangular garden has an area of 12 feet by 5 feet. A gravel path of equal width is to be built around the garden. How wide can the path be if there is enough gravel for 138 square feet?

11. _____

Objective 4 Solve applied problems using quadratic functions as models.
For extra help, see Examples 5 and 6 on page 526 of your text, the Section Lecture video for Section 9.4, and Exercise Solution Clip 43.

Solve each problem. When appropriate, round answers to the nearest tenth.

12. A population of microorganisms grows according to the function $p(x) = 100 + .2x + .5x^2$, where x is given in hours. How many hours does it take to reach a population of 250 microorganisms?

12. _____

13. An object is thrown downward from a tower 280 feet high. The distance the object has fallen at time t in seconds is given by $s(t) = 16t^2 + 68t$. How long will it take the object to fall 100 feet?

13. _____

14.   If Pablo throws an object upward from a height of 32   **14.** _____
feet with an initial velocity of 48 feet per second, then
its height h (in feet) after t seconds is given by the
formula $h = -16t^2 + 48t + 32$. At what times will it be
50 feet above the ground?

15.   George and Albert have found that the profit (in   **15.** _____
dollars) from their cigar shop is given by the formula
$P = -10x^2 + 100x + 300$, where x is the number of
units of cigars sold daily. How many units should be
sold for a profit of $460?

Chapter 9 QUADRATIC EQUATIONS, INEQUALITIES, AND FUNCTIONS

9.5   Graphs of Quadratic Functions

Learning Objectives
1   Graph a quadratic function.
2   Graph parabolas with horizontal and vertical shifts.
3   Use the coefficient of x^2 to predict the shape and direction in which a parabola opens.
4   Find a quadratic function to model data.

Key Terms

Use the vocabulary terms listed below to complete each statement in exercises 1–4.

parabola   vertex   axis   quadratic function

1.   The vertical (or horizontal) line through the vertex of a vertical (or horizontal) parabola is its _____.

2.   The point on a parabola that has the least y-value (if the parabola opens up) or the greatest y-value (if the parabola opens down) is called the _____ of the parabola.

3.   A function defined by $f(x) = ax^2 + bx + c$, for real numbers a, b, and c, with $a \neq 0$, is a _____.

4.   The graph of a quadratic function is a _____.

Guided Examples

Review these examples for Objective 2:

1.   Graph $g(x) = x^2 + 2$. Give the vertex, axis, domain, and range.

The graph of $g(x)$ has the same shape as that of $f(x) = x^2$ but is shifted 2 units up with vertex (0, 2). Every function value is 2 more than the corresponding function value of $f(x) = x^2$.

Now Try:

1.   Graph $f(x) = x^2 - 1$. Give the vertex, axis, domain, and range.

Vertex   _____

Axis   _____

Domain   _____

Range   _____

x	$f(x) = x^2$	$g(x) = x^2 + 2$
−2	4	6
−1	1	3
0	0	2
1	1	3
2	4	6

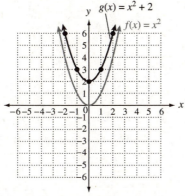

The vertex is (0, 2). The axis is $x = 0$. The domain is $(-\infty, \infty)$. The range is $[2, \infty)$.

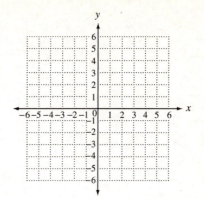

2. Graph $g(x) = (x - 1)^2$. Give the vertex, axis, domain, and range.

The graph of $g(x)$ has the same shape as that of $f(x) = x^2$ but is shifted 1 unit right with vertex (1, 0).

x	$f(x) = x^2$	$g(x) = (x - 1)^2$
−2	4	9
−1	1	4
0	0	1
1	1	0
2	4	1
3	9	4

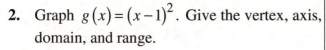

2. Graph $f(x) = (x + 3)^2$. Give the vertex, axis, domain, and range.

Vertex _____

Axis _____

Domain _____

Range _____

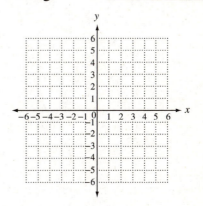

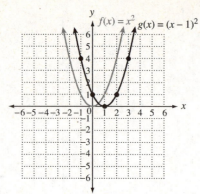

The vertex is (1, 0). The axis is $x = 1$. The domain is $(-\infty, \infty)$. The range is $[0, \infty)$.

3. Graph $g(x) = (x - 1)^2 - 2$. Give the vertex, axis, domain, and range.

The graph of $g(x)$ has the same shape as that of $f(x) = x^2$ but is shifted 1 unit right (since $x - 1 = 0$ if $x = 1$) and 2 units down (because of the -2).

x	$g(x) = (x - 1)^2 - 2$
-2	7
-1	2
0	-1
1	-2
2	-1
3	2
4	7

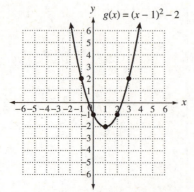

The vertex is (1, -2). The axis is $x = 1$. The domain is $(-\infty, \infty)$. The range is $[-2, \infty)$.

3. Graph $f(x) = (x + 2)^2 - 1$. Give the vertex, axis, domain, and range.

Vertex   _____

Axis   _____

Domain   _____

Range   _____

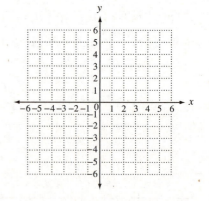

Name: Date:

Instructor: Section:

Review these examples for Objective 3:

4. Graph $g(x) = -2x^2$. Give the vertex, axis, domain, and range.

   The graph of $g(x)$ has the same shape as that of $f(x) = x^2$, but is narrower and opens downward.

x	$g(x) = -2x^2$
−2	−8
−1	−2
0	0
1	−2
2	−8

   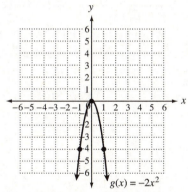

   The vertex is (0, 0). The axis is $x = 0$. The domain is $(-\infty, \infty)$. The range is $(-\infty, 0]$.

Now Try:

4. Graph $f(x) = -\frac{1}{4}x^2$. Give the vertex, axis, domain, and range.

   Vertex   _____

   Axis   _____

   Domain   _____

   Range   _____

   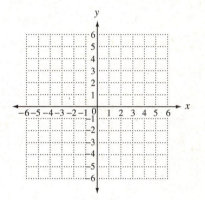

5. Graph $g(x) = -\frac{1}{2}(x+1)^2 - 2$. Give the vertex, axis, domain, and range.

   The parabola open down because $a < 0$ and is wider than the graph of $f(x) = x^2$. The parabola has vertex $(-1, -2)$.

x	$g(x) = -\frac{1}{2}(x+1)^2 - 2$
−3	−4
−2	−2.5
−1	−2
0	−2.5
1	−4

5. Graph $f(x) = 3(x-1)^2 + 1$. Give the vertex, axis, domain, and range.

   Vertex   _____

   Axis   _____

   Domain   _____

   Range   _____

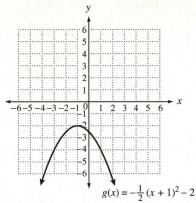

$$g(x) = -\frac{1}{2}(x+1)^2 - 2$$

The vertex is $(-1, -2)$. The axis is $x = -1$.
The domain is $(-\infty, \infty)$. The range is
$(-\infty, -2]$.

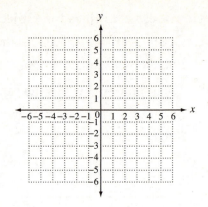

Review this example for Objective 4:

6. The number of ice cream cones sold by an ice cream parlor from 2003–2009 is shown in the following table.

Year	Years since 2003, x	Number of cones sold
2003	0	1775
2004	1	4194
2005	2	5063
2006	3	5161
2007	4	4663
2008	5	4639
2009	6	3710

a. Use the ordered pairs $(x,$ number of cones sold) to make a scatter diagram of the data.

b. Use the graph to determine whether the coefficient a of x^2 in a quadratic model should be positive or negative.

c. Determine a quadratic function that models these data by using a system of equations. Use the ordered pairs $(0, 1775)$, $(3, 5161)$, and $(6, 3710)$. Round the values of a, b, and c in your model to the nearest tenth, as necessary.

Now Try:

6. The table lists the average price of a Major League Baseball ticket.

Year	Years since 1990, x	Price
1991	1	$9.14
1994	4	$10.60
1997	7	$12.49
2000	10	$16.81
2004	14	$19.82
2010	20	$26.74

Source: www.ticketnews.com

a. Use the ordered pairs $(x,$ price) to make a scatter diagram of the data.

d. Use your model from part c to estimate the number of cones sold in 2010.

a.

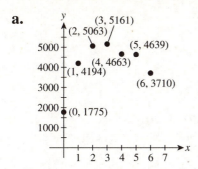

b. It appears that the parabola opens down, so the coefficient a is negative.

c. Using the chosen ordered pairs, we substitute the x- and y-values into the quadratic form $y = ax^2 + bx + c$ to obtain three equations.

$$a(0)^2 + b(0) + c = 1775 \quad (1)$$
$$a(3)^2 + b(3) + c = 5161 \quad (2)$$
$$a(6)^2 + b(6) + c = 3710 \quad (3)$$

Equation (1) simplifies to $c = 1775$, so substitute 1775 for c in equation (2) and (3).

$$9a + 3b + 1775 = 5161 \quad (2)$$
$$36a + 6b + 1775 = 3710 \quad (3)$$

Subtracting 1775 from each side gives

$$9a + 3b = 3386 \quad (2)$$
$$36a + 6b = 1935 \quad (3)$$

Multiply (2) by -2, then add the two equations and solve for a.

$$-18a - 6b = -6772 \quad -2 \times (2)$$
$$\underline{36a + 6b = 1935 \quad (3)}$$
$$18a = -4837$$
$$a \approx -268.7$$

Substitute this value for a into equation (2) and solve for b.

$$9(-268.7) + 3b = 3386$$
$$-2418.3 + 3b = 3386$$
$$3b = 5804.3$$
$$b \approx 1934.8$$

Therefore, the model is

$$y = -268.7x^2 + 1934.8x + 1775.$$

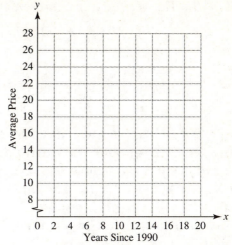

Years Since 1990

b. Use the graph to determine whether the coefficient a of x^2 in a quadratic model should be positive or negative.

_____

c. Determine a quadratic function that models these data by using a system of equations. Use the ordered pairs (1, 9.14), (10, 16.81), and (20, 26.74). Round the values of a, b, and c in your model to the nearest hundredth, as necessary.

_____

d. Use your model from part c to estimate the cost of tickets in 2015.

_____

d. The year 2010 is represented by $x = 7$.

$y = -268.7(7)^2 + 1934.8(7) + 1775.$

≈ 2152

According to the model, about 2152 ice cream cones were sold in 2010.

Objective 1 Graph a quadratic function.

Objective 2 Graph parabolas with horizontal and vertical shifts.

For extra help, see Example 1–3 on pages 532–534 of your text, the Section Lecture video for Section 9.5, and Exercise Solution Clips 21, 23, and 27.

Graph each parabola. Give the vertex, axis, domain, and range.

1. $f(x) = x^2 + 3$

1. vertex _____

    axis _____

    domain _____

    range _____

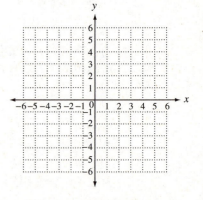

2. $f(x) = (x-3)^2$

2. vertex _____

axis _____

domain _____

range _____

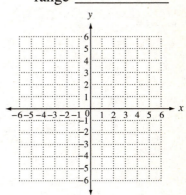

3. $f(x) = (x+3)^2 - 1$

3. vertex _____

axis _____

domain _____

range _____

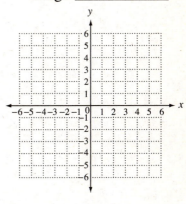

4. $f(x) = (x+2)^2 + 3$

4. vertex _____

   axis _____

   domain _____

   range _____

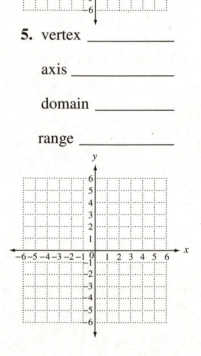

5. $f(x) = (x-1)^2 + 2$

5. vertex _____

   axis _____

   domain _____

   range _____

Objective 3 Use the coefficient of x^2 to predict the shape and direction in which a parabola opens.

For extra help, see Examples 4 and 5 on page 534–535 of your text, the Section Lecture video for Section 9.5, and Exercise Solution Clips 29 and 33.

For each quadratic function, tell whether the graph opens up or down and whether the graph is wider, narrower, or the same shape as the graph of $f(x) = x^2$. Then give the vertex.

6. $f(x) = -\dfrac{4}{3}x^2 - 1$ 6. _____

7. $f(x) = \dfrac{2}{3}(x-1)^2$ 7. _____

8. $f(x) = -\dfrac{1}{3}(x+3)^2 - 4$ 8. _____

9. $f(x) = \dfrac{5}{4}(x-1)^2 + 7$ 9. _____

10. $f(x) = 4 - x^2$ 10. _____

Objective 4 Find a quadratic function to model data.

For extra help, see Example 6 on pages 535–536 of your text and the Section Lecture video for Section 9.5.

Tell whether a linear or quadratic function would be a more appropriate model for each set of graphed data. If linear, tell whether the slope should be positive or negative. If quadratic, tell whether the coefficient a of x^2 should be positive or negative.

11.   11. _____

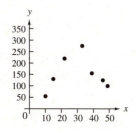

12.

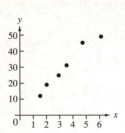

12. _____

13.

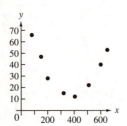

13. _____

Solve each problem.

14. The number of publicly traded companies filing for bankruptcy for selected years between 1990 and 2000 are shown in the table, with 0 representing 1990, 2 representing 1992, etc.

Year	Number of Bankruptcies
0	115
2	91
4	70
6	84
8	120
10	176

Source: Lial, Margaret L., John Hornsby, Terry McGinnis, _Intermediate Algebra_ Eighth Edition. Boston: Pearson Education, 2006.

a. Use the ordered pairs to make a scatter diagram of the data.

b. Use the scatter diagram to decide whether a linear or quadratic function would better model the data. If linear, is the slope positive or negative? If quadratic, should the coefficient a of x^2 be positive or negative?

14.a.

b. _____

c. _____

d. _____

c. Use the ordered pairs (0, 115), (4, 70), and (8, 120) to find a function that models the data. Round the values of a, b, and c to three decimal places, if necessary.

d. Use the model from part (c) to approximate the number of company bankruptcy filings in 2002. Round the answer to the nearest whole number.

15. The number of music DVDs released from 2000 through 2007, with 0 representing 2000, 1 representing 2001, etc., is shown in the table below.

Year	Number of DVDs (thousands)
0	4.0
1	5.6
2	7.4
3	10.4
4	12.2
5	13.9
6	13.6
7	12.1

Source: http://files.dvdnote.com/pdfs/press_kit.pdf.

a. Use the ordered pairs to make a scatter diagram of the data.

b. Use the scatter diagram to decide whether a linear or quadratic function would better model the data. If linear, is the slope positive or negative? If quadratic, should the coefficient a of x^2 be positive or negative?

15.a.

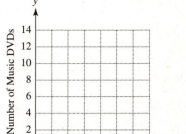

b. _____

c. _____

d. _____

c. Use the ordered pairs (0, 4.0), (5, 13.9), and (7, 12.1) to find a function that models the data. Round the values of *a*, *b*, and *c* to three decimal places, if necessary.

d. Use the model from part (c) to predict the number of music DVDs in 2010.

Chapter 9 QUADRATIC EQUATIONS, INEQUALITIES, AND FUNCTIONS

9.6 More About Parabolas and Their Applications

Learning Objectives
1 Find the vertex of a vertical parabola.
2 Graph a quadratic function.
3 Use the discriminant to find the number of x-intercepts of a parabola with a vertical axis.
4 Use quadratic functions to solve problems involving maximum or minimum value.
5 Graph parabolas with horizontal axes.

Key Terms

Use the vocabulary terms listed below to complete each statement in exercises 1–2.

   discriminant   **vertex**

1. The _____ of a quadratic function is found by using the formula $b^2 - 4ac$.

2. The graph of a quadratic equation defined by $f(x) = ax^2 + bx + c \ (a \neq 0)$ has _____ $\left(\frac{-b}{2a}, f\left(\frac{-b}{2a}\right)\right)$.

Guided Examples

Review these examples for Objective 1:
1. Find the vertex of the graph of
   $f(x) = x^2 + 6x + 10$.

   We can express $x^2 + 6x + 10$ in the form $(x - h)^2 + k$ by completing the square on $x^2 + 6x$. Because we want to keep $f(x)$ alone on one side of the equation, we add and subtract the appropriate number on just one side.

   $f(x) = x^2 + 6x + 10$
   $\qquad = \left(x^2 + 6x + 9 - 9\right) + 10$

   $\qquad \left[\frac{1}{2}(6)\right]^2 = 3^2 = 9$
   Add and subtract 9.

Now Try:
1. Find the vertex of the graph of
   $f(x) = x^2 - 6x + 4$.

   _____

$$f(x) = \left(x^2 + 6x + 9\right) + 10 - 9$$

Bring -9 outside the parentheses.

$$= (x+3)^2 + 1 \quad \text{Factor; combine like terms.}$$

The vertex of the parabola is $(-3, 1)$.

2. Find the vertex of the graph of
   $$f(x) = 3x^2 + 6x + 10.$$

   Because the x^2-term has a coefficient other than 1, we factor that coefficient out of the first two terms before completing the square.

   $$f(x) = 3x^2 + 6x + 10.$$
   $$= 3\left(x^2 + 2x\right) + 10$$
   $$= 3\left(x^2 + 2x + 1 - 1\right) + 10$$
   $$\left[\tfrac{1}{2}(1)\right]^2 = 1^2 = 1$$

   Add and subtract 1.

   $$= 3\left(x^2 + 2x + 1\right) + 3(-1) + 10$$

   Bring 1 outside the parentheses.
   Use the distributive property.

   $$= 3(x+1)^2 + 7$$

   Factor; combine like terms.

   The vertex is $(-1, 7)$.

3. Use the vertex formula to find the vertex of the graph of $f(x) = -4x^2 + 5x + 3$.

   The x-coordinate of the vertex is given by $\frac{-b}{2a}$.

   $$\frac{-b}{2a} = \frac{-5}{2(-4)} = \frac{5}{8} \qquad a = -4,\, b = 5,\, c = 3$$

   The y-coordinate is $f\left(\frac{-b}{2a}\right)$.

   $$f\left(\tfrac{5}{8}\right) = -4\left(\tfrac{5}{8}\right)^2 + 5\left(\tfrac{5}{8}\right) + 3 = \tfrac{73}{16}$$

   The vertex is $\left(\tfrac{5}{8}, \tfrac{73}{16}\right)$.

2. Find the vertex of the graph of
   $$f(x) = 2x^2 - 6x + 5.$$

   _____

3. Use the vertex formula to find the vertex of the graph of
   $$f(x) = -2x^2 + 4x - 1.$$

   _____

Review this example for Objective 2:

4. Graph the quadratic function defined by

   $f(x) = x^2 - 3x + 2$. Give the vertex, axis, domain, and range.

   Step 1: From the equation, $a = 1$, so the graph opens up.

   Step 2: The x-coordinate of the vertex is $\frac{3}{2}$.

   The y-coordinate of the vertex is

   $f\left(\frac{3}{2}\right) = \left(\frac{3}{2}\right)^2 - 3\left(\frac{3}{2}\right) + 2 = -\frac{1}{4}$. The vertex is

   $\left(\frac{3}{2}, -\frac{1}{4}\right)$.

   Step 3: Find any intercepts. Since the vertex is in quadrant IV and the graph opens up, there will be two x-intercepts. Let $f(x) = 0$ and solve.

   $x^2 - 3x + 2 = 0$

   $(x - 2)(x - 1) = 0$ Factor.

   $x - 2 = 0$ or $x - 1 = 0$ Zero-factor property

        $x = 2$ $x = 1$

   The x-intercepts are $(1, 0)$ and $(2, 0)$.
   Find the y-intercept by evaluating $f(0)$.

   $f(0) = 0^2 - 3(0) + 2 = 2$

   The y-intercept is $(0, 2)$.

   Step 4: Plot the points found so far and additional points as needed using symmetry about the axis, $x = \frac{3}{2}$.

   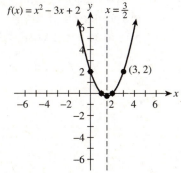

   The domain is $(-\infty, \infty)$ and the range is

   $\left[-\frac{1}{4}, \infty\right)$.

Now Try:

4. Graph the quadratic function defined by $f(x) = x^2 + 4x + 5$. Give the vertex, axis, domain, and range.

   Vertex   _____

   Axis   _____

   Domain   _____

   Range   _____

   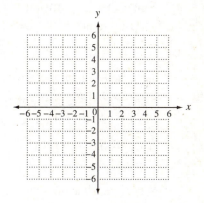

Review this example for Objective 3:

5. Find the discriminant and use it to determine the number of x-intercepts of the graph of each quadratic function.

    a. $f(x) = 3x^2 + 3x + 2$

    b. $f(x) = 4x^2 + 12x + 9$

    c. $f(x) = 6x^2 - 3x - 1$

    a. $f(x) = 3x^2 + 3x + 2$
        $b^2 - 4ac = 3^2 - 4(3)(2) = -15$
        Since the discriminant is negative, the graph has no x-intercepts.

    b. $f(x) = 4x^2 + 12x + 9$
        $b^2 - 4ac = 12^2 - 4(4)(9) = 0$
        Since the discriminant is zero, the graph has only one x-intercept, its vertex.

    c. $f(x) = 6x^2 - 3x - 1$
        $b^2 - 4ac = (-3)^2 - 4(6)(-1) = 33$
        Since the discriminant is positive, the graph has two x-intercepts.

Now Try:

5. Find the discriminant and use it to determine the number of x-intercepts of the graph of each quadratic function.

    a. $f(x) = x^2 + 5x + 4$

    _____

    b. $f(x) = 5x^2 - 5x + 2$

    _____

    c. $f(x) = 9x^2 - 24x + 16$

    _____

Review these examples for Objective 4:

6. A farmer has 1000 yards of fencing to enclose a rectangular field. What is the largest area that the farmer can enclose? What are the dimensions of the field when the area is maximized?

    If the length of the field is represented by l and the width of the field is represented by w, the perimeter is given by $2l + 2w = 1000$ or $l + w = 500$ or $w = 500 - l$.

    The area is given by $A = lw$ or

    $A(l) = l(500 - l) = 500l - l^2 = -l^2 + 500l$.

    This is a quadratic equation, so its maximum occurs at the vertex of its graph. The x-coordinate is given by $\frac{-b}{2a} = \frac{-500}{2(-1)} = 250$.

    The y-coordinate is

    $A(250) = -250^2 + 500(250) = 62,500$.

Now Try:

6. A farmer has 1000 yards of fencing to enclose a rectangular field next to a building. What is the largest area that the farmer can enclose? What are the dimensions of the field when the area is maximized?

    _____

If $l = 250$, then $w = 500 - 250 = 250$. Therefore, the maximum area is 62,500 sq yd when the length of the field is 250 yd and the width is 250 yd.

7. An object is launched directly upward at 64 feet per second from a platform 80 feet high. Its height above the ground is given by $s(t) = -16t^2 + 64t + 80,$ where t is the number of seconds after launch. What will be the object's maximum height? When will it reach this height?

For this function $a = -16$, $b = 64$, and $c = 80$. The vertex formula gives $t = \frac{-b}{2a} = \frac{-64}{2(-16)} = 2$. This indicates that the maximum height is reached at 2 seconds. Now calculate $s(2)$ to find the maximum height.

$s(2) = -16(2)^2 + 64(2) + 80 = 144$

Thus, it takes 2 seconds to reach the maximum height of 144 feet above the ground.

7. An object is launched directly upward at 48 feet per second from a platform 250 feet high. Its height above the ground is given by $s(t) = -16t^2 + 48t + 250$ where t is the number of seconds after launch. What will be the object's maximum height? When will it reach this height?

_____

Review these examples for Objective 5:

8. Graph $x = (y+1)^2 - 2$. Give the vertex, axis, domain, and range.

This graph has its vertex at $(-2, -1)$ since the roles of x and y are interchanged. It opens to the right since $a > 0$ and has the same shape as $y = x^2$.

x	y
2	−3
−1	−2
−2	−1
−1	0
2	1

Now Try:

8. Graph $x = (y-2)^2 + 1$. Give the vertex, axis, domain, and range.

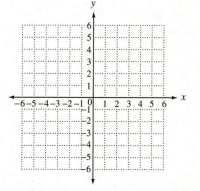

Vertex _____

Axis _____

The axis is $y = -1$. The domain is $[-2, \infty)$ and the range is $(-\infty, \infty)$.

9. Graph $x = 3y^2 + 6y - 2$. Give the vertex, axis, domain, and range.

We must complete the square in order to write the equation in $x = (y - k)^2 + h$ form.

$x = 3y^2 + 6y - 2$

$\quad = 3\left(y^2 + 2y\right) - 2$ Factor.

$\quad = 3\left(y^2 + 2y + 1 - 1\right) - 2$

> Complete the square within parentheses. Add and subtract 1.

$\quad = 3\left(y^2 + 2y + 1\right) - 2 + 3(-1)$

> Distributive property

$\quad = 3\left(y + 1\right)^2 - 5$ Factor; simplify.

The vertex is $(-5, -1)$. The axis is $y = -1$.

x	y
-2	-2
-5	-1
-2	0

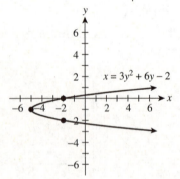

The domain is $[-5, \infty)$ and the range is $(-\infty, \infty)$.

Domain   _____

Range   _____

9. Graph $x = 2y^2 + 4y + 3$. Give the vertex, axis, domain, and range.

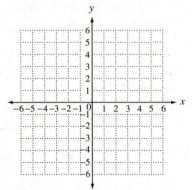

Vertex   _____

Axis   _____

Domain   _____

Range   _____

Objective 1 Find the vertex of a parabola
Objective 3 Use the discriminant to find the number of *x*-intercepts of a parabola with
**   a vertical axis.**

For extra help, see Examples 1–3 and 5 on pages 541–542 and page 544 of your text, the Section Lecture video for Section 9.6, and Exercise Solution Clips 5, 7, 9, and 11.

Find the vertex of each parabola. For each equation, decide whether the graph opens up, down, to the left, or to the right, and whether it is wider, narrower, or the same shape as the graph of $y = x^2$. *If it is a parabola with a vertical axis, find the discriminant and use it to determine the number of x-intercepts.*

1. $f(x) = x^2 - 2x + 4$ 1. _____

2. $f(x) = 5x^2 - 4x + 1$ 2. _____

3. $f(x) = -2x^2 - 4x + 1$ 3. _____

4. $f(x) = -\frac{1}{4}x^2 - 3x - 9$ 4. _____

5. $x = -2y^2 + 4y - 1$ 5. _____

6. $x = y^2 + 4y - 4$ 6. _____

Name: Date:
Instructor: Section:

Objective 2 Graph a quadratic function.

For extra help, see Example 4 on page 543 of your text, the Section Lecture video for Section 9.6, and Exercise Solution Clip 23.

Graph each parabola. Give the vertex, axis, domain, and range.

7. $f(x) = x^2 + 8x + 20$

7. vertex_____

axis _____

domain_____

range_____

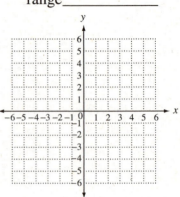

8. $f(x) = -\frac{1}{2}x^2 - 2x + 1$

8. vertex_____

axis _____

domain_____

range_____

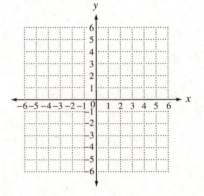

9. $f(x) = \dfrac{5}{4}x^2 + 5x + 3$

9. vertex_____

axis _____

domain_____

range_____

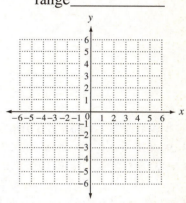

Objective 4 Use quadratic functions to solve problems involving maximum or minimum value.

For extra help, see Examples 6 and 7 on pages 545–546 of your text, the Section Lecture video for Section 9.6, and Exercise Solution Clips 35 and 37.

Solve each problem.

10. The perimeter of a rectangle is 16 m. What length will produce the maximum area? What is the maximum area?

10. _____

11. Jean sells ceramic pots. She has weekly costs of $C(x) = x^2 - 100x + 2700,$ where x is the number of pots she sells each week. How many pots should she sell to minimize her costs? What is the minimum cost?

11. _____

12. A projectile is fired upward so that its distance (in feet)   12. _____
above the ground t seconds after firing is given by
$s(t) = -16t^2 + 80t + 156$. Find the maximum height it
reaches and the number of seconds to reach that height.

Objective 5 Graph parabolas with horizontal axes.
For extra help, see Examples 4, 8, and 9 on pages 543 and 547 of your text, the Section
Lecture video for Section 9.6, and Exercise Solution Clips 23, 27, and 31.

Graph each parabola. Give the vertex, axis, domain, and range.

13. $x = -y^2 + 6y - 9$

13. vertex_____

   axis _____

   domain_____

   range_____

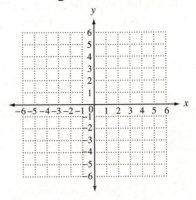

14. $3x = y^2 - 6y + 6$

14. vertex_____

axis _____

domain_____

range_____

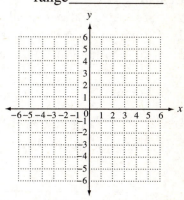

15. $x = 3y^2 - 6y + 4$

15. vertex_____

axis _____

domain_____

range_____

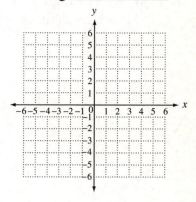

Chapter 9 QUADRATIC EQUATIONS, INEQUALITIES, AND FUNCTIONS

9.7   Polynomial and Rational Inequalities

Learning Objectives
1   Solve quadratic inequalities.
2   Solve polynomial inequalities of degree 3 or greater.
3   Solve rational inequalities.

Key Terms

Use the vocabulary terms listed below to complete each statement in exercises 1–2.

quadratic inequality   rational inequality

1.   A _____ can be written in the form $ax^2 + bx + c < 0$ (or with $\leq$ or $>$ or $\geq$), where a, b, and c are real numbers, with $a \neq 0$.

2.   An inequality that involves rational expressions is called a

_____.

Guided Examples

Review these examples for Objective 1:
1.   Use the graph to solve each quadratic inequality.

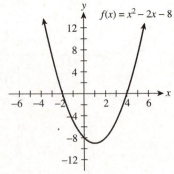

a. $x^2 - 2x - 8 > 0$

b. $x^2 - 2x - 8 < 0$

Note that the x-intercepts are $(-2, 0)$ and $(4, 0)$.

a.   From the graph, we see that the y-values are greater than 0 when the x-values are less than -2 or greater than 4.

Now Try:
1.   Use the graph to solve each quadratic inequality.

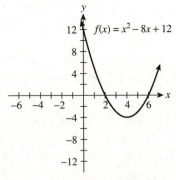

a. $x^2 - 8x + 12 > 0$ _____

b. $x^2 - 8x + 12 < 0$ _____

Thus, the solution set of $x^2 - 2x - 8 > 0$ is $(-\infty, -2) \cup (4, \infty)$.

b. From the graph, we see that the y-values are less than 0 when the x-values are greater than −2 and less than 4. Thus, the solution set of $x^2 - 2x - 8 < 0$ is $(-2, 4)$.

2. Solve and graph the solution set.

$x^2 + 5x + 4 \geq 0$

Solve the quadratic equation $x^2 + 5x + 4 = 0$ by factoring.

$(x+1)(x+4) = 0$

$x + 1 = 0$ or $x + 4 = 0$

$\quad x = -1 \qquad\qquad x = -4$

The numbers −4 and −1 divide a number line into intervals A, B, and C, as shown below.

$$
\begin{array}{ccc}
A & B & C
\end{array}
$$
$$-4 \qquad -1\ 0$$

Since the numbers −4 and −1 are the only numbers that make the quadratic expression $x^2 + 5x + 4$ equal to 0, all other numbers make the expression either positive or negative. If one number in an interval satisfies the inequality, then all the numbers in that interval will satisfy the inequality.

Choose any number in interval A as a test number; we will choose −5.

$$x^2 + 5x + 4 \geq 0$$
$$\overset{?}{}$$
$$(-5)^2 + 5(-5) + 4 \geq 0 \quad \text{Substitute } -5 \text{ for } x.$$
$$4 \geq 0 \ \checkmark$$

Because −5 satisfies the inequality, all numbers from interval A are solutions.

Now try −2 from interval B.

$$x^2 + 5x + 4 \geq 0$$
$$\overset{?}{}$$
$$(-2)^2 + 5(-2) + 4 \geq 0 \quad \text{Substitute } -2 \text{ for } x.$$
$$-2 \geq 0 \quad \text{False}$$

The numbers in interval B are not solutions.

2. Solve and graph the solution set.

$x^2 - x - 2 < 0$

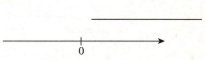

Finally, try 0 from interval C.

$$x^2 + 5x + 4 \geq 0$$
$$\overset{?}{(0)^2 + 5(0) + 4 \geq 0} \quad \text{Substitute 0 for } x.$$
$$4 \geq 0 \quad \checkmark$$

Because 0 satisfies the inequality, all number from interval C are solutions.

Because the inequality is greater than or equal to zero, we include the endpoints of the intervals in the solution set. Thus, the solution set is $(-\infty, -4] \cup [-1, \infty)$.

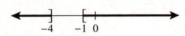

3. Solve each inequality.

    a. $(2k+5)^2 \leq -1$ **b.** $(2k+5)^2 \geq -1$

    a. Because $(2k+5)^2$ is never negative, there is no solution. The solution set is $\varnothing$.

    b. Because $(2k+5)^2$ is never negative, it is always greater than -1. The solution set is $(-\infty, \infty)$.

3. Solve each inequality.

    a. $(4m+1)^2 \leq -3$ _____

    b. $(4m+1)^2 \geq -3$ _____

Review this example for Objective 2:

4. Solve and graph the solution set.
    $$(x+1)(x-2)(x+4) \leq 0$$

    Set the factored polynomial equal to 0, then use the zero-factor property.

    $$x+1=0 \quad \text{or} \quad x-2=0 \quad \text{or} \quad x+4=0$$
    $$x=-1 \qquad\quad x=2 \qquad\quad x=-4$$

    Locate -4, -1, and 2 on a number line to determine the intervals A, B, C, and D.

    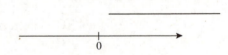

    Substitute a test number from each interval in the original inequality to determine which intervals satisfy the inequality.

Now Try:

4. Solve and graph the solution set.
    $$(2x-1)(2x+3)(3x+1) \leq 0$$

    _____

Interval	Test number	Test of inequality	True or False?
A	−5	−28 ≤ 0	True
B	−2	8 ≤ 0	False
C	0	−8 ≤ 0	True
D	5	162 ≤ 0	False

The numbers in intervals A and C are in the solution set. The three endpoints are included in the solution set since the inequality symbol, ≤, includes equality.

Thus, the solution set is $(-\infty, -4] \cup [-1, 2]$.

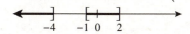

Review these examples for Objective 3:

5. Solve and graph the solution set of $\dfrac{7}{x-1} < 1$.

Write the inequality so that 0 is on one side.

$$\frac{7}{x-1} - 1 < 0 \quad \text{Subtract 1}$$

$$\frac{7}{x-1} - \frac{x-1}{x-1} < 0 \quad \text{The LCD is } x-1.$$

$$\frac{7-x+1}{x-1} < 0 \quad \text{Write as a single fraction.}$$

$$\frac{8-x}{x-1} < 0 \quad \text{Combine like terms.}$$

The sign of $\frac{8-x}{x-1}$ will change from positive to negative or negative to positive only at those numbers that make the numerator or denominator 0. These two numbers, 1 and 8, divide a number line into three intervals.

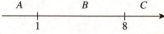

Test a number in each interval.

Interval	Test number	Test of inequality	True or False?
A	0	−8 < 0	True
B	2	6 < 0	False
C	10	$-\frac{2}{9} < 0$	True

Now Try:

5. Solve and graph the solution set of $\dfrac{y}{y+1} > 3$.

Name: **Date:**

Instructor: **Section:**

The solution set is $(-\infty, 1) \cup (8, \infty)$. This interval does not include 1 because it would make the denominator of the original inequality 0. The number 8 is not included because the inequality symbol, $<$, does not include equality.

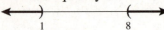

6. Solve and graph the solution set of $\dfrac{x+1}{x-5} \geq 3$.

Write the inequality so that 0 is on one side.

$$\dfrac{x+1}{x-5} - 3 \geq 0 \quad \text{Subtract 3.}$$

$$\dfrac{x+1}{x-5} - \dfrac{3(x-5)}{x-5} \geq 0 \quad \text{The LCD is } x-5.$$

$$\dfrac{x+1-3x+15}{x-5} \geq 0 \quad \text{Write as a single fraction.}$$

$$\dfrac{-2x+16}{x-5} \geq 0 \quad \text{Combine like terms.}$$

The sign of $\frac{-2x+16}{x-5}$ will change from positive to negative or negative to positive only at those numbers that make the numerator or denominator 0. These two numbers, 5 and 8, divide a number line into three intervals.

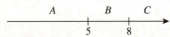

Test a number in each interval.

Interval	Test number	Test of inequality	True or False?
A	0	$-\dfrac{1}{5} \geq 3$	False
B	6	$7 \geq 3$	True
C	10	$\dfrac{11}{5} \geq 3$	False

6. Solve and graph the solution set of $\dfrac{z+2}{z-3} \leq 2$.

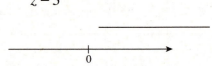

The solution set is $(5, 8]$. This interval does not include 5 because it would make the denominator of the original inequality 0. The number 8 is included because the inequality symbol, $\geq$, does includes equality.

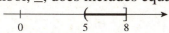

Objective 1   Solve quadratic inequalities.

For extra help, see Examples 1–3 on page 552–554 of your text, the Section Lecture video for Section 9.7, and Exercise Solution Clip 1, 9, and 23.

Solve each inequality, and graph the solution set.

1. $k^2 + 7k + 12 > 0$

1. _____

2. $a^2 - 3a - 18 \leq 0$

2. _____

3. $2y^2 + 5y < 3$

3. _____

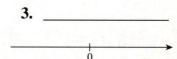

4. $15x^2 + 2 \geq 11x$

4. _____

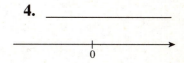

Solve each inequality.

5. $(3z-2)^2 < 0$ 5. _____

6. $(4m+1)^2 \geq 0$ 6. _____

Objective 2 Solve polynomial inequalities of degree 3 or greater.
For extra help, see Example 4 on page 555 of your text, the Section Lecture video for Section 9.7, and Exercise Solution Clip 27.

Solve each inequality and graph the solution set.

7. $(k+5)(k-1)(k+3) \leq 0$ 7. _____

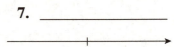

8. $(p-6)(p-4)(p-2) > 0$ 8. _____

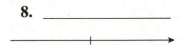

9. $(4q-3)(2q-7)(3q-10) \geq 0$ 9. _____

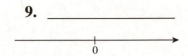

Objective 3 Solve rational inequalities.
For extra help, see Examples 5 and 6 on pages 556–557 of your text, the Section Lecture video for Section 9.7, and Exercise Solution Clips 35 and 41.

Solve each inequality and graph the solution set.

10. $\dfrac{r}{r-2} < 4$

10. _____

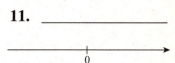

11. $\dfrac{6}{3r-2} \geq 1$

11. _____

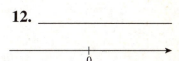

12. $\dfrac{-5}{2x-3} \leq 2$

12. _____

13. $\dfrac{2p-1}{3p+1} < 1$

13. _____

14. $\dfrac{4x}{2x+3} > 2$

14. _____

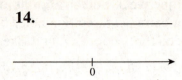

15. $\dfrac{3x-2}{x+2} \le 3$

15. _____

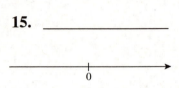

Chapter 10 INVERSE, EXPONENTIAL, AND LOGARITHMIC FUNCTIONS

10.1 Inverse Functions

Learning Objectives

1 Decide whether a function is one-to-one and, if it is, find its inverse.
2 Use the horizontal line test to determine whether a function is one-to-one.
3 Find the equation of the inverse of a function.
4 Graph f^{-1} from the graph of f.

Key Terms

Use the vocabulary terms listed below to complete each statement in exercises 1–2.

one-to-one function   **inverse of a function**

1. A(n) _____ is a function in which each x-value corresponds to only one y-value and each y-value corresponds to just one x-value.

2. If f is a one-to-one function, then the _____ f is the set of all ordered pairs of the form (y, x) where (x, y) belongs to f.

Guided Examples

Review this example for Objective 1:

1. Decide whether each function is one-to-one. If it is, find the inverse.
    a. $F = \{(-3, -1), (-2, 0), (-1, 1), (0, 2)\}$
    b. $G = \{(2, 1), (-1, 1), (0, 0), (1, 1)\}$

    c.
State	Number of National Parks
AK	8
AZ	3
CA	8
CO	4
FL	3
HA	2
UT	5

    a. Every x-value in F corresponds to only one y-value, and every y-value corresponds to only one x-value, so F is a one-to-one function.

Now Try:

1. Decide whether each function is one-to-one. If it is, find the inverse.
    a. $F = \{(2, 4), (-1, 1), (0, 0)$ $(1, 1), (2, 6)\}$ _____

    b. $G = \{(3, 2), (-3, -2), (2, 3),$ $(-2, -3)\}$ _____

    c.
State	Number of representatives
AK	1
AZ	8
CA	53
FL	25
NY	29
DE	1

    _____

b. Every *x*-value in *G* corresponds to only one *y*-value. However, the *y*-value 1 corresponds to two *x*-values, so *G* is not a one-to-one function.

c. Let *N* be the function defined in the table, with the states forming the domain and the number of national parks forming the range. Then, *N* is not one-to-one, because two different states have the same number of national parks.

Review this example for Objective 2:

2. Use the horizontal line test to determine whether each graph is the graph of a one-to-one function.

a.

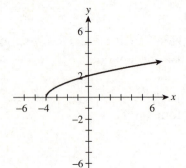

b.

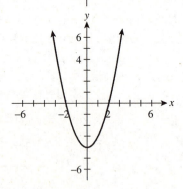

Now Try:

2. Use the horizontal line test to determine whether each graph is the graph of a one-to-one function.

a.

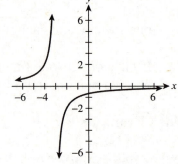

b.

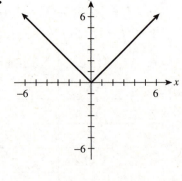

---

a.

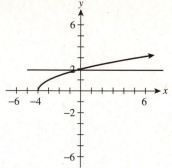

Every horizontal line will intersect the graph in exactly one point. The function is one-to-one.

b.

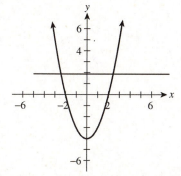

Because a horizontal line intersects the graph in more than one point, the function is not one-to-one.

Review this example for Objective 3:

3. Decide whether each equation defines a one-to-one function. If so, find the equation that defines the inverse.

  a. $f(x) = 3x - 5$

  b. $f(x) = 2x^2 + 3$

  c. $f(x) = x^3 + 1$

  a. The graph of $y = 3x - 5$ is a nonvertical line, so by the horizontal line test, f is a one-to-one function. To find the inverse, let $y = f(x)$, interchange x and y, then solve for y.

Now Try:

3. Decide whether each equation defines a one-ton-one function. If so, find the equation that defines the inverse.

  a. $f(x) = \sqrt{x-1},\ x \geq 1$

  _____

  b. $f(x) = 2x^3 - 3$

  _____

  c. $f(x) = -\frac{3}{2}x^2$

  _____

$$y = 3x - 5$$
$$x = 3y - 5 \quad \text{Interchange } x \text{ and } y.$$
$$x + 5 = 3y$$
$$\frac{x+5}{3} = y$$
$$f^{-1}(x) = \frac{x+5}{3} = \frac{x}{3} + \frac{5}{3} = \frac{1}{3}x + \frac{5}{3}$$

b. The graph of $y = 2x^2 + 3$ is a vertical parabola, so by the horizontal line test, f is not a one-to-one function and does not have an inverse.

c. The graph of $y = x^3 + 1$ is a cubing function. The function is one-to-one and has an inverse.

$$y = x^3 + 1$$
$$x = y^3 + 1 \quad \text{Interchange } x \text{ and } y.$$
$$x - 1 = y^3$$
$$\sqrt[3]{x-1} = y$$
$$f^{-1}(x) = \sqrt[3]{x-1}$$

Review this example for Objective 4:	Now Try:
4. Use the given graph to graph the inverse of f.	**4.** Use the given graph to graph the inverse of f.

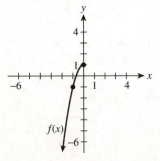

We can find the graph of f^{-1} from the graph of f by locating the mirror image of each point in f with respect to the line $y = x$.

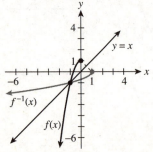

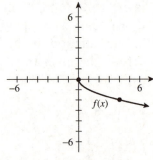

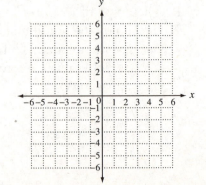

Name:

Date:

Instructor:

Section:

Objective 1 Decide whether a function is one-to-one and, if it is, find its inverse.

For extra help, see Example 1 on page 573 of your text, the Section Lecture video for Section 10.1, and Exercise Solution Clip 1.

Decide whether or not the function is one-to-one. If it is, find its inverse.

1. $\{(-3, 1), (-2, 2), (-1, 3), (0, 4)\}$

1. _____

2. $\{(1, 0), (2, 0), (3, 5), (4, 1)\}$

2. _____

3. $\{(0, 0), (1, 1), (-1, -1), (2, 2), (-2, -2)\}$

3. _____

4. $\{(4, 0), (2, 3), (0, 0), (3, 5)\}$

4. _____

Objective 2 Use the horizontal line test to determine whether a function is one-to-one.

For extra help, see Example 2 on page 574 of your text, the Section Lecture video for Section 10.1, and Exercise Solution Clip 7.

Use the horizontal line test to determine whether the graph is the graph of a one-to-one function.

5.

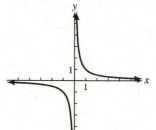

5. _____

6.

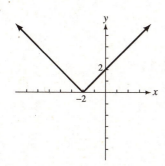

6. _____

Name: Date:
Instructor: Section:

7.

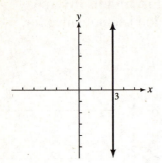

7. _____

8.

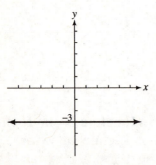

8. _____

Objective 3 Find the equation of the inverse of a function.
For extra help, see Example 3 on pages 574–575 of your text, the Section Lecture video for Section 10.1, and Exercise Solution Clip 13.

Decide whether the equation defines a one-to-one function. If so, find the equation of the inverse.

9. $f(x) = 2x - 5$

9. _____

10. $f(x) = x^2 - 1$

10. _____

11. $f(x) = x^3 - 1$

11. _____

12. $f(x) = \dfrac{3}{x-1}$

12. _____

Objective 4 Graph f^{-1} from the graph of f.

For extra help, see Example 4 on page 576 of your text, the Section Lecture video for Section 10.1, and Exercise Solution Clip 25.

The graphs of some functions are given below. For functions that are one-to-one, copy the graphs and graph the inverses with dashed curves on the same axes as the functions.

13.

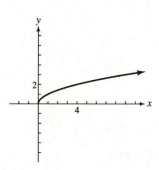

13.

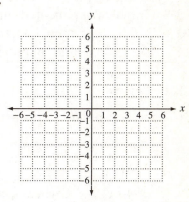

14.

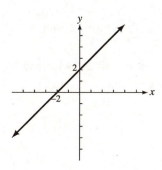

14.

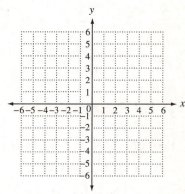

15.

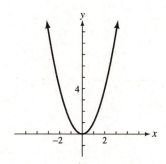

15.

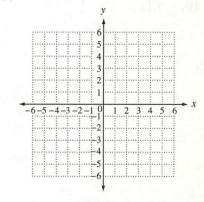

Chapter 10 INVERSE, EXPONENTIAL, AND LOGARITHMIC FUNCTIONS

10.2   Exponential Functions

Learning Objectives
1   Define an exponential function.
2   Graph an exponential function.
3   Solve exponential equations of the form $a^x = a^k$ for x.
4   Use exponential functions in applications involving growth or decay.

Key Terms

Use the vocabulary terms listed below to complete each statement in exercises 1–3.

exponential function   asymptote   exponential equation

1.   A(n) _____ is a function defined by an expression of the form
     $f(x) = a^x$, where $a > 0$ and $a \neq 1$ for all real numbers x.

2.   A(n) _____ is an equation that has a variable as an exponent.

3.   A line that a graph more and more closely approaches as the graph gets farther away
     from the origin is called a(n) _____ of the graph.

Guided Examples

Review these examples for Objective 2:

1.   Graph $f(x) = 6^x$.

     Create a table of values, then plot the points
     and draw a smooth curve through them.

x	$f(x) = 6^x$
-2	$\dfrac{1}{36}$
-1	$\dfrac{1}{6}$
0	1
1	6
2	36
3	216

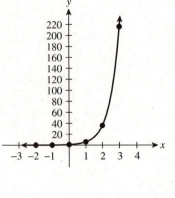

Now Try:

1.   Graph $f(x) = 3^x$.

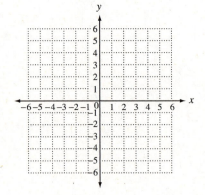

2. Graph $f(x) = \left(\frac{1}{6}\right)^x$.

Create a table of values, then plot the points
and draw a smooth curve through them.

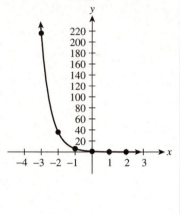

x	$f(x) = \left(\frac{1}{6}\right)^x$
-3	216
-2	36
-1	6
0	1
1	$\frac{1}{6}$
2	$\frac{1}{36}$

2. Graph $f(x) = \left(\frac{1}{3}\right)^x$.

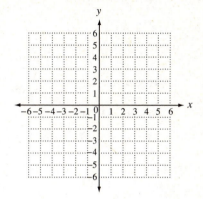

3. Graph $f(x) = 3^{2x-1}$.

Create a table of values, then plot the points
and draw a smooth curve through them.

x	$2x - 1$	$f(x) = 3^{2x-1}$
-1	-3	$\frac{1}{27}$
0	-1	$\frac{1}{3}$
1	1	3
2	3	27

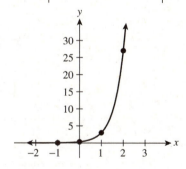

3. Graph $f(x) = 2^{1-x}$.

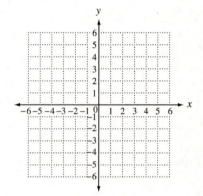

Review these examples for Objective 3:

4. Solve the equation $16^x = 64$.

$$16^x = 64$$

$$\left(2^4\right)^x = 2^6 \qquad \text{Write with the same base.}$$

$$2^{4x} = 2^6 \qquad \text{Power rule for exponents}$$

$$4x = 6 \qquad \text{If } a^x = a^y, \text{ then } x = y.$$

$$x = \frac{6}{4} = \frac{3}{2} \qquad \text{Solve for } x; \text{ simplify.}$$

Check:

$$16^{3/2} = \left(16^{1/2}\right)^3 = 4^3 = 64 \ \checkmark$$

The solution set is $\left\{\frac{3}{2}\right\}$.

5. Solve each equation.

  a. $16^{x-2} = 64^x$ **b.** $4^x = \frac{1}{64}$

  c. $\left(\frac{2}{5}\right)^x = \frac{125}{8}$

  a. $\quad 16^{x-2} = 64^x$

$$\left(2^4\right)^{x-2} = \left(2^6\right)^x \quad \text{Write with the same base.}$$

$$2^{4x-8} = 2^{6x} \quad \text{Power rule for exponents.}$$

$$4x - 8 = 6x \qquad \text{If } a^x = a^y, \text{ then } x = y.$$

$$-8 = 2x \qquad \text{Solve for } x.$$

$$-4 = x$$

  The solution set is $\{-4\}$.

  b. $4^x = \frac{1}{64}$

$$4^x = \frac{1}{4^3} \qquad 64 = 4^3$$

$$4^x = 4^{-3} \qquad \text{Write with the same base.}$$

$$x = -3 \qquad \text{Set exponents equal.}$$

Now Try:

4. Solve the equation $25^x = 125$.

_____

5. Solve each equation.

  a. $4^{x-1} = 8^x$ _____

  b. $3^x = \frac{1}{243}$ _____

  c. $\left(\frac{3}{2}\right)^x = \frac{16}{81}$ _____

c. $\left(\frac{2}{5}\right)^x = \frac{125}{8}$

$\left(\frac{2}{5}\right)^x = \left(\frac{8}{125}\right)^{-1}$

$\left(\frac{2}{5}\right)^x = \left[\left(\frac{2}{5}\right)^3\right]^{-1}$ Write with the same base.

$\left(\frac{2}{5}\right)^x = \left(\frac{2}{5}\right)^{-3}$ Power rule for exponents.

$x = -3$ Set exponents equal.

Review these examples for Objective 4:

6. Suppose the number of bacteria present in a certain culture after t minutes is given by the equation $Q(t) = Q_o\left(2^{0.05t}\right)$, where Q_0 represents the initial number of bacteria. If 5000 bacteria are present after 20 minutes, how many bacteria were present initially?

$Q(20) = 5000$ and $t = 20$, so we have

$Q(t) = Q_0\left(2^{0.05 \cdot 20}\right)$

$5000 = Q_0\left(2^1\right)$

$Q_0 = \frac{5000}{2} = 2500$

There were 2500 bacteria present initially.

7. The amount of radioactive material in a sample is given by the function

$A(t) = 90\left(\frac{1}{2}\right)^{t/18}$, where $A(t)$ is the amount present, in grams, t days after the initial measurement.

a. How many grams will be present after 3 days? Round to the nearest hundredth.

b. In how many days will there be 45 grams left in the sample?

a. $A(3) = 90\left(\frac{1}{2}\right)^{3/18}$ Let $t = 3$.

≈ 80.18 Use a calculator.

About 80.18 grams will be left in the sample.

Now Try:

6. The population of Evergreen Park is now 16,000. The population t years from now is given by the formula $P = 16,000\left(2^{t/10}\right)$. Using the model, what will be the population 40 years from now?

_____

7. An industrial city in Ohio has found that its population is declining according to the equation $y = 70,000(2)^{-0.01x}$, where x is the time in years from 1910.

a. According to the model, what will the city's population be in the year 2020?

_____

b. When will the city's population be 35,000?

_____

---

b. $45 = 90\left(\frac{1}{2}\right)^{t/18}$ $A(t) = 45.$

$\frac{1}{2} = \left(\frac{1}{2}\right)^{t/18}$ Divide by 90.

$\left(\frac{1}{2}\right)^{1} = \left(\frac{1}{2}\right)^{t/18}$ $\frac{1}{2} = \left(\frac{1}{2}\right)^{1}$

$1 = \frac{t}{18}$ Set exponents equal.

$18 = t$ Solve for t.

There will be 45 grams left in the sample in 18 days.

Objective 1 Define an exponential function.
For extra help, see page 580 of your text and the Section Lecture video for Section 10.2.

Decide whether or not the function defines an exponential function.

1. a. $f(x) = 3^{2x}$

  b. $f(x) = 2x^3$

  c. $f(x) = x + 2$

  d. $f(x) = (-2)^x$

1. a._____

   b. _____

   c._____

   d. _____

Objective 2 Graph an exponential function.
For extra help, see Examples 1–3 on pages 581–582 of your text, the Section Lecture video for Section 10.2, and Exercise Solution Clips 5, 7, and 11.

Graph each exponential function.

2. $f(x) = 2^x$

2.

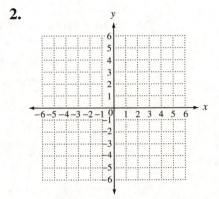

3. $f(x) = 2^{-x}$

3.

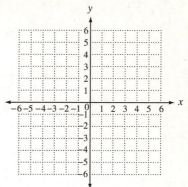

4. $f(x) = \left(\frac{1}{2}\right)^{x^2}$

4.

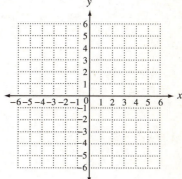

5. $f(x) = 2^{|x|}$

5.

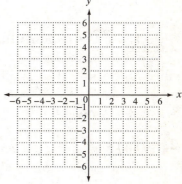

Objective 3 Solve exponential equations of the form $a^x = a^k$ for x.
For extra help, see Examples 4 and 5 on pages 583–584 of your text, the Section Lecture video for Section 10.2, and Exercise Solution Clips 17 and 19.

Solve each equation.

6. $27^k = 9$

6. _____

7. $25^{1-t} = 5$ 7. _____

8. $4^{k+2} = 32$ 8. _____

9. $8^{2x+1} = 4^{4x}$ 9. _____

10. $9^{x+1} = 27^{2x}$ 10. _____

Objective 4 Use exponential functions in applications involving growth or decay.
For extra help, see Examples 6 and 7 on pages 584–585 of your text, the Section Lecture video for Section 10.2, and Exercise Solution Clip 39.

Solve each problem.

11. The population of Canadian geese that spend the   11. _____
summer at Gemini Lake each year has been growing
according to $f(x) = 56(2)^{0.2x}$, where x is the time in
years from 1990. Find the number of geese in 2010.

12. The diameter in inches of a tree during a certain period
grew according to $f(x) = 2.5\left(9^{0.05x}\right)$, where x was the   12. _____
number of years after the start of this growth period.
Find the diameter of the tree after 10 years.

13. A culture of a certain kind of bacteria grows according to $f(x) = 7750(2)^{0.75x}$, where x is the number of hours after 12 noon. Find the number of bacteria in the culture at 12 noon.

13. _____

14. A sample of a radioactive substance decays according to $f(x) = 100\left(10^{-0.2x}\right)$, where y is the mass of the substance in grams and x is the time in hours after the original measurement. Find the mass of the substance after 10 hours.

14. _____

15. When a bactericide is placed in a certain culture of bacteria, the number of bacteria decreases according to $f(x) = 3200(4)^{-0.1x}$, where x is the time in hours. Find the number of bacteria in the culture in 20 hours.

15. _____

Chapter 10 INVERSE, EXPONENTIAL, AND LOGARITHMIC FUNCTIONS

10.3   Logarithmic Functions

Learning Objectives
1   Define a logarithm.
2   Convert between exponential and logarithmic forms.
3   Solve logarithmic equations of the form $\log_a b = k$ for a, b, or k.
4   Define and graph logarithmic functions.
5   Use logarithmic functions in applications involving growth or decay.

Key Terms

Use the vocabulary terms listed below to complete each statement in exercises 1–3.

logarithm   **logarithmic equation**   **logarithmic function with base** *a*

1.   If a and x are positive numbers with $a \neq 1$, then $f(x) = \log_a x$ defines the

_____.

2.   The _____ of a positive number is the exponent indicating the power to which it is necessary to raise a given number (the base) to give the original number.

3.   An equation with a logarithm in at least one term is a _____.

Guided Examples

Review this example for Objective 2:
1.   Fill in the blanks with the equivalent forms.

	Exponential Form	Logarithmic Form
a.	$5^3 = 125$	
b.	$81^{-1/4} = \frac{1}{3}$	
c.		$\log_{16} 4 = \frac{1}{2}$
d.		$\log_2\left(\frac{1}{64}\right) = -6$

a. $\log_5 125 = 3$

b. $\log_{81}\left(\frac{1}{3}\right) = -\frac{1}{4}$

Now Try:
1.   Fill in the blanks with the equivalent forms.

	Exponential Form	Logarithmic Form
a.	$8^2 = 64$	
b.	$81^{3/4} = 27$	
c.		$\log_{16} \frac{1}{4} = -\frac{1}{2}$
d.		$\log_{1/2} 4 = -2$

c. $16^{1/2} = 4$

d. $2^{-6} = \frac{1}{64}$

Review these examples for Objective 3:	**Now Try:**

2. Solve each equation.

  a. $\log_{3/2} x = -2$ **b.** $\log_5 (3x+1) = 2$

  c. $\log_x 6 = 2$ **d.** $\log_{64} \sqrt[4]{8} = x$

  a. By the definition of logarithm,
  $\log_{3/2} x = -2$ is equivalent to
  $x = \left(\frac{3}{2}\right)^{-2}$. So, $x = \left(\frac{3}{2}\right)^{-2} = \left(\frac{2}{3}\right)^2 = \frac{4}{9}$.
  The solution set is $\left\{\frac{4}{9}\right\}$.

  b. $\log_5 (3x+1) = 2$

  $3x+1 = 5^2$ Write in exponential form.

  $3x = 24$ Apply the exponent; subtract 1.

  $x = 8$ Divide by 3.

  The solution set is $\{8\}$.

  c. $\log_x 6 = 2$

  $x^2 = 6$ Write in exponential form.

  $x = \pm\sqrt{6}$ Take square root.

  Only the principal square root satisfies the equation since the base must be a positive number. The solution set is $\left\{\sqrt{6}\right\}$.

  d. $\log_{64} \sqrt[4]{8} = x$

  $64^x = \sqrt[4]{8}$ Write in exponential form.

  $\left(8^2\right)^x = 8^{1/4}$ Write with the same base.

  $8^{2x} = 8^{1/4}$ Power rule for exponents

  $2x = \frac{1}{4}$ Set exponents equal.

  $x = \frac{1}{8}$ Divide by 2.

  The solution set is $\left\{\frac{1}{8}\right\}$.

Now Try:

2. Solve each equation.

  a. $\log_4 x = -3$ _____

  b. $\log_9 (2x+1) = 2$ _____

  c. $\log_x 12 = 2$ _____

  d. $\log_{81} \sqrt[3]{9} = x$ _____

3. Evaluate each logarithm.

   a. $\log_8 8$ **b.** $\log_{\sqrt{5}} \sqrt{5}$

   c. $\log_{64} 1$ **c.** $\log_{0.2} 1$

   a. $\log_8 8 = 1$ **b.** $\log_{\sqrt{5}} \sqrt{5} = 1$

   c. $\log_{64} 1 = 0$ **d.** $\log_{0.2} 1 = 0$

3. Evaluate each logarithm.

   a. $\log_4 4$ _____

   b. $\log_{\sqrt{6}} \sqrt{6}$ _____

   c. $\log_{100} 1$ _____

   d. $\log_{1/3} 1$ _____

Review these examples for Objective 4:

4. Graph $f(x) = \log_5 x$.

Begin by writing $y = \log_5 x$ in exponential form. Then, create a table of values, plot the points and draw a smooth curve through them.

$x = 5^y$	y
$\frac{1}{5}$	-1
1	0
5	1
25	2

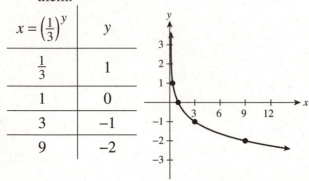

5. Graph $f(x) = \log_{1/3} x$.

Begin by writing $y = \log_{1/3} x$ in exponential form. Then, create a table of values, plot the points and draw a smooth curve through them.

$x = \left(\frac{1}{3}\right)^y$	y
$\frac{1}{3}$	1
1	0
3	-1
9	-2

Now Try:

4. Graph $f(x) = \log_3 x$.

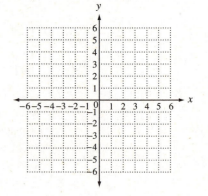

5. Graph $f(x) = \log_{1/4} x$.

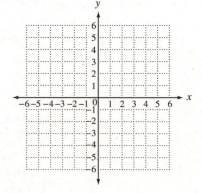

Review this example for Objective 5:

6. A company analyst found that total sales in thousands of dollars after a major advertising campaign are given by $S(x) = 100\log_2(x+2)$, where x is time in weeks after the campaign was introduced.

   a. Find the amount of sales at the beginning of the advertising campaign.

   b. Find the amount of sales 2 weeks after the advertising campaign.

   a. At the beginning of the campaign, $x = 0$, so we have
   $$S(0) = 100\log_2(0+2) = 100\log_2 2$$
   $$= 100 \cdot 1 = 100$$
   At the beginning of the campaign, sales were $100,000.

   b. Two weeks after the campaign, $x = 2$, so we have
   $$S(2) = 100\log_2(2+2) = 100\log_2 4$$
   $$= 100 \cdot 2 = 200$$
   Two weeks after the campaign, sales were $200,000.

Now Try:

6. The number of fish in an aquarium is given by $f = 5\log_6(3t+6)$, where t is time in months. Find the difference in the number of fish present between $t = 0$ and $t = 10$.

   _____

Objective 1 Define a logarithm.
Objective 2 Convert between exponential and logarithmic forms
For extra help, see Example 1 on page 589 of your text, the Section Lecture video for Section 10.3, and Exercise Solution Clips 3 and 27.

1. Evaluate each of the following.

   a. $\log_2 8$ b. $\log_8 64$

   c. $\log_4\left(\frac{1}{4}\right)$ d. $\log_3 \sqrt{3}$

   e. $\log_{1/2} 4$ f. $\log_{81} 3$

1. a. _____

   b. _____

   c. _____

   d. _____

   e. _____

   f. _____

2. Write in logarithmic form.

    a. $3^2 = 9$ **b.** $10^{-2} = \frac{1}{100}$

    c. $2^{-7} = \frac{1}{128}$ **d.** $5^{1/3} = \sqrt[3]{5}$

2. a. _____

    b. _____

    c. _____

    d. _____

3. Write in exponential form.

    a. $\log_3 27 = 3$ **b.** $\log_{16} 2 = \frac{1}{4}$

    c. $\log_4 \frac{1}{16} = -2$ **d.** $\log_{10} 0.001 = -3$

3. a. _____

    b. _____

    c. _____

    d. _____

Objective 3 Solving logarithmic equations of the form $\log_a b = k$ for a, b, or k.
For extra help, see Examples 2 and 3 on pages 589–590 of your text, the Section Lecture video for Section 10.3, and Exercise Solution Clip 29.

Solve each equation.

4. $\log_3 x = -1$

4. _____

5. $\log_x 4 = -2$

5. _____

6. $\log_4 (3x + 1) = 2$

6. _____

7. $\log_{49} \sqrt[4]{7} = x$

7. _____

Objective 4 Define and graph logarithmic functions.

For extra help, see Examples 4 and 5 on pages 590–591 of your text, the Section Lecture video for Section 10.3, and Exercise Solution Clips 49 and 51.

Graph each logarithmic function.

8. $y = \log_2 x$

8.

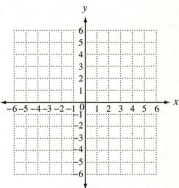

9. $y = \log_{1/2} x$

9.

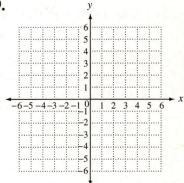

10. $y = \log_2 2x$

10.

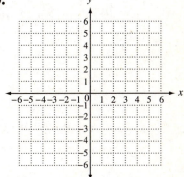

11. $y = \log_4 x$

11.

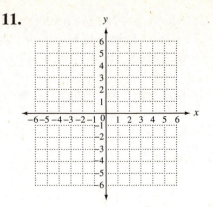

Objective 5 Use logarithmic functions in applications involving growth or decay.
For extra help, see Example 6 on page 592 of your text, the Section Lecture video for
Section 10.3, and Exercise Solution Clip 61.

Solve each problem.

12. After black squirrels were introduced to Williams Park,
their population grew according to $y = 14\log_5 (x+5)$,
where x is the number of months after the squirrels were
introduced. Find the number of squirrels after 20
months.

12. _____

13. A manufacturer receives revenue y given in dollars for
selling x units of an item according to
$y = 200\log_3 (x+1)$. Find the revenue for selling 26
units.

13. _____

14. Under certain conditions, the velocity v of the wind in
centimeters per second is given by $v = 300\log_2 \left(\frac{10x}{7}\right)$,
where x is the height in centimeters above the ground.
Find the wind velocity at 11.2 centimeters above the
ground.

14. _____

15.   A decibel is a measure of the loudness of a sound. A   **15.** _____
very faint sound is assigned an intensity of I_0 ; then
another sound is given an intensity I found in terms of
I_o , the faint sound. The decibel rating of the sound is

given in decibels by $d = 10 \ \log_{10} \dfrac{I}{I_0}$. Find the decibel

rating of rock music that has intensity
$I = 100{,}000{,}000{,}000 I_0$.

Chapter 10 INVERSE, EXPONENTIAL, AND LOGARITHMIC FUNCTIONS

10.4   Properties of Logarithms

Learning Objectives
1   Use the product rule for logarithms.
2   Use the quotient rule for logarithms.
3   Use the power rule for logarithms.
4   Use properties to write alternative forms of logarithmic expressions

---

Key Terms

Use the vocabulary terms listed below to complete each statement in exercises 1–4.

    product rule for logarithms   **quotient rule for logarithms**

    power rule for logarithms   **special properties**

1.   The equations $b^{\log_b x} = x$, $x > 0$ and $\log_b b^x = x$ are referred to as the _____ of logarithms.

2.   The equation $\log_b \frac{x}{y} = \log_b x - \log_b y$ is referred to as the _____.

3.   The equation $\log_b xy = \log_b x + \log_b y$ is referred to as the _____.

4.   The equation $\log_b x^r = r\log_b x$ is referred to as the _____.

Guided Examples

Review this example for Objective 1:

1.   Use the product rule to rewrite each logarithm. Assume $x > 0$.

    a. $\log_4(6 \cdot 11)$ **b.** $\log_3 8 + \log_3 2$

    c. $\log_2(2x)$ **d.** $\log_3 x^4$

    a. $\log_4(6 \cdot 11) = \log_4 6 + \log_4 11$

    b. $\log_3 8 + \log_3 2 = \log_3(8 \cdot 2) = \log_3 16$

    c. $\log_2(2x) = \log_2 2 + \log_2 x = 1 + \log_2 x$

Now Try:

1.   Use the product rule to rewrite each logarithm. Assume $x > 0$.

    a. $\log_6(5 \cdot 3)$ _____

    b. $\log_5 7 + \log_5 3$ _____

    c. $\log_4(4x)$ _____

    d. $\log_4 x^3$ _____

d. $\log_3 x^4 = \log_3(x \cdot x \cdot x \cdot x)$

$\qquad = \log_3 x + \log_3 x + \log_3 x + \log_3 x$

$\qquad = 4\log_3 x$

Review this example for Objective 2:

2. Use the quotient rule to rewrite each logarithm. Assume $x > 0$.

   a. $\log_4 \dfrac{8}{7}$ **b.** $\log_3 8 - \log_3 x$

   c. $\log_2 \dfrac{16}{11}$

   a. $\log_4 \dfrac{8}{7} = \log_4 8 - \log_4 7$

   b. $\log_3 8 - \log_3 x = \log_3 \dfrac{8}{x}$

   c. $\log_2 \dfrac{16}{11} = \log_2 16 - \log_2 11 = 4 - \log_2 11$

Now Try:

2. Use the quotient rule to rewrite each logarithm. Assume $x > 0$.

   a. $\log_5 \dfrac{4}{9}$ _____

   b. $\log_6 x - \log_6 3$ _____

   c. $\log_4 \dfrac{16}{11}$ _____

Review these examples for Objective 3:

3. Use the power rule to rewrite each logarithm. Assume $b > 0$, $x > 0$, and $b \neq 1$.

   a. $\log_4 3^5$ **b.** $\log_b x^3$

   c. $\log_b \sqrt{11}$ **d.** $\log_2 \sqrt[5]{x^4}$

   a. $\log_4 3^5 = 5\log_4 3$

   b. $\log_b x^3 = 3\log_b x$

   c. $\log_b \sqrt{11} = \log_b 11^{1/2} = \frac{1}{2}\log_b 11$

   d. $\log_2 \sqrt[5]{x^4} = \log_2 x^{4/5} = \frac{4}{5}\log_2 x$

4. Find each value.

   a. $\log_3 3^5$ **b.** $\log_4 64$

   c. $3^{\log_3 6}$

   a. $\log_3 3^5 = 5$

   b. $\log_4 64 = \log_4 4^3 = 3$

Now Try:

3. Use the power rule to rewrite each logarithm. Assume $b > 0$, $x > 0$, and $b \neq 1$.

   a. $\log_6 4^3$ _____

   b. $\log_b x^6$ _____

   c. $\log_b \sqrt{13}$ _____

   d. $\log_3 \sqrt[4]{x^3}$ _____

4. Find each value.

   a. $\log_6 6^7$ _____

   b. $\log_5 125$ _____

   c. $2^{\log_2 5}$ _____

---

c. $3^{\log_3 6} = 6$

Review these examples for Objective 4:

5. Use the properties of logarithms to rewrite each expression if possible. Assume that all variables represent positive real numbers.

   a. $\log_5 6x^3$ **b.** $\log_b \sqrt{3x}$

   c. $3\log_b x - \left(2\log_b y + \frac{1}{2}\log_b z\right)$

   d. $2\log_3 x + \log_3(x-1) - \frac{1}{2}\log_3(x+1)$

   e. $\log_2(2x+3y)$

   a. $\log_5 6x^3 = \log_5 6 + \log_5 x^3$ Product rule

                   $= \log_5 6 + 3\log_5 x$ Power rule

   b. $\log_b \sqrt{3x} = \log_b(3x)^{1/2}$

                 $= \frac{1}{2}\log_b(3x)$ Power rule

                 $= \frac{1}{2}\left(\log_b 3 + \log_b x\right)$

                           Product rule

   c. $3\log_b x - \left(2\log_b y + \frac{1}{2}\log_b z\right)$

        $= \log_b x^3 - \left(\log_b y^2 + \log_b z^{1/2}\right)$

                  Power rule

        $= \log_b \dfrac{x^3}{y^2\sqrt{z}}$ Product and quotient rules

   d. $2\log_3 x + \log_3(x-1) - \frac{1}{2}\log_3(x+1)$

        $= \log_3 x^2 + \log_3(x-1) - \log_3(x+1)^{1/2}$

                   Power rule

        $= \log_3\left(x^2(x-1)\right) - \log_3\sqrt{x+1}$

                   Product rule

        $= \log_3 \dfrac{x^3 - x^2}{\sqrt{x+1}}$ Multiply in numerator; quotient rule

   e. $\log_2(2x+3y)$ cannot be rewritten using the properties of logarithms.

Now Try:

5. Use the properties of logarithms to rewrite each expression if possible. Assume that all variables represent positive real numbers.

   a. $\log_6 36x^3$ _____

   b. $\log_b \sqrt{\dfrac{x}{3}}$ _____

   c. $\log_b x + 4\log_b y - \log_b z$

       _____

   d.

    $2\log_2 x + \log_2(x-1) - \frac{1}{3}\log_2\left(x^2+1\right)$

       _____

   e. $\log_3(3x-y)$ _____

6. Given that $\log_3 5 = 1.4650$ and $\log_3 4 = 1.2619$, evaluate the following.

    a. $\log_3 20$ **b.** $\log_3 0.8$

    c. $\log_3 25$

    a. $\log_3 20 = \log_3 (4 \cdot 5)$ Factor 20.

               $= \log_3 4 + \log_3 5$ Product rule

               $= 1.2619 + 1.4650$ Substitute.

               $= 2.7269$ Add.

    b. $\log_3 0.8 = \log_3 \frac{4}{5}$ $0.8 = \frac{4}{5}$

              $= \log_3 4 - \log_3 5$ Quotient rule

              $= 1.2619 - 1.4650$ Substitute.

              $= -0.2031$ Subtract.

    c. $\log_3 25 = \log_3 5^2$ $25 = 5^2$

             $= 2\log_3 5$ Power rule

             $= 2(1.4650)$ Substitute.

             $= 2.9300$ Multiply.

7. Decide whether each statement is *true* or *false*.

    a. $\log_2 32 - \log_2 16 = \log_2 16$

    b. $\log_3 (\log_4 64) = \dfrac{\log_{12} 144}{\log_6 36}$

    a. Evaluate each side.
    Left side:

$$\log_2 32 - \log_2 16 = \log_2 2^5 - \log_2 2^4$$

                          Write 32 and 16
                          as powers of 2.

                 $= 5 - 4$ $\log_a a^x = a$

                 $= 1$

    Right side:

$$\log_2 16 = \log_2 2^4 = 4$$

    The statement is false because $1 \neq 4$.

6. Given that $\log_3 7 = 1.7712$ and $\log_3 10 = 2.0959$, evaluate the following.

    a. $\log_3 70$ _____

    b. $\log_3 0.7$ _____

    c. $\log_3 49$ _____

7. Decide whether each statement is *true* or *false*.

    a. $\log_3 27 + \log_3 9 = \log_5 5$

        _____

    b. $(\log_2 8)(\log_2 4) = \log_2 32$

        _____

b. Evaluate each side.
   Left side:

$$\log_3(\log_4 64) = \log_3\left(\log_4 4^3\right)$$
$$= \log_3 3$$
$$= 1$$

   Right side:

$$\frac{\log_{12} 144}{\log_6 36} = \frac{\log_{12} 12^2}{\log_6 6^2} = \frac{2}{2} = 1$$

   The statement is true because $1 = 1$.

Objective 1 Use the product rule for logarithms.
For extra help, see Example 1 on page 596 of your text, the Section Lecture video for Section 10.4, and Exercise Solution Clip 7.

Use the product rule for logarithms to rewrite each logarithm

1. $\log_3 (6)(5)$ 1. _____

2. $\log_7 5m,\ m > 0$ 2. _____

3. $\log_{10} 7 + \log_{10} 9$ 3. _____

Objective 2 Use the quotient rule for logarithms.
For extra help, see Example 2 on page 597 of your text, the Section Lecture video for Section 10.4, and Exercise Solution Clip 9.

Use the quotient rule for logarithms to rewrite each logarithm.

4. $\log_2 \dfrac{7}{9}$ 4. _____

5. $\log_2 r - \log_2 s$ 5. _____

6. $\log_2 7q^4 - \log_2 5q^2$ 6. _____

Objective 3 Use the power rule for logarithms.
For extra help, see Examples 3 and 4 on pages 598–599 of your text, the Section Lecture
video for Section 10.4, and Exercise Solution Clip 11.

Use the power rule for logarithms to rewrite the logarithm.

7. $\log_m 2^7$

7. _____

8. $\log_2 \sqrt[3]{x^2}$

8. _____

9. $\log_5 125^{1/3}$

9. _____

Objective 4 Use properties to write alternative forms of logarithmic expressions.
For extra help, see Examples 5–7 on pages 599–601 of your text, the Section Lecture video
for Section 10.4, and Exercise Solution Clips 13, 33, and 45.

*Use the properties of logarithms to rewrite each expression if possible. Assume that all
variables represent positive real numbers.*

10. $\log_2 4p^3$

10. _____

11. $\log_5 \dfrac{7m^3}{8y}$

11. _____

12. $\log_7 \dfrac{8r^7}{3a^3}$

12. _____

13. $\log_{10} 4k^2 j - \log_{10} 3kj^2$

13. _____

14. $\log_4 8y + \log_4 3y - \log_4 6y^3$

14. _____

15. $\log_2 (x-1) + \log_2 (x+1) - \log_2 (x^2 - 1)$

15. _____

Chapter 10 INVERSE, EXPONENTIAL, AND LOGARITHMIC FUNCTIONS

10.5   Common and Natural Logarithms

Learning Objectives
1   Evaluate common logarithms using a calculator.
2   Use common logarithms in applications.
3   Evaluate natural logarithms using a calculator.
4   Use natural logarithms in applications.
5   Use the change-of-base rule.

Key Terms

Use the vocabulary terms listed below to complete each statement in exercises 1–3.

common logarithm   natural logarithm   universal constant

1.   The number e is called a(n) _____ because of its importance in many areas of mathematics.

2.   A(n) _____ is a logarithm to base 10.

3.   A(n) _____ is a logarithm to base e.

Guided Examples

Review this example for Objective 1:
1.   Using a calculator, evaluate each logarithm to four decimal places.

   a. $\log 436.2$ **b.** $\log 543,210$

   c. $\log 0.125$

   a. $\log 436.2 \approx 2.6397$

   b. $\log 543,210 \approx 5.7350$

   c. $\log 0.125 \approx -0.9031$

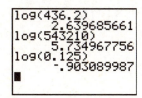

Now Try:
1.   Using a calculator, evaluate each logarithm to four decimal places.

   a. $\log 983.5$ _____

   b. $\log 79,315$ _____

   c. $\log 0.333$ _____

Review these examples for Objective 2:

2. Wetlands are classified as bogs, fens, marshes, and swamps, on the basis of pH values. A pHvalue between 6.0 and 7.5 indicates that the wetland is a "rich fen." When the pH is between 3.0 and 6.0, the wetland is a "poor fen," and if the pH falls to 3.0 or less, it is a "bog." Suppose that the hydronium ion concentration of a sample of water from a wetland is 5.4×10^{-4}. Find the pH value for the water and determine how the wetland should be classified.

$$pH = -\log\left(5.4 \times 10^{-4}\right) \quad \text{Definition of pH}$$
$$= -\left(\log 5.4 + \log 10^{-4}\right) \quad \text{Product rule}$$
$$= -(0.7324 - 4) \quad \begin{array}{l}\text{Use a calculator to}\\\text{find } \log 5.4.\end{array}$$
$$= 3.2676$$

Since the pH is between 3.0 and 6.0, the wetland is a poor fen.

3. Find the hydronium ion concentration of a solution with pH 5.4.

$$pH = -\log\left[H_3O^+\right] \quad \text{Definition of pH}$$
$$5.4 = -\log\left[H_3O^+\right]$$
$$\log\left[H_3O^+\right] = -5.4 \quad \text{Multiply by } -1.$$
$$H_3O^+ = 10^{-5.4} \quad \begin{array}{l}\text{Write in}\\\text{exponential form.}\end{array}$$
$$\approx 4.0 \times 10^{-6} \quad \text{Use a calculator.}$$

4. Find the decibel level to the nearest whole number of the sound with intensity I of $5.012 \times 10^{10} I_0$.

$$D = 10\log\left(\frac{I}{I_0}\right) = 10\log\left(\frac{5.012 \times 10^{10} I_0}{I_0}\right)$$
$$= 10\log\left(5.012 \times 10^{10}\right)$$
$$\approx 107 \text{ db}$$

Now Try:

2. Suppose that the hydronium ion concentration of a sample of water from a wetland is 6.2×10^{-8}. Find the pH value for the water and determine how the wetland should be classified.

_____

3. Find the hydronium ion concentration of a solution with pH 3.6.

_____

4. Find the decibel level to the nearest whole number of the sound with intensity I of $3.16 \times 10^8 I_0$.

_____

Review this example for Objective 3:

5. Using a calculator, evaluate each logarithm to four decimal places.

    a. $\ln 436.2$ **b.** $\ln 54.3$

    c. $\ln 0.125$

    a. $\ln 436.2 \approx 6.0781$

    b. $\ln 54.3 \approx 3.9945$

    c. $\log 0.125 \approx -2.0794$

    ```
    ln(436.2)
            6.078100854
    ln(54.3
            3.994524227
    ln(0.125)
            -2.079441542
    ```

Now Try:

5. Using a calculator, evaluate each logarithm to four decimal places.

    a. $\ln 98$ _____

    b. $\ln 793$ _____

    c. $\ln 0.333$ _____

Review this example for Objective 4:

6. The time t in years for an investment increasing at a rate of r percent (in decimal form) to double is given by

    $$t = \frac{\ln 2}{\ln(1+r)}.$$

    This is called the doubling time. Find the doubling time to the nearest tenth for an investment at 4%.

    $4\% = 0.04$, so

    $$t = \frac{\ln 2}{\ln(1+0.04)} = \frac{\ln 2}{\ln 1.04} \approx 17.7$$

    The doubling time for the investment is about 17.7 years.

Now Try:

6. Use the formula at the left to find the doubling time to the nearest tenth for an investment at 6%.

    _____

Review these examples for Objective 5:

7. Evaluate $\log_7 28$ to four decimal places.

    $$\log_7 28 = \frac{\log 28}{\log 7} = 1.7124$$

Now Try:

7. Evaluate $\log_5 180$ to four decimal places.

    _____

8. According to selected figures from the last two decades of the 20th century, the number of trillion cubic feet of dry natural gas consumed worldwide can be approximated by the function defined by

$f(x) = 51.47 + 6.044 \log_2 x$, where $x = 1$ represents 1980, $x = 1$ represents 1981, etc. Use this function to approximate consumption in 2000 to the nearest hundredth. (*Source*: *Intermediate Algebra* 11e by Lial/Hornsby/McGinnis)

The year 2000 is represented by $x = 21$.
$f(21) = 51.47 + 6.044 \log_2 21 \approx 78.02$

Natural gas consumption in 2000 was about 78.02 trillion ft^3.

8. Use the model at the left to find the natural gas consumption in 1990.

_____

Objective 1 Evaluating common logarithms using a calculator.

For extra help, see Example 1 on page 604 of your text, the Section Lecture video for Section 10.5, and Exercise Solution Clip 7.

Find each logarithm. Give approximations to four decimal places.

1. log 57.23

1. _____

2. log 0.0914

2. _____

3. log 280,037

3. _____

Objective 2 Use common logarithms in applications.

For extra help, see Examples 2–4 on pages 605–606 of your text, the Section Lecture video for Section 10.5, and Exercise Solution Clips 29, 37, and 41.

4. Find the pH of a solution with the given hydronium ion concentration. Round the answer to the nearest tenth.

   a. 4.3×10^{-9} b. 2.8×10^{-6}

4. a. _____

   b. _____

5. Find the hydronium ion concentration of a solution with the given pH value.

   a. 5.2   b. 1.3

5. a. _____

   b. _____

6. Find the decibel level to the nearest whole number of the   **6.** _____

sound with intensity I of $2.5 \times 10^{13} I_0$.

Objective 3 Evaluate natural logarithms using a calculator.

For extra help, see Example 5 on page 607 of your text, the Section Lecture video for Section 10.5, and Exercise Solution Clip 15.

Find each logarithm. Give approximations to four decimal places.

7. $\ln 0.12$ **7.** _____

8. $\ln 143$ **8.** _____

9. $\ln 6$ **9.** _____

Objective 4 Use natural logarithms in applications.

For extra help, see Example 6 on page 607 of your text, the Section Lecture video for Section 10.5, and Exercise Solution Clip 43.

The half-life of a radioactive substance is the time it takes for half of the material to decay. The amount A in pounds of substance remaining after t years is given by

$$-\frac{h \ln \frac{A}{C}}{\ln 2} = t,$$ *where C is the initial amount in pounds, and h is its half-life in years. Use the formula to solve problems 10 and 11.*

10. The half-life of radium-226 is 1620 years. How long, to   **10.** _____

the nearest year, will it take for 100 pounds to decay to 25 pounds?

11. The half-life of strontium-90 is 28.1 years. How long, to   **11.** _____

the nearest year, will it take for 200 pounds to decay to 150 pounds?

12. The formula $t = 16.625 \ln\left(\dfrac{x}{x-750}\right)$ approximates the number of years t of a home mortgage of \$150,000 at 6% in terms of the monthly payment x in dollars. If a homeowners monthly payment is \$900, in how many years will the mortgage be paid off? Round to the nearest whole year.

12. _____

Objective 5 Use the change-of-base rule.
For extra help, see Examples 7 and 8 on pages 608–609 of your text, the Section Lecture video for Section 10.5, and Exercise Solution Clips 49 and 61.

Use the change-of-base rule with either common or natural logarithms to find each logarithm to four decimal places.

13. $\log_{16} 27$

13. _____

14. $\log_6 3$

14. _____

15. $\log_{1/2} 5$

15. _____

Chapter 10 INVERSE, EXPONENTIAL, AND LOGARITHMIC FUNCTIONS

10.6   Exponential and Logarithmic Equations; Further Applications

Learning Objectives
1   Solve equations involving variables in the exponents.
2   Solve equations involving logarithms.
3   Solve applications of compound interest.
4   Solve applications involving base e exponential growth and decay.

Key Terms

Use the vocabulary terms listed below to complete each statement in exercises 1–2.

compound interest   continuous compounding

1.   The formula for _____ is $A = Pe^{rt}$.

2.   The formula for _____ is $A = P\left(1 + \frac{r}{n}\right)^{nt}$.

Guided Examples

Review these examples for Objective 1:

1.   Solve $4^x = 30$. Approximate the solution to three decimal places.

$$4^x = 30$$

$$\log 4^x = \log 30 \qquad \text{If } x = y, \text{ and } x > 0, \ y > 0,$$
$$\text{then } \log_b x = \log_b y.$$

$$x \log 4 = \log 30 \qquad \text{Power rule}$$

$$x = \frac{\log 30}{\log 4} \qquad \text{Divide by log 4.}$$

$$x \approx 2.453 \qquad \text{Use a calculator.}$$

Check

$$4^x = 4^{2.453} \approx 30 \ \checkmark$$

The solution set is $\{2.453\}$.

Now Try:

1.   Solve $3^x = 15$. Approximate the solution to three decimal places.

_____

2. Solve $e^{0.005x} = 9$. Approximate the solution to three decimal places.

$$e^{0.005x} = 9$$

$\ln e^{0.005x} = \ln 9$ If $x = y$, and $x > 0$, $y > 0$, then $\ln x = \ln y$.

$0.005x \ln e = \ln 9$ Power rule

$0.005x = \ln 9$ $\ln e = 1$

$$x = \frac{\ln 9}{0.005}$$ Divide by 0.005.

$x \approx 439.445$ Use a calculator.

The solution set is $\{439.445\}$.

2. Solve $e^{0.4x} = 15$. Approximate the solution to three decimal places.

_____

Review these examples for Objective 2:

3. Solve $\log_3 (x-1)^2 = 3$. Give the exact solution.

$$\log_3 (x-1)^2 = 3$$

$(x-1)^2 = 3^3$ Write in exponential form.

$(x-1)^2 = 27$

$x-1 = \pm\sqrt{27}$ Take the square root on each side.

$x-1 = \pm 3\sqrt{3}$ Simplify the square root.

$x = 1 \pm 3\sqrt{3}$ Add 1.

Since the domain of $\log_b x$ is $(0, \infty)$, we disregard the negative solution, $1 - 3\sqrt{3}$.

Check:

$$\log_3 (x-1)^2 = 3$$

$$\log_3 \left(1 + 3\sqrt{3} - 1\right)^2 \overset{?}{=} 3$$

$$\log_3 \left(3\sqrt{3}\right)^2 \overset{?}{=} 3$$

$$\log_3 27 \overset{?}{=} 3$$

$$3^3 \overset{?}{=} 27$$

$$27 = 27 \checkmark$$

The solution set is $\left\{1 + 3\sqrt{3}\right\}$.

Now Try:

3. Solve $\log_6 (x+1)^3 = 2$. Give the exact solution.

_____

4. Solve $\log_3(5x+42) - \log_3 x = \log_3 26$.

$$\log_3(5x+42) - \log_3 x = \log_3 26$$

$$\log_3 \frac{5x+42}{x} = \log_3 26$$

$$\frac{5x+42}{x} = 26$$

If $\log_b x = \log_b y$, then $x = y$.

$$5x + 42 = 26x \quad \text{Multiply by } x.$$

$$42 = 21x \quad \text{Subtract } 5x.$$

$$2 = x \quad \text{Divide by 21.}$$

Since we cannot take the logarithm of a nonpositive number, both $5x + 42$ and x must be positive. If $x = 2$, this condition is satisfied.

Check:

$$\log_3(5x+42) - \log_3 x = \log_3 26$$

$$\log_3(5\cdot 2+42) - \log_3 2 \overset{?}{=} \log_3 26$$

$$\log_3 52 - \log_3 2 \overset{?}{=} \log_3 26$$

$$\log_3 \frac{52}{2} \overset{?}{=} \log_3 26$$

$$\log_3 26 = \log_3 26 \ \checkmark$$

The solution set is $\{2\}$.

5. Solve $\log_2(x+7) + \log_2(x+3) = \log_2 77$.

$$\log_2(x+7) + \log_2(x+3) = \log_2 77$$

$$\log_2[(x+7)(x+3)] = \log_2 77$$

Product rule

$$(x+7)(x+3) = 77$$

If $\log_b x = \log_b y$, then $x = y$.

$$x^2 + 10x + 21 = 77 \quad \text{Multiply.}$$

$$x^2 + 10x - 56 = 0 \quad \text{Subtract 77.}$$

$$(x-4)(x+14) = 0 \quad \text{Factor.}$$

$$x - 4 = 0 \quad \text{or} \quad x + 14 = 0 \qquad \text{Zero-factor}$$

$$x = 4 \qquad\qquad x = -14 \quad \text{property}$$

The value -14 must be rejected since it lead to the logarithm of a negative number in the original equation. Check that the only solution is 4.

The solution set is $\{4\}$.

4. Solve

$$\log_6(2x+7) - \log_6 x = \log_6 16.$$

_____

5. Solve.

$$\log_4(4x-3) + \log_4 x = \log_4(2x-1)$$

_____

Review these examples for Objective 3:

6. How much money will there be in an account at the end of 5 years if $5000 is deposited at 4% compounded monthly? How much interest was earned?

Because interest is compounded monthly, $n = 12$. The other given values are $P = 5000$, $r = 0.04$, and $t = 5$.

$$A = P\left(1 + \frac{r}{n}\right)^{nt}$$

$$A = 5000\left(1 + \frac{0.04}{12}\right)^{12 \cdot 5}$$

$$A = 5000(1.0033)^{60}$$

$$A = 6104.98$$

There will be $6104.98 in the account at the end of 5 years. The amount of interest earned is $6104.98 - $5000 = $104.98.

7. Approximate the time it would take for money deposited in an account paying 5% interest compounded quarterly to double. Round to the nearest hundredth.

We want the number of years t for P dollars to grow to $2P$ dollars at a rate of 5% per year. In the compound interest formula, we substitute $2P$ for A, and let $r = 0.05$ and $n = 4$.

$$2P = P\left(1 + \frac{0.05}{4}\right)^{4t}$$

$$2 = 1.0125^{4t}$$

$$\log 2 = \log 1.025^{4t}$$

$$\log 2 = 4t \log 1.0125$$

$$t = \frac{\log 2}{4 \log 1.0125}$$

$$t \approx 13.95$$

It will take about 13.95 years for the investment to double.

Now Try:

6. How much money will there be in an account at the end of 5 years if $10,000 is deposited at 4% compounded quarterly? How much interest was earned?

_____

7. Approximate the time it would take for money deposited in an account paying 5% interest compounded monthly to double. Round to the nearest hundredth.

_____

8. Suppose that $5000 is invested at 4% interest for 3 years.

  a. How much will the investment be worth if it is compounded continuously?

  b. Approximate the amount of time it would take for the investment to double. Round to the nearest tenth.

  a. $A = Pe^{rt}$

     $A = 5000e^{0.04 \cdot 3}$

     $A = 5000e^{0.12}$

     $A = 5637.48$

     The investment will be worth $5637.48.

  b. $A = Pe^{rt}$

    $10{,}000 = 5000e^{0.04t}$

       $2 = e^{0.04t}$ Divide by 5000.

    $\ln 2 = \ln e^{0.04t}$ If $x = y$, then $\ln x = \ln y$.

    $\ln 2 = 0.04t$ $\ln e^k = k$

    $\dfrac{\ln 2}{0.04} = t$ Divide by 0.04.

       $t \approx 17.3$

    It will take about 17.3 years for the amount to double.

---

8. Suppose that $5000 is invested at 2% interest for 3 years.

  a. How much will the investment be worth if it is compounded continuously?

    _____

  b. Approximate the amount of time it would take for the investment to double. Round to the nearest tenth.

    _____

Review this example for Objective 4:

9. A sample of 500 g of lead-210 decays according to the function $y = y_0 e^{-0.032t}$, where t is the time in years, y is the amount of the sample at time t, and y_0 is the initial amount present at $t = 0$.

  a. How much lead will be left in the sample after 20 years? Round to the nearest tenth of a gram.

  b. Approximate the half-life of lead-210 to the nearest tenth.

  a. Let $t = 20$ and $y_0 = 500$.

    $y = 500e^{-0.032 \cdot 20} \approx 263.6$

    There will be about 263.6 grams after 20 years.

Now Try:

9. Cesium-137, a radioactive isotope used in radiation therapy, decays according to the function $y = y_0 e^{-0.0231t}$, where t is the time in years and y_0 is the initial amount present at $t = 0$.

  a. If an initial sample contains 36 mg of cesium-137, how much cesium-137 will be left in the sample after 50 years? Round to the nearest tenth

    _____

  b. Approximate the half-life of cesium-137 to the nearest tenth. _____

b. Let $y = \frac{1}{2}(500) = 250$.

$$250 = 500e^{-0.032t}$$

$$0.5 = e^{-0.032t}$$

$$\ln 0.5 = \ln e^{-0.032t}$$

$$\ln 0.5 = -0.032t$$

$$t = \frac{\ln 0.5}{-0.032} \approx 21.7$$

The half-life of lead-210 is about 21.7 years.

Objective 1 Solve equations involving variables in the exponents.
For extra help, see Examples 1 and 2 on pages 613–614 of your text, the Section Lecture video for Section 10.6, and Exercise Solution Clips 1 and 13.

Solve each equation. Give solutions to three decimal places.

1. $2^{3y-9} = 7$ 1. _____

2. $25^{x+2} = 125^{3-x}$ 2. _____

3. $4^{x-1} = 3^{2x}$ 3. _____

4. $e^{4-3x} = 11$ 4. _____

Objective 2 Solve equations involving logarithms.
For extra help, see Examples 3–5 on pages 614–615 of your text, the Section Lecture video for Section 10.6, and Exercise Solution Clips 31, 37, and 41.

Solve each equation. Give exact solutions.

5. $\log_2 (x+1) - \log_2 x = \log_2 5$

5. _____

6. $\log (-y) + \log 4 = \log (2y+5)$

6. _____

7. $\log_3 \left(x^2 - 10 \right) - \log_3 x = 1$

7. _____

8. $\ln (x+4) + \ln (x-2) = \ln 7$

8. _____

Objective 3 Solve applications of compound interest.
For extra help, see Examples 6–8 on pages 616–618 of your text, the Section Lecture video for Section 10.6, and Exercise Solution Clips 47 and 49.

Solve each problem.

9. How much will be in an account after 10 years if $25,000 is invested at 8% compounded quarterly?

9. _____

10. How much will be in an account after 5 years if $10,000   **10.** _____
is invested at 4.5% compounded continuously?

11. How long will it take an investment to triple if it is   **11.** _____
placed in an account paying 6% interest compounded
semiannually? Round to the nearest tenth.

12. How long will it take an investment to double if it is   **12.** _____
placed in an account paying 9% interest compounded
continuously? Round to the nearest tenth.

Objective 4 Solve applications involving base *e* exponential growth and decay.
For extra help, see Example 9 on page 618 of your text, the Section Lecture video for
Section 10.6, and Exercise Solution Clip 61.

Solve each problem.

13. Iodine-131 decays according to the function   **13. a.**_____
$y = y_0 e^{-0.0864t}$, where t is the time in days, y is the
amount of the sample at time t, and y_0 is the initial   **b.** _____
amount present at $t = 0$.
a. If an initial sample contains 20 mg of iodine-131,
how much iodine-131 will be left in the sample after
5 days? Round to the nearest tenth.

b. Approximate the half-life of iodine-131 to the nearest tenth.

14. The concentration of a drug in a person's system decreases according to the function $C(t) = 2e^{-0.2t}$, where $C(t)$ is given in mg and t is in hours. Approximate answers to the nearest hundredth.

    a. How much of the drug will be in the person's system after one hour?

14. a._____

   b. _____

    b. Approximate the time it will take for the concentration of the drug to be half of its original amount.

15. The number of bacteria in a sample can be modeled by $N(t) = N_0 e^{0.014t}$, where N_0 is the original amount of bacteria in the sample and t is given in hours.

    a. If 5000 bacteria are present initially, how many bacteria will be present after 3 hours?

15. a._____

   b. _____

    b. Approximate the time it will take for the number of bacteria to double. Round to the nearest tenth.

Chapter 11 NONLINEAR FUNCTIONS, CONIC SECTIONS, AND NONLINEAR SYSTEMS

11.1   Additional Graphs of Functions

Learning Objectives

1   Recognize the graphs of elementary functions defined by $|x|$, $\frac{1}{x}$, and $\sqrt{x}$, and graph their translations.

2   Recognize and graph step functions.

Key Terms

Use the vocabulary terms listed below to complete each statement in exercises 1–7.

squaring function	**absolute value function**	**reciprocal function**
asymptotes	**square root function**	**greatest integer function**
step function		

1.   A _____ is a function that looks like a series of steps.

2.   The function defined by $f(x) = [\![x]\!]$ is called the _____.

3.   Lines that a graph approaches without actually touching are called _____.

4.   The function defined by $f(x) = |x|$ with a graph that includes portions of two lines is called the _____.

5.   The function defined by $f(x) = \sqrt{x}$, with $x \geq 0$, is called the _____.

6.   The function defined by $f(x) = x^2$ is called the _____.

7.   The _____ is defined by $f(x) = \frac{1}{x}$.

Name: _____   Date: _____

Instructor: _____   Section: _____

Guided Examples

Review these examples for Objective 1:

1. Graph $f(x) = \dfrac{1}{x-2}$. Give the domain and range.

    The graph of $y = \dfrac{1}{x-2}$ is obtained by shifting the graph of $y = \dfrac{1}{x}$ two units to the right.

x	y	x	y
-4	$-\dfrac{1}{6}$	$\dfrac{3}{2}$	-2
-3	$-\dfrac{1}{5}$	$\dfrac{5}{2}$	2
-2	$-\dfrac{1}{4}$	3	1
-1	$-\dfrac{1}{3}$	4	$\dfrac{1}{2}$
0	$-\dfrac{1}{2}$	5	$\dfrac{1}{3}$
1	-1	6	$\dfrac{1}{4}$

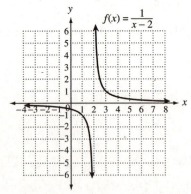

The domain is $(-\infty, 2) \cup (2, \infty)$. The range is $(-\infty, 0) \cup (0, \infty)$.

Now Try:

1. Graph $f(x) = |x + 4|$. Give the domain and range.

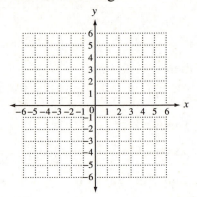

_____

2. Graph $f(x) = \sqrt{x} - 2$. Give the domain and range.

The graph of $y = \sqrt{x} - 2$ is obtained by shifting the graph of $y = \sqrt{x}$ two units down.

x	y
0	-2
1	-1
4	0

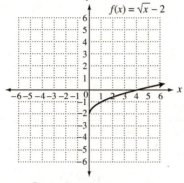

The domain is $[0, \infty)$. The range is $[-2, \infty)$.

3. Graph $f(x) = \dfrac{1}{x+2} - 1$. Give the domain and range.

The graph of $y = \dfrac{1}{x+2} - 1$ is obtained by shifting the graph of $y = \dfrac{1}{x}$ two units to the left and then one unit down.

x	y	x	y
-6	$-\dfrac{5}{4}$	-1	0
-5	$-\dfrac{4}{3}$	0	$-\dfrac{1}{2}$
-4	$-\dfrac{3}{2}$	1	$-\dfrac{2}{3}$
-3	-2	2	$-\dfrac{3}{4}$
$-\dfrac{5}{2}$	-3	3	$-\dfrac{4}{5}$
$-\dfrac{3}{2}$	1	4	$-\dfrac{5}{6}$

2. Graph $f(x) = |x| - 1$. Give the domain and range.

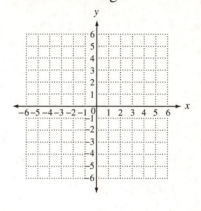

_____

3. Graph $f(x) = \sqrt{x-1} - 1$. Give the domain and range.

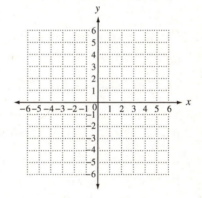

_____

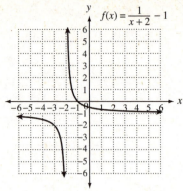

$$f(x) = \frac{1}{x+2} - 1$$

The domain is $(-\infty, -2) \cup (-2, \infty)$. The range is $(-\infty, -1) \cup (-1, \infty)$.

Review these examples for Objective 2:

4. Evaluate each expression.

   a. $[\![-5]\!]$ **b.** $[\![-6.5]\!]$

   c. $[\![3.5]\!]$ **d.** $\left[\!\left[\frac{9}{5}\right]\!\right]$

   a. $[\![-5]\!] = -5$ **b.** $[\![-6.5]\!] = -7$

   c. $[\![3.5]\!] = 3$ **d.** $\left[\!\left[\frac{9}{5}\right]\!\right] = 1$

5. Graph $f(x) = [\![x]\!] + 2$. Give the domain and range.

The graph of $y = [\![x]\!] + 2$ is obtained by shifting the graph of $y = [\![x]\!]$ two units up.

If $-3 \leq x < -2$, then $[\![x]\!] + 2 = -3 + 2 = -1$.

If $-2 \leq x < -1$, then $[\![x]\!] + 2 = -2 + 2 = 0$.

If $-1 \leq x < 0$, then $[\![x]\!] + 2 = -1 + 2 = 1$.

If $0 \leq x < 1$, then $[\![x]\!] + 2 = 0 + 2 = 2$, etc.

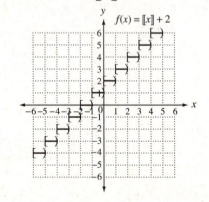

$$f(x) = [\![x]\!] + 2$$

Now Try:

4. Evaluate each expression.

   a. $[\![8]\!]$ _____

   b. $\left[\!\left[-\frac{5}{2}\right]\!\right]$ _____

   c. $[\![5.3]\!]$ _____

   d. $[\![-6.1]\!]$ _____

5. Graph $f(x) = [\![x - 2]\!]$. Give the domain and range.

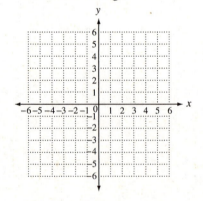

_____

The domain is $(-\infty, \infty)$. The range is $\{\ldots,$ $-2, -1, 0, 1, 2, \ldots\}$ or the set of integers.

6. In 2011, U.S. first class letter postage is 46¢ for the first ounce and 18¢ for each additional ounce. Let $p(x)$ = the cost of sending a letter that weighs x ounces. Graph the function over the interval (0, 6].

For x in the interval (0, 1], $y = \$0.46$.
For x in the interval (1, 2],
$y = \$0.46 + \$0.18 = \$0.64$.
For x in the interval (2, 3],
$y = \$0.64 + \$0.18 = \$0.82$.
For x in the interval (3, 4],
$y = \$0.82 + \$0.18 = \$1.00$.
For x in the interval (4, 5],
$y = \$1.00 + \$0.18 = \$1.18$.
For x in the interval (5, 6],
$y = \$1.18 + \$0.18 = \$1.36$.

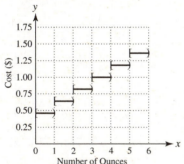

6. At the end of December, Mary's dad lent her $500 to buy an IPad. She agreed to repay the loan at $50 per month on the first of each month, starting in January. If $f(x)$ represents the amount to be still to be repaid in month x, graph the function over its entire domain.

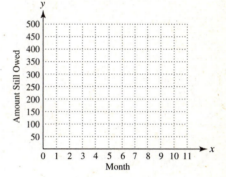

Name: Date:

Instructor: Section:

Objective 1   Recognize the graphs of elementary functions defined by $|x|$, $\frac{1}{x}$, and $\sqrt{x}$, and graph their translations.

For extra help, see Examples 1–3 on pages 637–638 of your text, the Section Lecture video for Section 11.1, and Exercise Solution Clips 11, 13, and 19.

Graph each function. Give the domain and range.

1. $f(x) = \sqrt{x+3}$

1.

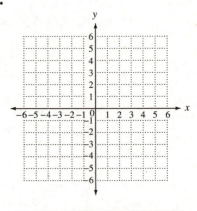

_____

2. $f(x) = |x-3| + 3$

2.

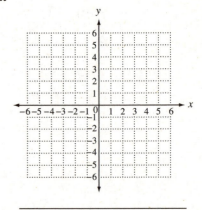

_____

3. $f(x) = \dfrac{1}{x+1}$

3.

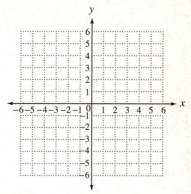

_____

4. $f(x) = \sqrt{x-4} + 4$

4.

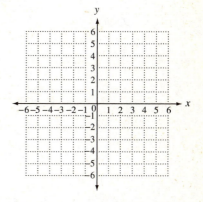

_____

5. $f(x) = |x+3| - 3$

5.

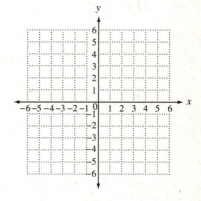

_____

6. $f(x) = \dfrac{1}{x-1} - 2$

6.

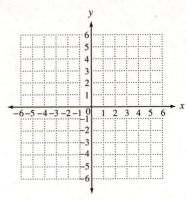

_____

7. $f(x) = \sqrt{x+3} - 1$

7.

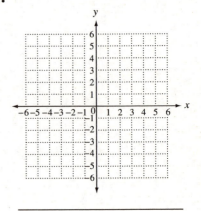

_____

8. $f(x) = \dfrac{1}{x} - 1$

8.

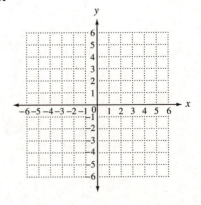

_____

9. $f(x) = |x| + 3$

9.

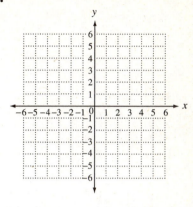

_____

10. $f(x) = \dfrac{1}{x+2} - 3$

10.

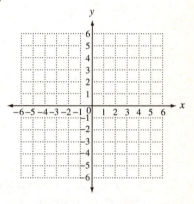

_____

Objective 2 Recognize and graph step functions.

For extra help, see Examples 4–6 on pages 638–639 of your text, the Section Lecture video for Section 11.1, and Exercise Solution Clips 29, 33, and 37.

Graph each function.

11. $f(x) = [\![x + 3]\!]$

11.

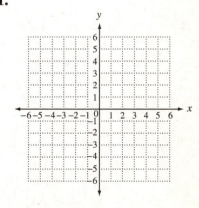

12. $f(x) = [\![x]\!] - 3$

12.

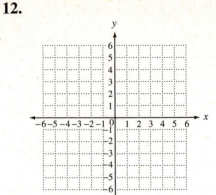

13. $f(x) = [\![x + 2]\!] - 3$

13.

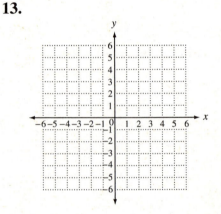

14.   The cost of parking a car in an hourly parking lot is $6.00 for the first hour and $4.00 for each additional hour or fraction of an hour. Graph the function f that models the cost of parking a car for x hours over the interval $(0, 6]$.

14.

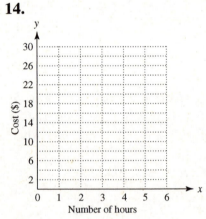

15.   In 1991, U.S. first class postage was 29¢ and each additional ounce cost 23¢. Let $p(x)$ = the cost of sending a letter that weighs x ounces. Graph the function over the interval $(0, 6]$.

15.

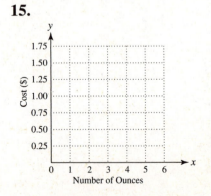

Chapter 11 NONLINEAR FUNCTIONS, CONIC SECTIONS, AND NONLINEAR SYSTEMS

11.2   The Circle and the Ellipse

Learning Objectives
1   Find an equation of a circle given the center and radius.
2   Determine the center and radius of a circle given its equation.
3   Recognize an equation of an ellipse.
4   Graph ellipses.

Key Terms

Use the vocabulary terms listed below to complete each statement in exercises 1–8.

conic sections   circle   center (of a circle)   radius

center-radius form   ellipse   foci   center (of an ellipse)

1.   A(n) _____ is the set of all points in a plane that lie a fixed distance from a fixed point.

2.   A(n) _____ is the set of all points in a plane the sum of whose distances from two fixed points is constant.

3.   Figures that result from the intersection of an infinite cone with a plane are called _____.

4.   A fixed point such that every point on a circle is a fixed distance from it is the _____.

5.   The distance from the center of a circle to a point on the circle is called the _____.

6.   The _____ of the equation of a circle with center (h, k) and radius r is $(x-h)^2 + (y-k)^2 = r^2$.

7.   _____ are fixed points used to determine the points that form a parabola, and ellipse, or a hyperbola.

8.   The _____ of the ellipse defined by $\dfrac{(x-3)^2}{9} + \dfrac{(y+2)^2}{16} = 1$ is $(3, -2)$.

Guided Examples

Review these examples for Objective 1:

1. Find an equation of the circle with radius 2 and center (0, 0) and graph it.

   If the point (x, y) is on the circle, then the distance from (x, y) to the center (0, 0) is 2.

   $$\sqrt{(x_2 - x_1)^2 + (y_2 - y_1)^2} = d \quad \text{Distance formula}$$

   $$\sqrt{(x - 0)^2 + (y - 0)^2} = 2 \quad \begin{array}{l} \text{Let } x_1 = 0, \ y_1 = 0, \\ \text{and } d = 2. \end{array}$$

   $$x^2 + y^2 = 4 \quad \text{Square each side.}$$

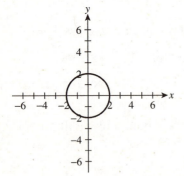

Now Try:

1. Find an equation of the circle with radius 5 and center (0, 0) and graph it.

   _____

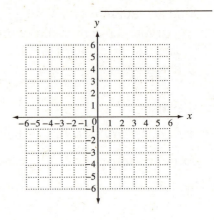

2. Find an equation of the circle with center (−3, 2) and radius 3 and graph it.

   $$\sqrt{(x_2 - x_1)^2 + (y_2 - y_1)^2} = d \quad \text{Distance formula}$$

   $$\sqrt{(x - (-3))^2 + (y - 2)^2} = 3 \quad \begin{array}{l} \text{Let } x_1 = -3, \\ y_1 = 2, \ d = 3. \end{array}$$

   $$(x + 3)^2 + (y - 2)^2 = 9 \quad \text{Square each side.}$$

   To graph the circle, plot the center (−3, 2), then move three units right, left, up, and down from the center, plotting the points (0, 2), (−3, 5), (−6, 2), and (−3, −1). Draw a smooth curve through the points.

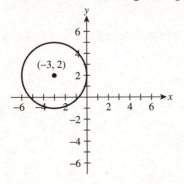

2. Find an equation of the circle with center (−5, 4) and radius 4 and graph it.

   _____

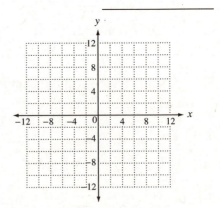

3. Find an equation of the circle with center at $(-2, -4)$ and radius $2\sqrt{5}$.

$$(x-h)^2 + (y-k)^2 = r^2 \quad \text{Center-radius form}$$

$$[x-(-2)]^2 + [y-(-4)]^2 = (2\sqrt{5})^2$$

$$h = -2, \; k = -4, \; r = 2\sqrt{5}$$

$$(x+2)^2 + (y+4)^2 = 20 \quad \text{Simplify.}$$

3. Find an equation of the circle with center at $(0, 3)$ and radius $\sqrt{2}$.

_____

Review this example for Objective 2:

4. Find the center and radius of the circle.

$$x^2 + y^2 + 8x + 4y - 29 = 0$$

To find the center and radius, complete the squares on x and y.

$$x^2 + y^2 + 8x + 4y - 29 = 0$$

$$x^2 + y^2 + 8x + 4y = 29$$

Transform so that the constant is on the right.

$$\left(x^2 + 8x \quad\right) + \left(y^2 + 4y \quad\right) = 29$$

$$\left[\tfrac{1}{2}(8)\right]^2 = 16 \quad \left[\tfrac{1}{2}(4)\right]^2 = 4$$

Square half the coefficient of each middle term.

$$\left(x^2 + 8x + 16\right) + \left(y^2 + 4y + 4\right) = 29 + 16 + 4$$

Complete the squares on x and y.

$$(x+4)^2 + (y+2)^2 = 49$$

Factor on the left; add on the right.

$$[x-(-4)]^2 + [y-(-2)]^2 = 7^2$$

Center-radius form

The center is $(-4, -2)$ and the radius is 7.

Now Try:

4. Find the center and radius of the circle.

$$x^2 + y^2 - 8x - 2y + 15 = 0$$

Center: _____

Radius: _____

Name: Date:
Instructor: Section:

Review these examples for Objective 4: | **Now Try:**

5. Graph $\dfrac{x^2}{25} + \dfrac{y^2}{9} = 1$.

$a^2 = 25$, so $a = 5$, and the x-intercepts are (5, 0) and (−5, 0). $b^2 = 9$, so $b = 3$, and the y-intercepts are (0, 3) and (0, −3).

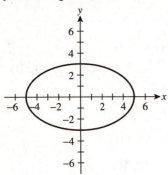

5. Graph $\dfrac{x^2}{16} + \dfrac{y^2}{9} = 1$

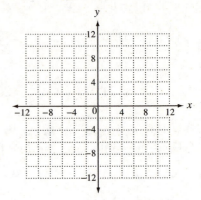

6. Graph $\dfrac{(x-1)^2}{9} + \dfrac{(y+2)^2}{4} = 1$.

The center is (1, −2), $a = 3$, and $b = 2$. The ellipse passes through the points (4, −2), (1, −4), (−2, −2), and (1, 0).

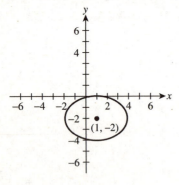

6. Graph $\dfrac{(x+1)^2}{4} + \dfrac{(y-2)^2}{9} = 1$.

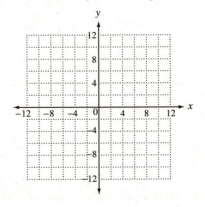

Objective 1 Find an equation of a circle given the center and radius.
For extra help, see Examples 1–3 on pages 642–643 of your text, the Section Lecture video for Section 11.2, and Exercise Solution Clips 1, 7, and 9.

Find the equation of a circle satisfying the given conditions. Then graph the circle.

1. Center: (2, −3), radius 5

1. _____

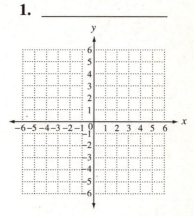

2. Center: (1, 4), radius 2

2. _____

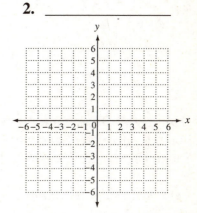

3. Center: (−2, −3), radius $\sqrt{3}$

3. _____

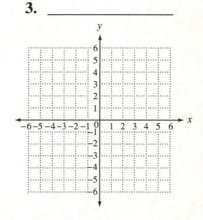

473

4. Center: (−5, 4), radius 4

4. _____

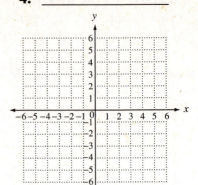

5. Center (0, 5), radius 3

5. _____

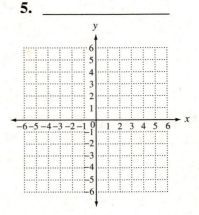

Objective 2 Determine the center and radius of a circle given its equation.
For extra help, see Example 4 on page 644 of your text, the Section Lecture video for
Section 11.2, and Exercise Solution Clip 11.

Find the center and radius of each circle.

6. $x^2 + y^2 - 4x + 8y + 11 = 0$

6. _____

7. $x^2 + y^2 - 2x - 6y - 15 = 0$

7. _____

8. $x^2 + y^2 + 3x + 2y - 2 = 0$

8. _____

9. $4x^2 + 4y^2 - 24x + 16y + 43 = 0$

9. _____

10. $3x^2 + 3y^2 + 12y + 30x = 21$

10. _____

Objective 3 Recognize the equation of an ellipse.
Objective 4 Graph ellipses.
For extra help, see Examples 5 and 6 on pages 645–646 of your text, the Section Lecture video for Section 11.2, and Exercise Solution Clips 29 and 37.

Graph each ellipse.

11. $\dfrac{x^2}{36} + \dfrac{y^2}{9} = 1$

11.

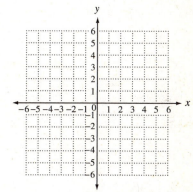

12. $\dfrac{x^2}{4} + \dfrac{y^2}{9} = 1$

12.

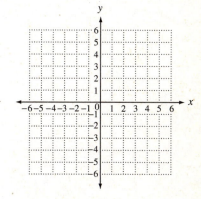

Name:

Date:

Instructor:

Section:

13. $\dfrac{(x-1)^2}{9}+\dfrac{y^2}{25}=1$

13.

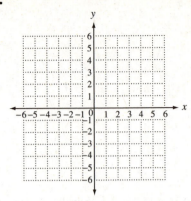

14. $\dfrac{(x+2)^2}{4}+\dfrac{(y-3)^2}{25}=1$

14.

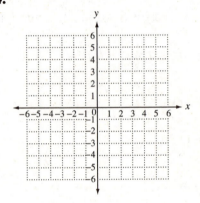

15. $\dfrac{(x+3)^2}{9}+\dfrac{(y+3)^2}{4}=1$

15.

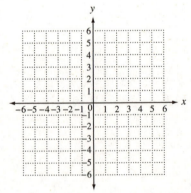

Chapter 11 NONLINEAR FUNCTIONS, CONIC SECTIONS, AND NONLINEAR SYSTEMS

11.3   The Hyperbola and Functions Defined by Radicals

Learning Objectives
1   Recognize the equation of a hyperbola.
2   Graph hyperbolas by using asymptotes.
3   Identify conic sections by their equations.
4   Graph certain square root functions.

Key Terms

Use the vocabulary terms listed below to complete each statement in exercises 1–5.

   hyperbola   transverse axis   asymptotes

   fundamental rectangle   generalized square root function

1.   A(n) _____ is the set of all points in a plane such that the absolute value of the difference of the distances from two fixed points is constant.

2.   Two intersecting lines that the branches of a hyperbola approach, but never reach, are its _____.

3.   The asymptotes of a hyperbola are the extended diagonals of its _____.

4.   The _____ of a hyperbola with x-intercepts $(a, 0)$ and $(-a, 0)$ lies on the x-axis.

5.   For an algebraic expression in x defined by u, with $u \geq 0$, a function of the form $f(x) = \sqrt{u}$, is a _____.

Guided Examples

Review these examples for Objective 2:

1. Graph $\dfrac{x^2}{9} - \dfrac{y^2}{16} = 1$.

Step 1: $a = 3$, $b = 4$. The x-intercepts are $(3, 0)$ and $(-3, 0)$.

Step 2: The four points $(3, 4)$, $(3, -4)$, $(-3, -4)$, and $(-3, 4)$ are the vertices of the fundamental rectangle.

Step 3: The equations of the asymptotes are $y = \pm\dfrac{4}{3}x$.

Step 4: Sketch the graph.

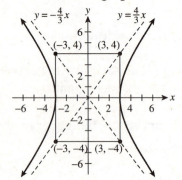

2. Graph $\dfrac{y^2}{25} - \dfrac{x^2}{16} = 1$.

$b = 5$, $a = 4$, so the y-intercepts are $(0, 5)$ and $(0, -5)$. The asymptotes are the extended diagonals of the rectangle with vertices at $(4, 5)$, $(4, -5)$, $(-4, -5)$, and $(-4, 5)$. Their equations are $y = \pm\dfrac{5}{4}x$.

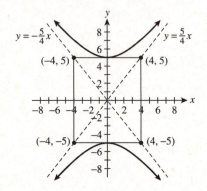

Now Try:

1. Graph $\dfrac{x^2}{16} - \dfrac{y^2}{4} = 1$.

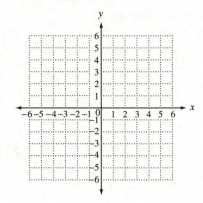

2. Graph $\dfrac{y^2}{4} - \dfrac{x^2}{9} = 1$.

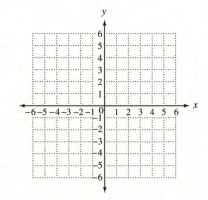

Review this example for Objective 3:

3. Identify the graph of each equation.

   a. $3x^2 + y = 36$ **b.** $4x^2 = 36 - 4y^2$

   c. $4x^2 = 36 + 4y^2$

   a. $3x^2 + y = 36$

      Only one of the variables is squared, so this is the vertical parabola

      $y = -3x^2 + 36$.

   b. $4x^2 = 36 - 4y^2$

      Both variables are squared, so the graph is either an ellipse or a hyperbola. Note that a circle is a special case of an ellipse. Rewrite the equation so that the x^2- and y^2-terms are on one side of the equation.

      $4x^2 + 4y^2 = 36$

        $x^2 + y^2 = 9$ Divide each side by 4.

      The graph of this equation is a circle with center (0, 0) and radius 3.

   c. $4x^2 = 36 + 4y^2$

      Again, both variables are squared, so the graph is either an ellipse or a hyperbola. Rewrite the equation so that the x^2- and y^2-terms are on one side of the equation.

      $4x^2 - 4y^2 = 36$

      $\dfrac{x^2}{9} - \dfrac{y^2}{9} = 1$ Divide each side by 36.

      The graph of this equation is a hyperbola.

Now Try:

3. Identify the graph of each equation.

   a. $5x^2 - 6y^2 = 30$ _____

   b. $4y^2 = 12 - 3x^2$ _____

   c. $2x - 9 = y^2$ _____

Review these examples for Objective 4:

4. Graph $f(x) = \sqrt{36 - x^2}$. Give the domain and range.

$$f(x) = \sqrt{36 - x^2}$$
$$y = \sqrt{36 - x^2} \quad \text{Replace } f(x) \text{ with } y.$$
$$y^2 = 36 - x^2 \quad \text{Square both sides.}$$
$$x^2 + y^2 = 36 \quad \text{Add } x^2.$$

This is the graph of a circle with center at $(0, 0)$ and radius 6. Since $f(x)$ represents a principal square root in the original equation, $f(x)$ must be nonnegative. This restricts the graph to the upper half of the circle.

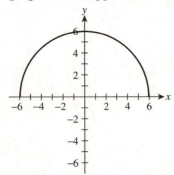

Domain: $[-6, 6]$; range: $[0, 6]$.

5. Graph $f(x) = -3\sqrt{1 - \dfrac{x^2}{4}}$. Give the domain and range.

$$f(x) = -3\sqrt{1 - \frac{x^2}{4}}$$
$$y^2 = 9\left(1 - \frac{x^2}{4}\right) \quad \begin{array}{l}\text{Replace } f(x) \text{ with } y.\\ \text{Square each side.}\end{array}$$
$$\frac{y^2}{9} = 1 - \frac{x^2}{4} \quad \text{Divide by 9.}$$
$$\frac{x^2}{4} + \frac{y^2}{9} = 1 \quad \text{Add } \frac{x^2}{4}.$$

This is the equation of an ellipse with x-intercepts $(2, 0)$ and $(-2, 0)$ and y-intercepts $(0, 3)$ and $(0, -3)$.

Now Try:

4. Graph $f(x) = \sqrt{9 - x^2}$. Give the domain and range.

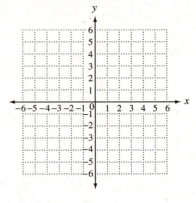

5. Graph $f(x) = \sqrt{9 - 9x^2}$. Give the domain and range.

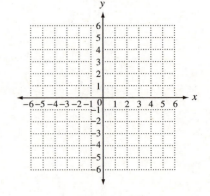

Since the original equation has a negative square root, y must be nonpositive, restricting the graph to the lower half of the ellipse.

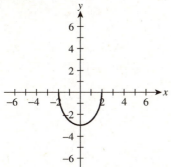

Domain: $[-2, 2]$; range: $[0, -3]$.

Objective 1 Recognize the equation of a hyperbola.
Objective 2 Graph hyperbolas by using asymptotes.
For extra help, see Examples 1 and 2 on page 651 of your text, the Section Lecture video for Section 11.3, and Exercise Solution Clips 5 and 7.

Graph each hyperbola.

1. $\dfrac{y^2}{25} - \dfrac{x^2}{16} = 1$

1.

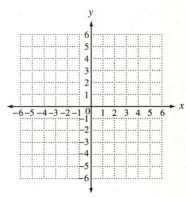

2. $\dfrac{x^2}{36} - \dfrac{y^2}{49} = 1$

2.

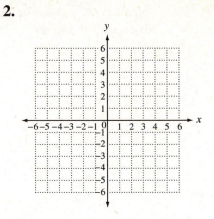

3. $\dfrac{y^2}{4} - \dfrac{x^2}{25} = 1$

3.

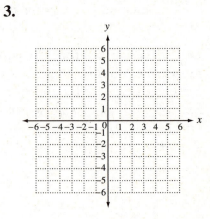

4. $\dfrac{x^2}{16} - y^2 = 1$

4.

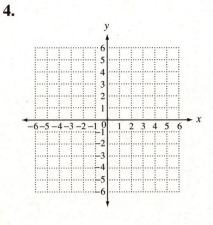

5. $\dfrac{x^2}{\frac{1}{4}} - \dfrac{y^2}{\frac{1}{9}} = 1$

5.

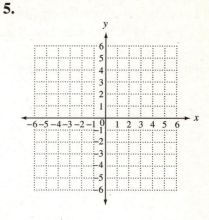

Objective 3 Identify conic sections by their equations.
For extra help, see Example 3 on page 653 of your text, the Section Lecture video for
Section 11.3, and Exercise Solution Clip 15.

Identify the graph of each equation as a parabola, circle, ellipse, or hyperbola.

6. $x^2 = y^2 + 9$

6. _____

7. $2x^2 + y^2 = 16$

7. _____

8. $2x + y^2 = 16$

8. _____

9. $25y^2 + 100 = 4x^2$

9. _____

10. $5x^2 = 25 - 5y^2$

10. _____

Name: Date:

Instructor: Section:

Objective 4 Graph certain square root functions.

For extra help, see Examples 4 and 5 on page 654 of your text, the Section Lecture video for Section 11.3, and Exercise Solution Clips 23 and 25.

Graph each generalized square root function. Give the domain and range.

11. $f(x) = -\sqrt{16 - x^2}$

11.

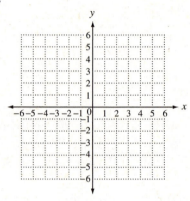

12. $f(x) = -\sqrt{1 + x}$

12.

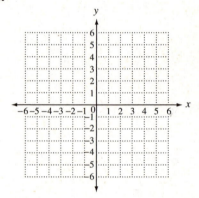

13. $f(x) = -\sqrt{6 - x}$

13.

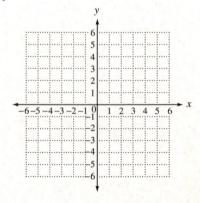

14. $f(x) = -3\sqrt{1 + \dfrac{x^2}{25}}$

14.

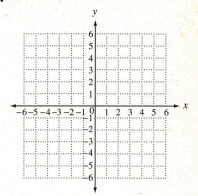

15. $f(x) = -5\sqrt{1 - \dfrac{x^2}{9}}$

15.

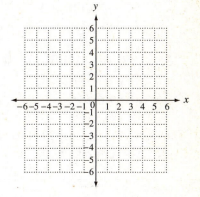

Chapter 11 NONLINEAR FUNCTIONS, CONIC SECTIONS, AND NONLINEAR SYSTEMS

11.4   Nonlinear Systems of Equations

Learning Objectives
1   Solve a nonlinear system by substitution.
2   Solve a nonlinear system by elimination.
3   Solve a nonlinear system that requires a combination of methods.

Key Terms

Use the vocabulary terms listed below to complete each statement in exercises 1–2.

nonlinear equation   **nonlinear system of equations**

1. An equation in which some terms have more than one variable or a variable of degree 2 or greater is a _____.

2. A system with at least one nonlinear equation is a _____.

Guided Examples

Review these examples for Objective 1:

1. Solve the system.
$$x^2 + 3y^2 = 3 \quad\quad (1)$$
$$x + y = -1 \quad\quad (2)$$

The graph of (1) is an ellipse and the graph of (2) is a line, so the graphs could intersect in zero, one, or two points.
Solve (2) for x, then substitute that expression into (1) and solve for y.
$$x + y = -1$$
$$x = -y - 1$$

$$x^2 + 3y^2 = 3 \quad (1)$$
$$(-y-1)^2 + 3y^2 = 3 \quad \text{Let } x = -y - 1.$$
$$y^2 + 2y + 1 + 3y^2 = 3 \quad \text{Expand } (-y-1)^2.$$
$$4y^2 + 2y - 2 = 0 \quad \text{Combine like terms. Subtract 3.}$$
$$2(2y-1)(y+1) = 0 \quad \text{Factor.}$$
$$(2y-1)(y+1) = 0 \quad \text{Divide by 2.}$$

Now Try:

1. Solve the system.
$$x^2 + y^2 = 17$$
$$x + y = -3$$

_____

$2y - 1 = 0$ or $y + 1 = 0$ Zero-factor

$y = \frac{1}{2}$ $y = -1$ property;

Solve for y.

Let $y = \frac{1}{2}$ in (2) to get $x = -\frac{3}{2}$.

If $y = -1$ in (1), then $x = 0$.

The solution set is $\left\{ \left(-\frac{3}{2}, \frac{1}{2} \right), (0, -1) \right\}$.

2. Solve the system.

$$xy = -10 \quad \text{(1)}$$
$$2x - y = 9 \quad \text{(2)}$$

The graph of (1) is a hyperbola and the graph of (2) is a line. There may be zero, one, or two points of intersection. Since neither equation has a squared term, solve either equation for one of the variables and then substitute the result into the other equation.

Solving (1) for y gives $y = -\frac{10}{x}$. Substituting into (2) gives

$$2x - \left(-\frac{10}{x} \right) = 9$$
$$2x + \frac{10}{x} = 9$$
$$2x^2 + 10 = 9x \quad \text{Multiply by } x.$$
$$2x^2 - 9x + 10 = 0 \quad \text{Subtract } 9x.$$
$$(2x - 5)(x - 2) = 0 \quad \text{Factor.}$$

$2x - 5 = 0$ or $x - 2 = 0$ Zero-factor

$x = \frac{5}{2}$ $x = 2$ property;

Solve for x.

Let $x = \frac{5}{2}$ in (1) to get $y = -4$.

If $x = 2$ in (1), then $y = -5$.

The solution set is $\left\{ (2, -5), \left(\frac{5}{2}, -4 \right) \right\}$.

2. Solve the system.

$$xy = 1$$
$$x + y = 2$$

_____

Review this example for Objective 2:

3. Solve the system.

    $$3x^2 + y^2 = 35 \quad \text{(1)}$$
    $$2x^2 - y^2 = 15 \quad \text{(2)}$$

    The graph of (1) is an ellipse and the graph of (2) is a hyperbola. There may be zero, one, two, three, or four points of intersection. Adding the two equations will eliminate y.

    $$3x^2 + y^2 = 35 \quad \text{(1)}$$
    $$\underline{2x^2 - y^2 = 15} \quad \text{(2)}$$
    $$5x^2 = 50 \quad \text{Add.}$$
    $$x^2 = 10 \quad \text{Divide by 5.}$$
    $$x = \pm\sqrt{10} \quad \text{Square root property}$$

    Substitute these values for x in (1) and solve for y.

    $$3\left(\sqrt{10}\right)^2 + y^2 = 35 \quad \text{Let } x = \sqrt{10}.$$
    $$30 + y^2 = 35$$
    $$y^2 = 5$$
    $$y = \pm\sqrt{5}$$

    $$3\left(-\sqrt{10}\right)^2 + y^2 = 35 \quad \text{Let } x = -\sqrt{10}.$$
    $$30 + y^2 = 35$$
    $$y^2 = 5$$
    $$y = \pm\sqrt{5}$$

    The solution set is $\left\{\left(\sqrt{10}, -\sqrt{5}\right), \left(\sqrt{10}, \sqrt{5}\right), \left(-\sqrt{10}, \sqrt{5}\right), \left(-\sqrt{10}, -\sqrt{5}\right)\right\}$.

Review this example for Objective 3:

4. Solve the system.

    $$x^2 + 5xy - y^2 = 13 \quad \text{(1)}$$
    $$x^2 - y^2 = 3 \quad \text{(2)}$$

    We will use the elimination method in combination with the substitution method.

Now Try:

3. Solve the system.

    $$x^2 - 2y = 8$$
    $$x^2 + y^2 = 16$$

    _____

Now Try:

4. Solve the system.

    $$5x^2 - xy + 5y^2 = 89$$
    $$x^2 + y^2 = 17$$

    _____

$$x^2 + 5xy - y^2 = 13$$
$$-x^2 + \qquad y^2 = -3 \quad \text{Multiply (2) by } -1.$$
$$5xy = 10 \quad \text{Add.}$$
$$y = \frac{2}{x} \quad \text{Solve for } y. \quad (3)$$

Now substitute $\frac{2}{x}$ for y in (2) and solve for x.

$$x^2 - \left(\frac{2}{x}\right)^2 = 3$$
$$x^2 - \frac{4}{x^2} = 3 \quad \text{Square } \frac{2}{x}.$$
$$x^4 - 4 = 3x^2 \quad \text{Multiply by } x^2, \; x \geq 0.$$
$$x^4 - 3x^2 - 4 = 0 \quad \text{Subtract } 3x^2.$$
$$(x-2)(x+2)(x^2+1) = 0 \quad \text{Factor.}$$

Using the zero-factor property, we have
$x = 2, x = -2, x = i, x = -i$.
Substitute each of these values into (3) and
solve for y.
If $x = 2$, then $y = 1$.
If $x = -2$, then $y = -1$.

If $x = i$, then $y = \frac{2}{i} = \frac{2}{i} \cdot \frac{-i}{-i} = \frac{-2i}{-i^2} = -2i$.

If $x = -i$, then $y = \frac{2}{-i} = \frac{2}{-i} \cdot \frac{i}{i} = \frac{2i}{-i^2} = 2i$.

It is important to check all answers in the
original equations because it is possible to
obtain extraneous solutions.
The solution set is $\{(2, 1), (-2, -1), (i, -2i),$
$(-i, 2i)\}$.

Objective 1 Solve a nonlinear system by substitution.
For extra help, see Examples 1 and 2 on pages 658−659 of your text, the Section Lecture
video for Section 11.4, and Exercise Solution Clips 17 and 19.

Solve each system by the substitution method.

1. $x^2 + y^2 = 17$
$\qquad 2x = y + 9$

1. _____

2. $2x^2 - y^2 = -1$
$\quad\ \ 2x\ + y\ = 7$

2. _____

3. $4x^2 + 3y^2 = 7$
$\quad\ \ 2x - 5y\ = -7$

3. _____

4. $\quad xy = -6$
$\quad\ x + y = 1$

4. _____

5. $\quad xy = 24$
$\quad\ \ y = 2x + 2$

5. _____

Objective 2 Solve a nonlinear system by elimination.
For extra help, see Example 3 on pages 659–660 of your text, the Section Lecture video for Section 11.4, and Exercise Solution Clip 33.

Solve each system by the elimination method.

6. $5x^2 - y^2 = 55$
$\quad\ \ 2x^2 + y^2 = 57$

6. _____

7. $x^2 + 2y^2 = 11$
$2x^2 - y^2 = 17$

7. _____

8. $3x^2 + 2y^2 = 30$
$2x^2 + y^2 = 17$

8. _____

9. $5x^2 + y^2 = 6$
$2x^2 - 3y^2 = -1$

9. _____

10. $4x^2 - 3y^2 = -8$
$2x^2 + y^2 = 5$

10. _____

Objective 3 Solve a nonlinear system that requires a combination of methods.
For extra help, see Example 4 on pages 660–661 of your text, the Section Lecture video for
Section 11.4, and Exercise Solution Clip 39.

Solve each system.

11. $4x^2 - 2xy + 4y^2 = 64$
$x^2 \qquad + \ y^2 = 13$

11. _____

12. $x^2 + 2xy + 3y^2 = 6$
$x^2 + 4xy + 3y^2 = 8$

12. _____

13. $3x^2 - 4xy + 2y^2 = \ 59$
$-3x^2 + 5xy - 2y^2 = -65$

13. _____

14. $x^2 + 3xy + 2y^2 = 12$
$-x^2 + 8xy - 2y^2 = 10$

14. _____

15. $x^2 + 5xy - y^2 = 20$

     $x^2 - 2xy - y^2 = -8$

15. _____

Chapter 11 NONLINEAR FUNCTIONS, CONIC SECTIONS, AND NONLINEAR SYSTEMS

11.5   Second-Degree Inequalities and Systems of Inequalities

Learning Objectives
1   Graph second-degree inequalities.
2   Graph the solution set of a system of inequalities.

Key Terms

Use the vocabulary terms listed below to complete each statement in exercises 1–2.

second-degree inequality   system of inequalities

1.   A _____ is an inequality with at least one variable of degree 2 and no variable with degree greater than 2.

2.   A _____ consists of two or more inequalities to be solved at the same time.

Guided Examples

Review these examples for Objective 1:

1.   Graph $x^2 + y^2 \geq 4$.

The boundary of the inequality $x^2 + y^2 \geq 4$ is the graph of the equation $x^2 + y^2 = 4$, a circle with radius 2 and center at the origin. To determine if the inequality will include the points outside or inside the circle, substitute any test point not on the circle into the original inequality.

$$x^2 + y^2 \geq 4$$

$$0^2 + 0^2 \overset{?}{\geq} 4$$

$$0 \geq 4 \quad \text{False}$$

Since a false statement results, shade the region that does not include (0, 0). Thus, the original inequality includes the points outside the circle, along with the boundary.

Now Try:

1.   Graph $x^2 + y^2 \leq 9$.

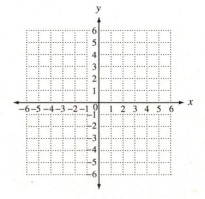

Copyright © 2012 Pearson Education, Inc. Publishing as Addison-Wesley.

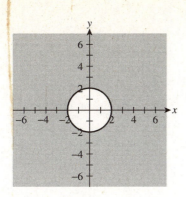

2. Graph $y > (x-1)^2 - 2$.

The boundary is a parabola that opens up with vertex $(1, -2)$.

$$y > (x-1)^2 - 2$$

$$0 \overset{?}{>} (0-1)^2 - 2 \quad \text{Use } (0, 0) \text{ as a test point.}$$

$$0 > -1 \qquad \text{True}$$

Because the final inequality is true, shade the region that includes $(0, 0)$. Thus, the original inequality includes those points inside the parabola, but does not include the parabola.

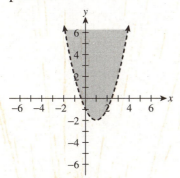

3. Graph $16x^2 \geq -9y^2 + 144$.

$$16x^2 \geq -9y^2 + 144$$

$$16x^2 + 9y^2 \geq 144 \quad \text{Add } 9y^2.$$

$$\frac{x^2}{9} + \frac{y^2}{16} \geq 1 \qquad \text{Divide by 144.}$$

Thus, the boundary is an ellipse with $a = 3$ and $b = 4$. Using $(0, 0)$ as a test point gives

$$\frac{0^2}{9} + \frac{0^2}{16} \geq 1 \Rightarrow 0 \geq 1, \text{ which is false.}$$

2. Graph $y > -(x-2)^2 + 1$.

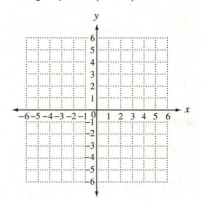

3. Graph $16x^2 < 9y^2 + 144$.

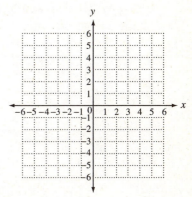

Therefore, we shade the region outside the ellipse.

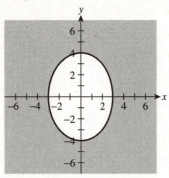

Review these examples for Objective 2:

4. Graph the solution set of the system.

$$x^2 + y^2 \leq 16$$
$$y > x$$

Begin by graphing $x^2 + y^2 \leq 16$. The boundary line is a circle centered at the origin with radius 4. The test point $(0, 0)$ leads to a true statement, so we shade inside the circle.

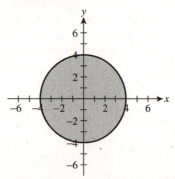

The boundary of the solution set of $y > x$ is a dashed line passing through $(0, 0)$. Using the test point $(0, 1)$ leads to a true statement, so shade above the line.

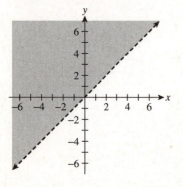

Now Try:

4. Graph the solution set of the system.

$$x^2 + y^2 < 9$$
$$y < x^2 - 3$$

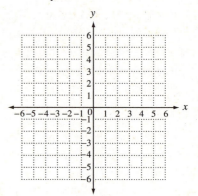

The graph of the solution set of the system is the intersection of the graphs of the two inequalities.

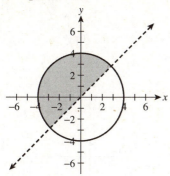

5. Graph the solution set of the system.

$$x + y < 2$$
$$y \le 4x + 3$$
$$x \ge 0$$

The graph of $x + y < 2$ consists of all points that lie below (to the right) of the dashed line $x + y = 2$. The graph of $y \le 4x + 3$ is the region that lies above (to the right) of the line $y = 4x + 3$. Finally, the graph of $x \ge 0$ is the region to the right of the y-axis. The graph of the system is the intersection of the three regions.

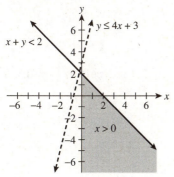

5. Graph the solution set of the system.

$$4x - y > 2$$
$$y > -x - 2$$
$$x \le 3$$

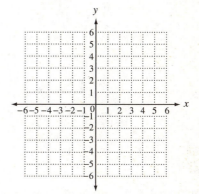

6. Graph the solution set of the system.

$$25x^2 + 9y^2 < 225$$
$$y \le -x^2 + 4$$
$$y < -x$$

The graph of $25x^2 + 9y^2 < 225$ is a dashed ellipse with $a = 3$ and $y = 5$. To satisfy the inequality, a point must lie inside the ellipse.

The graph of $y \le -x^2 + 4$ is a parabola with vertex (0, 4) opening downward. The inequality includes the points on the boundary along with the points inside the parabola. The graph of $y < -x$ includes all points below the line $y = -x$. Therefore, the graph of the system is the shaded region, which lies inside the ellipse and the parabola, and below the line.

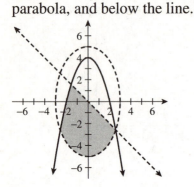

6. Graph the solution set of the system.

$$y \ge (x+2)^2 - 5$$
$$2x + y < -5$$
$$(x+3)^2 + (y-1)^2 < 9$$

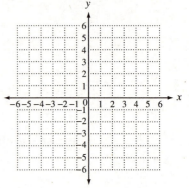

Objective 1 Graph second-degree inequalities.
For extra help, see Examples 1–3 on pages 664–665 of your text, the Section Lecture video for Section 11.3, and Exercise Solution Clips 9 and 15.

Graph each nonlinear inequality.

1. $x \ge y^2$

1.

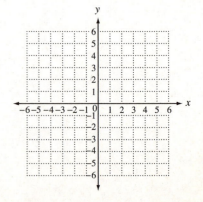

2. $25y^2 \leq 100 - 4x^2$

2.

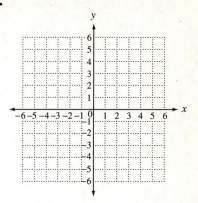

3. $x^2 + 4y^2 > 4$

3.

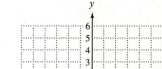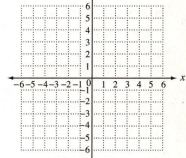

4. $x^2 + y^2 \geq 16$

4.

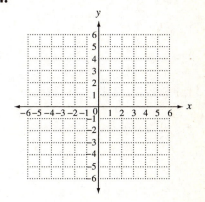

5. $4y^2 \geq 196 + 49x^2$

5.

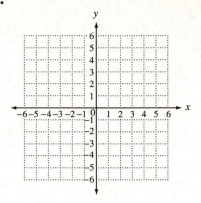

6. $9x^2 < 4y^2 + 36$

6.

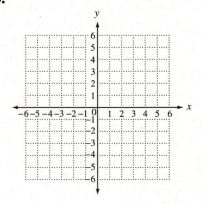

7. $y < x^2 - 4x + 5$

7.

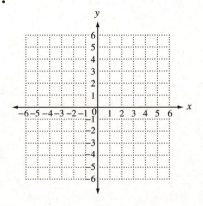

8. $(x-2)^2 + (y+3)^2 < 25$

8.

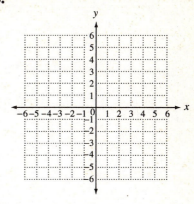

Objective 2 Graphing the solution set of a system of inequalities.
For extra help, see Examples 4−6 on pages 665−666 of your text, the Section Lecture video for Section 11.3, and Exercise Solution Clips 27, 29, and 31.

Graph each system of inequalities.

9. $x - 2y \geq -6$
$x + 4y \geq 12$

9.

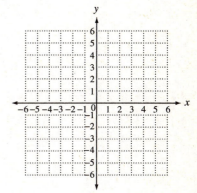

10. $x^2 + y^2 \leq 25$
$3x - 5y > -15$

10.

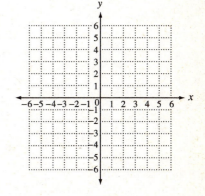

11. $9x^2 + 16y^2 < 144$

        $y^2 - \ \ x^2 > 4$

11.

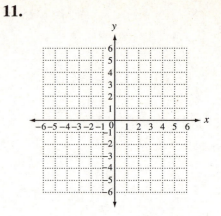

12. $4x^2 + 16y^2 \le 64$

        $16x^2 + \ 4y^2 \ge 64$

12.

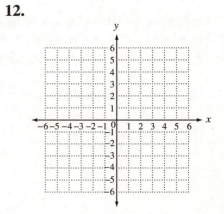

13. $x^2 > 9 - y^2$

        $x \le 0$

        $y \ge 0$

13.

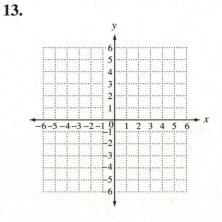

14.
$$y \le x^2 - 4x + 4$$
$$y > x^2 - 4x + 2$$
$$(x-2)^2 + (y+1)^2 < 9$$

14.

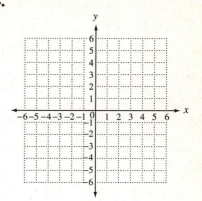

15. $x^2 + y^2 \le 25$
$x^2 - y^2 > 4$
$x + y > 3$

15.

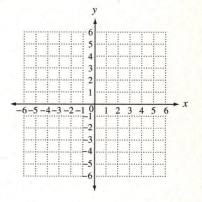

Chapter 12 SEQUENCES AND SERIES

12.1   Sequences and Series

Learning Objectives
1   Find the terms of a sequence, given the general term.
2   Find the general term of a sequence.
3   Use sequences to solve applied problems
4   Use summation notation to evaluate a series.
5   Write a series with summation notation.
6   Find the arithmetic mean (average) of a group of numbers.

Key Terms

Use the vocabulary terms listed below to complete each statement in exercises 1–8.

infinite sequence	**finite sequence**	**terms of a sequence**
general term	**series**	**summation notation**
index of summation	**arithmetic mean (average)**	

1.   A(n) _____ is a function whose domain is the set of positive integers.

2.   A _____ is the sum of the terms of a sequence.

3.   _____ is a compact way of writing a series using the general term of the corresponding sequence.

4.   When using summation notation $\displaystyle\sum_{i}^{n} f(i)$, the letter i is called the

       _____.

5.   The expression a_n which defines a sequence, is called the _____ of the sequence.

6.   The _____ of a group of numbers is given by the formula $\dfrac{\displaystyle\sum_{i=1}^{n} x_i}{n}$.

7.   The function values $a_1, a_2, a_3, \ldots$, written in order, are the _____.

8.   A(n) _____ is a function with domain of the form $\{1, 2, 3, \ldots, n\}$, where n is a positive integer.

505

Guided Examples

Review this example for Objective 1:

1. Given an infinite sequence with $a_n = 3n - 2$, find a_4.

   $a_4 = 3(4) - 2 = 10$

Now Try:

1. Given an infinite sequence with $a_n = \dfrac{n+2}{5n}$ find a_6.

   _____

Review this example for Objective 2:

2. Find an expression for the general term a_n of the sequence $1, \frac{1}{2}, \frac{1}{3}, \frac{1}{4}, \frac{1}{5}, \ldots$

   Notice that the denominators are consecutive integers, so the general term is $a_n = \dfrac{1}{n}$.

Now Try:

2. Find an expression for the general term a_n of the sequence $2, -4, 8, -16, 32, \ldots$.

   _____

Review this example for Objective 3:

3. Suppose a copier loses $\frac{1}{4}$ of its value each year; that is, at the end of any given year, the value is $\frac{3}{4}$ of its value at the beginning of the year. If a new copier costs $4800, what is its value at the end of 4 years?

   At the end of the first year, the value of the copier is $\frac{3}{4}(\$4800) = \$3600$.

   At the end of the second year, the value of the copier is $\frac{3}{4}(\$3600) = \$2700$.

   At the end of the third year, the value of the copier is $\frac{3}{4}(\$2700) = \$2025$.

   At the end of the fourth year, the value of the copier is $\frac{3}{4}(\$2025) = \$1518.75$.

Now Try:

3. Carla borrows $6000 and agrees to pay $500 monthly plus interest of 2% on the unpaid balance from the beginning of the first month. Find the payments for the first four months and the remaining debt at the end of that period.

   Month 1 _____

   Month 2 _____

   Month 3 _____

   Month 4 _____

   Remaining balance _____

Review this example for Objective 4:

4. Write out the terms and evaluate each series.

   a. $\displaystyle\sum_{i=1}^{6}(2i-6)$ **b.** $\displaystyle\sum_{i=1}^{5}i(i+2)$

   c. $\displaystyle\sum_{i=2}^{5}2i^2$

   a. $\displaystyle\sum_{i=1}^{6}(2i-6)$
   $$= [2(1)-6]+[2(2)-6]+[2(3)-6]$$
   $$+[2(4)-6]+[2(5)-6]+[2(6)-6]$$
   $$= -4+(-2)+0+2+4+6$$
   $$= 6$$

   b. $\displaystyle\sum_{i=1}^{5}i(i+2)$
   $$= 1(1+2)+2(2+2)+3(3+2)$$
   $$+4(4+2)+5(5+2)$$
   $$= 3+8+15+24+35$$
   $$= 85$$

   c. $\displaystyle\sum_{i=2}^{5}2i^2 = 2(2)^2+2(3)^2+2(4)^2+2(5)^2$
   $$= 8+18+32+50$$
   $$= 108$$

Now Try:

4. Write out the terms and evaluate each series.

   a. $\displaystyle\sum_{i=1}^{7}(4i-3)$ _____

   b. $\displaystyle\sum_{i=2}^{6}(i+3)(i-4)$ _____

   c. $\displaystyle\sum_{i=1}^{4}2^i$ _____

Review this example for Objective 5:

5. Write each sum with summation notation.

   a. $3+\frac{3}{2}+1+\frac{3}{4}+\frac{3}{5}$

   b. $3+9+27+81+243$

   a. First, find a general term. If we rewrite the series as $\frac{3}{1}+\frac{3}{2}+\frac{3}{3}+\frac{3}{4}+\frac{3}{5}$, then it appears that $a_n=\frac{3}{n}$. Thus, the sum can be written as $\displaystyle\sum_{i=1}^{5}\frac{3}{i}$.

Now Try:

5. Write each sum with summation notation.

   a. $4+8+12+16+20+24$

   _____

   b. $\frac{1}{2}+\frac{2}{3}+\frac{3}{4}+\frac{4}{5}+\frac{5}{6}$

   _____

b. Each term is a power of 3, so $a_n = 3^n$.

Thus, the sum can be written as $\displaystyle\sum_{i=1}^{5} 3^i$.

Review this example for Objective 6:	Now Try:

Review this example for Objective 6:

6. The amount of rain that fell in Philadelphia from March 2009 through August 2009 is given in the table.

Month	Amount of Precipitation
March	7.55 in.
April	3.85 in.
May	2.61 in.
June	2.15 in.
July	8.33 in.
August	1.08 in.

Source: Franklin Institute

What was the average amount of rain each month for this six-month period?

$$\overline{x} = \frac{\displaystyle\sum_{i=1}^{6} x_i}{6}$$

$$= \frac{7.55 + 3.85 + 2.61 + 2.15 + 8.33 + 1.08}{6}$$

$$= \frac{25.57}{6} \approx 4.26$$

The average amount of rain was about 4.26 inches per month during the six-month period.

Now Try:

6. The amount of rain that fell in Seattle during October from 2004 through 2010 is given in the table.

Month	Amount of Precipitation
2004	2.81 in.
2005	3.01 in.
2006	1.55 in.
2007	3.32 in.
2008	2.17 in.
2009	5.54 in.
2010	5.24 in.

Source: National Weather Service

What is the average amount of rain that fell during October for the seven-year period?

_____

Objective 1 Find the terms of a sequence, given the general term.

For extra help, see Example 1 on page 678 of your text, the Section Lecture video for Section 12.1, and Exercise Solution Clip 1.

Write out the first five terms of each sequence.

1. $a_n = (-1)^n$

1. _____

2. $a_n = \dfrac{1+n}{n}$

2. _____

Find the indicated term for the sequence.

3. $a_n = \dfrac{3n-2}{5n+2}; \; a_8$

3. _____

Objective 2 Find the general term of a sequence.
For extra help, see Example 2 on page 679 of your text, the Section Lecture video for
Section 12.1, and Exercise Solution Clip 19.

Find a general term a_n for the given terms of each sequence.

4. 3, 5, 7, 9, 11, …

4. _____

5. $\sqrt{3}, \; 3, \; 3\sqrt{3}, \; 9, \; 9\sqrt{3}, \; …$

5. _____

6. $-\dfrac{1}{5}, \dfrac{1}{10}, -\dfrac{1}{15}, \dfrac{1}{20}, -\dfrac{1}{25}, \cdots$

6. _____

Objective 3 Use sequences to solve applied problems.
For extra help, see Example 3 on page 679 of your text, the Section Lecture video for
Section 12.1, and Exercise Solution Clip 27.

Solve each applied problem by writing the first few terms of a sequence.

7. A colony of bacteria doubles in weight every hour. If the
colony weighs 2 grams at the beginning of an
experiment, find the weight after 3 hours.

7.

8. Ms. Burley is offered a new job with a salary of $35,000
per year and a 2% raise at the end of each year. Write a
sequence showing her salary for each of the first five
years. (Round to the nearest dollar.)

8.

Objective 4 Use summation notation to evaluate a series.
For extra help, see Example 4 on pages 680–681 of your text, the Section Lecture video for Section 12.1, and Exercise Solution Clip 31.

Write out each series and evaluate it.

9. $\displaystyle\sum_{i=1}^{4}(2i+3)$ 9. _____

10. $\displaystyle\sum_{i=1}^{4}\left(-\frac{1}{2i}\right)$ 10. _____

Objective 5 Write a series with summation notation.
For extra help, see Example 5 on page 681 of your text, the Section Lecture video for Section 12.1, and Exercise Solution Clip 39.

Write each series with summation notation.

11. $1+4+9+16+25$ 11. _____

12. $\frac{1}{5}+\frac{1}{8}+\frac{1}{11}+\frac{1}{14}+\frac{1}{17}$ 12. _____

13. $1+2x+3x^2+4x^3+5x^4$ 13. _____

Objective 6 Find the arithmetic mean (average) of a group of numbers.
For extra help, see Example 6 on page 682 of your text and the Section Lecture video for Section 12.1.

Find the arithmetic mean for each collection of numbers.

14. $2, 8, 18, 32, -5, -7$ 14. _____

15. $\frac{1}{3}, -\frac{1}{2}, \frac{1}{4}, \frac{7}{6}$ 15. _____

Chapter 12 SEQUENCES AND SERIES

12.2   Arithmetic Sequences

Learning Objectives
1   Find the common difference of an arithmetic sequence.
2   Find the general term of an arithmetic sequence.
3   Use an arithmetic sequence in an application.
4   Find any specified term or the number of terms of an arithmetic sequence.
5   Find the sum of a specified number of terms of an arithmetic sequence.

Key Terms

Use the vocabulary terms listed below to complete each statement in exercises 1–2.

arithmetic sequence (arithmetic progression)   **common difference**

1.   A(n) _____ is a sequence in which each term after the first differs from the preceding term by a constant amount.

2.   The _____ d is the difference between any two adjacent terms of an arithmetic sequence.

Guided Examples

Review these examples for Objective 1:

1.   Determine the common difference d for the arithmetic sequence.

     35, 32, 29, 26, …

     Since the sequence is arithmetic, d is the difference between any two adjacent terms. We arbitrarily choose the terms 29 and 26. $d = 26 - 29 = -3$

2.   Write the first five terms of the arithmetic sequence with first term 9 and common difference 5.

     The second term is found by adding 5 to the first term 9, getting 14. For the next term, add 5 to 14, and so on. The first five terms are 9, 14, 19, 24, 29.

Now Try:

1.   Determine the common difference d for the arithmetic sequence.

     −13, −11, −9, −7, …

     _____

2.   Write the first five terms of the arithmetic sequence with first term 9 and common difference −5.

     _____

Review this example for Objective 2:

3. Determine the general term of the arithmetic sequence.

$$-3, -7, -11, -15, -19, \ldots$$

Then use the general term to find a_{20}.

The first term is $a_1 = -3$.
$d = -11 - (-7) = -4$
Now find a_n.

$a_n = a_1 + (n-1)d$	Formula for a_n.
$a_n = -3 + (n-1)(-4)$	$a_1 = -3$, $d = -4$
$a_n = -3 - 4n + 4$	Distributive property
$a_n = -4n + 1$	Combine like terms.
$a_{20} = -4(20) + 1 = -79$	Let $n = 20$.

Now Try:

3. Determine the general term of the arithmetic sequence.

$$35, 39, 43, 47, 51, \ldots$$

Then use the general term to find a_{20}.

_____

Review this example for Objective 3:

4. After knee surgery, your trainer tells you to return to your jogging program slowly. He suggests jogging for 12 minutes each day for the first week. Each week thereafter, he suggests that you increase that time by 6 minutes per day. How long will you be jogging in the 6th week?

After n weeks, you will be jogging
$a_n = 12 + 6(n-1)$ minutes. To find how long you will be jogging in the 6th week, find a_6.
$a_6 = 12 + 6(6-1) = 42$

You will be jogging 42 minutes in the 6th week.

Now Try:

4. Mary accepts a teaching job that pays $32,530 the first year with a guarantee of a $1,030 per year raise. What will her salary be in her 10th year on the job?

_____

Review these examples for Objective 4:

5. Evaluate the indicated term for each arithmetic sequence.

   a. $a_1 = 28$, $d = 10$, a_{12}

   b. $a_4 = 24$, $a_{10} = -6$, a_{20}

   a. $a_n = a_1 + (n-1)d$ Formula for a_n.

   $a_{12} = 28 + (12-1)(10)$ $a_1 = 28$, $n = 12$, $d = 10$

   $a_{12} = 138$

Now Try:

5. Evaluate the indicated term for each arithmetic sequence.

   a. $a_1 = 9$, $d = 6$, a_{15}

   _____

   b. $a_3 = 7$, $a_{10} = -14$, a_{20}

   _____

b. Any term can be found if a_1 and d are known. Use the formula for a_n.

$a_4 = a_1 + (4-1)d$	$a_{10} = a_1 + (10-1)d$
$a_4 = a_1 + 3d$	$a_{10} = a_1 + 9d$
$24 = a_1 + 3d$	$-6 = a_1 + 9d$

This gives a system of two equations in two variables.

$$24 = a_1 + 3d \quad (1)$$
$$-6 = a_1 + 9d \quad (2)$$

Multiply (2) by -1, then add the resulting equation to (1) to eliminate a_1.

$$24 = a_1 + 3d$$
$$\underline{6 = -a_1 - 9d}$$
$$30 = -6d$$
$$-5 = d$$

Now find a_1.

$$24 = a_1 + 3(-5)$$
$$24 = a_1 - 15$$
$$39 = a_1$$

Thus, $a_n = 39 - 5(n-1) = 44 - 5n$.

$$a_{20} = 44 - 5(20) = -56$$

6. Find the number of terms in the sequence 7, 10, 13, …, 55.

$a_1 = 7$, $a_n = 55$, $d = 10 - 7 = 3$

$a_n = a_1 + (n-1)d$ Formula for a_n.

$55 = 7 + (n-1)3$ $a_1 = 7$, $a_n = 55$, $d = 3$

$55 = 7 + 3n - 3$ Distributive property

$55 = 4 + 3n$ Combine like terms.

$51 = 3n$ Subtract 4.

$17 = n$ Divide by 3.

There are 17 terms in the sequence.

6. Find the number of terms in the sequence
$$3, \frac{9}{2}, 6, \frac{15}{2}, 9, \ldots, \frac{51}{2}$$

_____

Review these examples for Objective 5:

7. Find the sum of the first eight terms of the sequence in which $a_n = 7 - 2n$.

Begin by evaluating a_1 and a_8.

$$a_1 = 7 - 2(1) = 5$$
$$a_8 = 7 - 2(8) = -9$$

Now Try:

7. Find the sum of the first six terms of the sequence in which $a_n = 3n - 4$.

_____

Now evaluate the sum using $a_1 = 5$, $a_8 = -9$, and $n = 8$.

$$S_n = \frac{n}{2}(a_1 + a_n)$$

$$S_8 = \frac{8}{2}(5 + (-9)) = 4(-4) = -16$$

8. Find the sum of the first 15 terms of the arithmetic sequence having first term 5 and common difference -4.

$$S_n = \frac{n}{2}[2a_1 + (n-1)d]$$

$$S_{15} = \frac{15}{2}[2(5) + (15-1)(-4)]$$

$$\qquad a_1 = 5,\ d = -4,\ n = 15$$

$$= \frac{15}{2}(-46)$$

$$= -345$$

9. Evaluate $\displaystyle\sum_{i=1}^{15}(3i + 2)$.

This is the sum of the first 15 terms of the arithmetic sequence having $a_n = 3n + 2$.

$$S_n = \frac{n}{2}(a_1 + a_n)$$

$$S_{15} = \frac{15}{2}[(3(1) + 2) + (15(1) + 2)]$$

$$\qquad a_1 = 3(1) + 2 = 5,\ a_{15} = 15(1) + 2 = 17$$

$$= \frac{15}{2}(5 + 17)$$

$$= \frac{15}{2}(22)$$

$$= 165$$

8. Find the sum of the first 13 terms of the arithmetic sequence having first term 2 and common difference 10.

_____

9. Evaluate $\displaystyle\sum_{i=1}^{10}(7i - 2)$.

_____

Objective 1 Find the common difference of an arithmetic sequence.

For extra help, see Examples 1 and 2 on page 685 of your text, the Section Lecture video for Section 12.2, and Exercise Solution Clip 7.

Find the common difference for the arithmetic sequence.

1. $-9, -5, -1, 3, \ldots$

1. _____

2. $\frac{1}{4}, -\frac{1}{4}, -\frac{3}{4}, -\frac{5}{4}, \ldots$

2. _____

3. Write the first five terms of the arithmetic sequence with first term 6 and common difference −3.

3. _____

Objective 2 Find the general term of an arithmetic sequence.

For extra help, see Example 3 on pages 685−686 of your text, the Section Lecture video for Section 12.2, and Exercise Solution Clip 11.

Use the formula for a_n to find the general term of each arithmetic sequence.

4. $a_1 = -7$, $d = -5$

4. _____

5. $\dfrac{1}{2}, \dfrac{5}{6}, \dfrac{7}{6}, \dfrac{3}{2}, \dfrac{11}{6}, \ldots$

5. _____

6. 0.8, 0.5, 0.2, …, −29.8, …

6. _____

Objective 3 Use an arithmetic sequence in an application.

For extra help, see Example 4 on page 686 of your text, the Section Lecture video for Section 12.2, and Exercise Solution Clip 41.

Solve each problem.

7. Ben's father has started a savings fund for Ben's college education. He makes an initial deposit of $5000 and each month contributes an additional $100. How much money will be in the account after 60 months? (Disregard any interest.)

7. _____

8. Lisa is starting a swimming program. She plans to swim 50 laps per day the first week and add 5 laps per day each week until she is swimming 125 laps. In which week will this occur?

8. _____

9. The starting salary at a supermarket is $8.75 per hour. Every six months an employee receives a raise of $0.75 per hour. What will the employee's salary be after two years?

9. _____

Objective 4 Find any specified term or the number of terms of an arithmetic sequence.
For extra help, see Examples 5 and 6 on pages 686–687 of your text, the Section Lecture video for Section 12.2, and Exercise Solution Clip 17.

Evaluate the indicated term for each arithmetic sequence.

10. $a_1 = -5$, $d = 9$; a_{12} 10. _____

11. $a_1 = -\frac{1}{2}$, $d = \frac{3}{2}$; a_{15} 11. _____

12. Find the number of terms in the sequence   12. _____
     19, 26, 33, …, 96

Objective 5 Find the sum of a specified number of terms of an arithmetic sequence.
For extra help, see Examples 7–9 on pages 688–689 of your text, the Section Lecture video for Section 12.2, and Exercise Solution Clips 29, 33, and 35.

13. Evaluate S_{10} for the arithmetic sequence defined by   13. _____
     $a_n = 2 - 3n$.

14. Evaluate S_8 for the arithmetic sequence with $a_1 = 3$ and
     $d = 2$.   14. _____

15. Evaluate $\displaystyle\sum_{i=1}^{10} (2i + 3)$.   15. _____

Chapter 12 SEQUENCES AND SERIES

12.3   Geometric Sequences

Learning Objectives

1   Find the common ratio of a geometric sequence.
2   Find the general term of a geometric sequence.
3   Find any specified term of a geometric sequence.
4   Find the sum of a specified number of terms of a geometric sequence.
5   Apply the formula for the future value of an ordinary annuity.
6   Find the sum of an infinite number of terms of certain geometric sequences.

Key Terms

Use the vocabulary terms listed below to complete each statement in exercises 1–7.

geometric sequence (geometric progression)	**common ratio**
annuity **ordinary annuity**	**payment period**
future value of an annuity	**term of an annuity**

1.   A(n) _____ r is the constant multiplier between adjacent terms in a geometric sequence.

2.   A(n) _____ is a sequence of equal payments made at equal periods of time.

3.   A(n) _____ is a sequence in which each term after the first is a constant multiple of the preceding term.

4.   The _____ is the sum of the compound amounts of all the payments, compounded to the end of the term.

5.   If the payments for an annuity are made at the end of the period, and if the frequency of payments is the same as the frequency of compounding, the annuity is called a(n) _____.

6.   The time from the beginning of the first payment period to the end of the last is called the _____.

7.   The time between payments of an annuity is called the _____.

Name: _____   Date:
Instructor: _____   Section:

Guided Examples

Review this example for Objective 1:

1. Determine r for the geometric sequence.
4, −8, 16, −32, 64, …

   To find r, choose any two successive terms and divide the second one by the first. We choose the third and fourth terms of the sequence.

   $$r = \frac{a_4}{a_3} = \frac{-32}{16} = -2$$

Now Try:

1. Determine r for the geometric sequence.
$$3, \frac{3}{4}, \frac{3}{16}, \frac{3}{64}, \dots$$

   _____

Review this example for Objective 2:

2. Determine the general term of the sequence.
−3, −15, −75, −375, …

   The first term is $a_1 = -3$ and the common ratio is $r = 5$.

   $$a_n = a_1 r^{n-1} = -3(5)^{n-1}$$

Now Try:

2. Determine the general term of the sequence.
$$\sqrt{3}, \ 3, \ 3\sqrt{3}, \ 9, \ 9\sqrt{3}, \ 27, \ \dots$$

   _____

Review these examples for Objective 3:

3. Evaluate the indicated term for each geometric sequence.

   a. $a_1 = 64, r = -\frac{1}{2}; a_6$

   b. 4, 12, 36, 108, …; a_9

   a. $a_n = a_1 r^{n-1}$

   $a_6 = 64\left(-\frac{1}{2}\right)^{6-1}$ $a_1 = 64, r = -\frac{1}{2}, n = 6$

   $\quad = -2$

   b. $a_n = a_1 r^{n-1}$

   $a_9 = 4(3)^{9-1}$ $a_1 = 4, r = 3, n = 9$

   $\quad = 26,244$

Now Try:

3. Evaluate the indicated term for each geometric sequence.

   a. $a_1 = -1, \ r = 3; \ a_7$

   _____

   b. $\frac{4}{3}, \frac{2}{3}, \frac{1}{3}, \frac{1}{6}, \dots; a_{10}$

   _____

4. Write the first five terms of the geometric sequence whose first term is −3 and whose common ratio is $\frac{2}{3}$.

   Use $a_n = a_1 r^{n-1}$ with $a_1 = -3$, $r = \frac{2}{3}$, and

   $n = 1, 2, 3, 4, 5$.

4. Write the first five terms of the geometric sequence whose first term is 0.8 and whose common ratio is −5.

   _____

$a_1 = -3, \; a_2 = -3\left(\frac{2}{3}\right) = -2,$

$a_3 = -3\left(\frac{2}{3}\right)^2 = -\frac{4}{3}, \; a_4 = -3\left(\frac{2}{3}\right)^3 = -\frac{8}{9},$

$a_5 = -3\left(\frac{2}{3}\right)^4 = -\frac{16}{27}$

Review these examples for Objective 4:

5. Evaluate the sum of the first seven terms of the geometric sequence with first term 36 and common ratio −6.

$$S_n = \frac{a_1\left(1 - r^n\right)}{1 - r}$$

$$S_7 = \frac{36\left(1 - (-6)^7\right)}{1 - (-6)} \qquad a_1 = 36, \; r = -6, \; n = 7$$

$$= \frac{36(1 - (-279,936))}{7}$$

$$= \frac{36(279,937)}{7}$$

$$= 1,439,676$$

6. Evaluate $\displaystyle\sum_{i=1}^{8} 4(3)^i.$

Since the series is in the form $\displaystyle\sum_{i=1}^{n} a \cdot b^i$, it represents the sum of the first n terms of the geometric sequence with $a_1 = a \cdot b^1$ and $r = b$. $a_1 = 4 \cdot (3)^1 = 12$ and $r = 3$.

$$S_n = \frac{a_1\left(1 - r^n\right)}{1 - r}$$

$$S_8 = \frac{12\left(1 - 3^8\right)}{1 - 3} \qquad a_1 = 12, \; r = 3, \; n = 8$$

$$= \frac{12(1 - 6561)}{-2}$$

$$= 39,360$$

Now Try:

5. Evaluate the sum of the first six terms of the geometric sequence with first term $\frac{2}{3}$ and common ratio 3.

_____

6. Evaluate $\displaystyle\sum_{i=1}^{5} \frac{1}{2}\left(4^i\right).$

_____

Review this example for Objective 5:

7. **a.** Matt decides to start a college fund for his son. He plans on depositing $1500 at the end of each year into an account paying 4.5% per year, compounded annually. How much will be in the account after 18 years?

   b. How much will be in the account after 18 years if Matt deposits $1500 at the end of each year into an account paying 3.5% per year, compounded quarterly?

   a. The payments form an ordinary annuity with $R = 1500$, $n = 18$, and $i = 0.045$.

   $$S = R\left[\frac{(1+i)^n - 1}{i}\right]$$

   $$S = 1500\left[\frac{(1+0.045)^{18} - 1}{0.045}\right]$$

   $$\approx 40,282.63$$

   b. The payments form an ordinary annuity with $R = 1500$, $n = 4(18)$, and $i = \frac{0.035}{4}$.

   $$S = R\left[\frac{(1+i)^n - 1}{i}\right]$$

   $$S = 1500\left[\frac{\left(1+\frac{0.035}{4}\right)^{4 \cdot 18} - 1}{\frac{0.035}{4}}\right]$$

   $$\approx 149,566.71$$

Now Try:

7. To save for retirement, Sara decides to deposit $2000 into an IRA at the end of each year. How much will be in the account at the end of 30 years if it is invested at

   a. 10% per year compounded annually? _____

   b. 5% per year compounded semiannually? _____

Review these examples for Objective 6:

8. Evaluate the sum of the terms of the infinite geometric sequence with $a_1 = -3$ and $r = \frac{2}{3}$.

   $$S = \frac{a_1}{1-r} \quad \text{Infinite sum formula}$$

   $$= \frac{-3}{1-\frac{2}{3}} \quad a_1 = -3, \, r = \frac{2}{3}$$

   $$= \frac{-3}{\frac{1}{3}}$$

   $$= -9$$

Now Try:

8. Evaluate the sum of the terms of the infinite geometric sequence with $a_1 = 5$ and $r = -\frac{1}{5}$.

   _____

9. Evaluate $\displaystyle\sum_{i=1}^{\infty} 3\left(\frac{1}{4}\right)^i$.

This is the infinite geometric series with

$a_1 = \frac{3}{4}$ and $r = \frac{1}{4}$. Since $|r| < 1$, find the sum

using the formula $S = \dfrac{a_1}{1-r}$.

$$S = \frac{\frac{3}{4}}{1-\frac{1}{4}} = 1$$

9. Evaluate $\displaystyle\sum_{i=1}^{\infty} 3^i$.

_____

Objective 1 Find the common ratio of a geometric sequence.

For extra help, see Example 1 on page 692 of your text, the Section Lecture video for Section 12.3, and Exercise Solution Clip 1.

Find the common ratio of the geometric sequence.

1. $10,\ 10\sqrt{2},\ 20,\ 20\sqrt{2},\ 40,\ \ldots$

1. _____

2. $-\dfrac{1}{2},\ \dfrac{1}{4},\ -\dfrac{1}{8},\ \dfrac{1}{16},\ -\dfrac{1}{32},\ \ldots$

2. _____

Objective 2 Finding the general term of a geometric sequence.

For extra help, see Example 2 on page 692 of your text and the Section Lecture video for Section 12.3.

Determine the general term of the geometric sequence.

3. $6, 12, 24, 48, 96, \ldots$

3. _____

4. $-\dfrac{2}{3},\ \dfrac{2}{9},\ -\dfrac{2}{27},\ \dfrac{2}{81},\ \ldots$

4. _____

5. $-\dfrac{4}{5},\ -\dfrac{4}{25},\ -\dfrac{4}{125},\ -\dfrac{4}{625},\ \ldots$

5. _____

Objective 3 Find any specified term of a geometric sequence.
For extra help, see Examples 3 and 4 on page 693 of your text, the Section Lecture video for
Section 12.3, and Exercise Solution Clips 15 and 21.

Evaluate the indicated term for each geometric sequence.

6. $a_1 = -4$, $r = 2$; a_7

6. _____

7. $a_1 = \frac{2}{3}$, $r = -3$; a_6

7. _____

8. Write the first five terms of the geometric sequence
    whose first term is $\frac{1}{2}$ and whose common ratio is 5.

8. _____

Objective 4 Find the sum of a specified number of terms of a geometric sequence.
For extra help, see Examples 5 and 6 on page 694 of your text, the Section Lecture video for
Section 12.3, and Exercise Solution Clips 25 and 29.

9. Evaluate the sum of the first seven terms of the
    geometric sequence with first term 2 and common ratio
    3.

9. _____

Evaluate each sum.

10. $\sum_{i=1}^{12} 2^i$

10. _____

11. $\sum_{i=1}^{7} 3(-2)^i$

11. _____

Name: Date:
Instructor: Section:

Objective 5 Apply the formula for the future value of an ordinary annuity.
For extra help, see Example 7 on page 695 of your text and the Section Lecture video for
Section 12.3.

Solve each problem.

12. Julio is a professional wrestler who believes that his
    wrestling career will last ten years. To prepare for
    retirement, he deposits $14,000 at the end of each year
    for ten years into an account paying 7% compounded
    annually. How much will he have on deposit after ten
    years?

12. _____

13. Gordon wants to buy a new car. If he puts $1,100 at the
    end of each year into an account paying 4.9%
    compounded annually for 3 years, how much will he
    have to spend on a car?

13. _____

Objective 6 Find the sum of an infinite number of terms of certain geometric sequences.
For extra help, see Examples 8 and 9 on page 697 of your text, the Section Lecture video for
Section 12.3, and Exercise Solution Clip 37.

Find the sum, if possible, of the infinite geometric sequence.

14. $a_1 = 2, \ r = -\frac{1}{3}$

14. _____

15. $\displaystyle\sum_{i=1}^{\infty} -5\left(-\frac{5}{3}\right)^{i}$

15. _____

Chapter 12 SEQUENCES AND SERIES

12.4 The Binomial Theorem

Learning Objectives
1 Expand a binomial raised to a power.
2 Find any specified term of the expansion of a binomial.

Key Terms

Use the vocabulary terms listed below to complete each statement in exercises 1–2.

Pascal's triangle   **binomial theorem (general binomial expansion)**

1. The _____ is a formula used to expand a binomial raised to a power.

2. Writing the coefficients of the terms of binomial expansions in a triangular pattern gives _____.

Guided Examples

Review these examples for Objective 1:

1. Evaluate 8!

    $8! = 8 \cdot 7 \cdot 6 \cdot 5 \cdot 4 \cdot 3 \cdot 2 \cdot 1 = 40,320$

2. Find the value of each expression.

    a. $\dfrac{8!}{6!2!}$ b. $\dfrac{8!}{4!4!}$

    c. $\dfrac{8!}{8!0!}$ d. $\dfrac{8!}{7!1!}$

    a. $\dfrac{8!}{6!2!} = \dfrac{8 \cdot 7 \cdot 6 \cdot 5 \cdot 4 \cdot 3 \cdot 2 \cdot 1}{(6 \cdot 5 \cdot 4 \cdot 3 \cdot 2 \cdot 1)(2 \cdot 1)} = \dfrac{8 \cdot 7}{2 \cdot 1} = 28$

    b. $\dfrac{8!}{4!4!} = \dfrac{8 \cdot 7 \cdot 6 \cdot 5 \cdot 4 \cdot 3 \cdot 2 \cdot 1}{(4 \cdot 3 \cdot 2 \cdot 1)(4 \cdot 3 \cdot 2 \cdot 1)}$

    $= \dfrac{8 \cdot 7 \cdot 6 \cdot 5}{4 \cdot 3 \cdot 2 \cdot 1} = 70$

    c. $\dfrac{8!}{8!0!} = \dfrac{8 \cdot 7 \cdot 6 \cdot 5 \cdot 4 \cdot 3 \cdot 2 \cdot 1}{(8 \cdot 7 \cdot 6 \cdot 5 \cdot 4 \cdot 3 \cdot 2 \cdot 1)(1)} = 1$

    d. $\dfrac{8!}{7!1!} = \dfrac{8 \cdot 7 \cdot 6 \cdot 5 \cdot 4 \cdot 3 \cdot 2 \cdot 1}{(7 \cdot 6 \cdot 5 \cdot 4 \cdot 3 \cdot 2 \cdot 1)(2 \cdot 1)} = \dfrac{8}{1} = 8$

Now Try:

1. Evaluate 10!

    _____

2. Find the value of each expression.

    a. $\dfrac{9!}{9!0!}$ _____

    b. $\dfrac{9!}{7!2!}$ _____

    c. $\dfrac{9!}{8!1!}$ _____

    d. $\dfrac{9!}{5!4!}$ _____

3. Evaluate $_9C_6$.

$$_9C_6 = \frac{9!}{6!(9-6)!} = \frac{9!}{6!3!}$$
$$= \frac{9 \cdot 8 \cdot 7 \cdot 6 \cdot 5 \cdot 4 \cdot 3 \cdot 2 \cdot 1}{(6 \cdot 5 \cdot 4 \cdot 3 \cdot 2 \cdot 1)(3 \cdot 2 \cdot 1)}$$
$$= 84$$

3. Evaluate $_{10}C_8$.

_____

4. Expand $(2x+5y)^4$.

$$(2x+5y)^4$$
$$= (2x)^4 + \frac{4!}{1!3!}(2x)^3(5y) + \frac{4!}{2!2!}(2x)^2(5y)^2$$
$$+ \frac{4!}{3!1!}(2x)(5y)^3 + (5y)^4$$
$$= (2x)^4 + 4(2x)^3(5y) + 6(2x)^2(5y)^2$$
$$+ 4(2x)(5y)^3 + (5y)^4$$
$$= 16x^4 + 160x^3y + 600x^2y^2$$
$$+ 1000xy^3 + 625y^4$$

4. Expand $(3r-2s)^4$.

_____

5. Expand $\left(\frac{1}{2} - a^2\right)^4$.

$$\left(\frac{1}{2} - a^2\right)^4$$
$$= \left(\frac{1}{2}\right)^4 + \frac{4!}{1!3!}\left(\frac{1}{2}\right)^3(-a^2) + \frac{4!}{2!2!}\left(\frac{1}{2}\right)^2(-a^2)^2$$
$$+ \frac{4!}{3!1!}\left(\frac{1}{2}\right)(-a^2)^3 + (-a^2)^4$$
$$= \left(\frac{1}{2}\right)^4 + 4\left(\frac{1}{2}\right)^3(-a^2) + 6\left(\frac{1}{2}\right)^2(-a^2)^2$$
$$+ 4\left(\frac{1}{2}\right)(-a^2)^3 + (-a^2)^4$$
$$= a^8 - 2a^6 + \frac{3}{2}a^4 - \frac{1}{2}a^2 + \frac{1}{16}$$

5. Expand $\left(2 - \frac{x}{2}\right)^5$.

_____

Review this example for Objective 2:

6. Find the sixth term of the expansion of $(a - 2b)^9$.

In the sixth term of $(x - y)^9$, x has an exponent of $9 - 5 = 4$ and y has an exponent of $9 - 4 = 5$. Letting $n = 9$, $x = a$, $y = -2b$, and $r = 6$, we have

$$\frac{9!}{4!5!}a^4(-2b)^5 = \frac{9 \cdot 8 \cdot 7 \cdot 6}{4 \cdot 3 \cdot 2 \cdot 1}a^4(-2b)^5$$
$$= 126a^4\left(-32b^5\right)$$
$$= -4032a^4b^5$$

Now Try:

6. Find the sixth term of the expansion of $\left(m^3 - 3r\right)^7$.

_____

Objective 1 Expand a binomial raised to a power.
For extra help, see Examples 1–5 on pages 702–704 of your text, the Section Lecture video for Section 12.4, and Exercise Solution Clips 1, 5, 11, 19, and 21.

Evaluate each expression.

1. $7!$

1. _____

2. $\dfrac{6!}{1!5!}$

2. _____

3. $\dfrac{10!}{0!10!}$

3. _____

4. $_9C_4$

4. _____

5. $_{15}C_3$

5. _____

Use the binomial theorem to expand each expression.

6. $\left(8m^3 - 3n^2\right)^3$

6. _____

7. $\left(\frac{1}{2}z - \frac{1}{3}x\right)^3$

7. _____

8. $\left(5a - 2b^2\right)^4$

8. _____

9. $(2y + 3z)^5$

9. _____

10. $\left(y^2 - x^2\right)^6$

10. _____

Objective 2 Find any specified term of the expansion of a binomial.
For extra help, see Example 6 on page 705 of your text, the Section Lecture video for
Section 12.4, and Exercise Solution Clip 33.

Find the indicated term of each binomial expansion.

11. Third term of $(m - 3)^5$

11. _____

12. Fourth term of $(4x - 3y)^5$

12. _____

13. Fifth term of $(3r - 1)^6$

13. _____

14. Sixth term of $\left(\dfrac{x}{2} + 2\right)^7$

15. Eighth term of $\left(3k - p^2\right)^9$

Chapter 1
Review of the Real Number System

1.1   Basic Concepts
Key Terms

1. variable   2.   graph

3. elements

4. additive inverses

5. infinite set   6.   equation

7. interval   8.   inequality

9. number line   10.   finite set

11. coordinate

12. set-builder notation

13. signed number

14. absolute value

15. three-part inequality

16. empty set   17.   set

18. interval notation

Now Try

1. $\{1, 3, 5, 7\}$

2. Answers may vary. One answer is $\{x \mid x$ is a multiple of 2 between 0 and 8, inclusive$\}$.

3.a. $\{0\}$ b. $\{-9, 0\}$

c. $\left\{\sqrt{3}, \pi\right\}$

d. $\left\{-9, -\frac{4}{3}, 0, 0.\overline{3}, \sqrt{3}, \pi\right\}$

4.a. True

b. False; no irrational numbers are rational numbers.

5.a. 2   b. -2

c. -4

6. Philadelphia from 2000–2009

7.a. True   b.   False

8.a. False   b.   False

9. $(-\infty, -10)$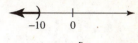

10. $[1, \infty)$

11. $[-3, 0]$

Objective 1

1. True

Objective 2

3.

Objective 3

5.a False   b.   False

Objective 5

7. 5

Objective 6

9.a False   b.   True

11. False

Objective 7

13. $[3, \infty)$;

15.a. $[-2, 4]$ b. $[-3, 5)$

1.2   Operations on Real Numbers
Key Terms

1. product   2.   sum

3. quotient   4.   reciprocals

5. difference

Now Try

1.a. -9 b. -20.8

c. $-\frac{15}{22}$

2.a. -5 b. -2.9

c. $-\frac{2}{15}$

3.a. 46　　　　b.　−16.1

c.　$-\frac{2}{5}$

4.a. −13　　　b.　−21

5. 5

6.a 112　　　b.　−13.12

c.　$-\frac{7}{12}$

7.a. 6　　　　b.　$-\frac{1}{6}$

c. −3

Objective 1

1.　−12.4　　　3.　$-\frac{4}{3}$

Objective 2

5.　−3.78

Objective 3

7.　7　　　　9.　12

Objective 4

11.　105

Objective 5

13.　$-\frac{1}{4}$　　　15.　−3

1.3　Exponents, Roots, and Order of Operations

Key Terms

1. principal square root

2. factors

3. algebraic expression

4. base

5. exponential expression

6. exponent　　　7.　square root

8. negative square root

Now Try

1.a. $\left(\frac{2}{3}\right)^5$　　　b.　x^6

2.a. $\frac{64}{27}$　　　b.　$\frac{256}{81}$

c. $-\frac{256}{81}$

3.a. 100　　　b.　$-\frac{2}{7}$

c. not a real number

4.　−31　　　5.　47

6.　$\frac{16}{3}$

7.a. $\frac{4}{3}$　　　b.　$-\frac{2}{127}$

Objective 1

1.　−8　　　3.　0.25

Objective 2

5.　not a real number

Objective 3

7.　$-\frac{1}{2}$　　　9.　20

11.　−1

Objective 4

13.　8　　　15.　$-\frac{34}{13}$

1.4　Properties of Real Numbers

Key Terms

1. combine like terms　2.　unlike terms

3. identity element for multiplication

4. like terms

5. numerical coefficient

6. identity element for addition

7. term

Now Try

1.a. $45x - 72y$　　　b.　$-9y$

c. cannot be rewritten

2.a. $-10x$　　　b.　$2a + 3b - c$

3.　$-4x - 30$

4.a. −12   b. $5p + 18$

Objective 1

1. $-3z + 21$

3. cannot be simplified

Objective 2

5. $-12y$

Objective 3

7. −7   9. 1

Objective 4

11. $5w$ 13. $-3d - 8$

Objective 5

15. 0

Chapter 2
Linear Equations, Inequalities, And Applications

2.1 Linear Equations in One Variable

Key Terms

1. linear equation in one variable

2. conditional equation

3. contradiction   4. solution set

5. equivalent equations

6. identity   7. solution

Now Try

1.a. equation   b. expression

2. $-\frac{8}{3}$ 3. $\left\{-\frac{9}{7}\right\}$

4. {3}   5. {300}

6.a. {all real numbers}; identity

b. $\varnothing$; contradiction

c. {−3}; conditional equation

Objective 1

1. equation

Objective 2

3. No

Objective 3

5. $\left\{-\frac{2}{7}\right\}$

Objective 4

7. $\{0\}$

Objective 5

9. {3}   11. {10}

Objective 6

13. Contradiction; $\varnothing$

15. Identity; {all real numbers}

2.2 Formulas and Percent

Key Terms

1. formula   2. percent

3. mathematical model

Now Try

1. $w = \dfrac{V}{lh}$ 2. $h = \dfrac{2A}{B+b}$

3. $B = \dfrac{2A}{h} - b$ or $B = \dfrac{2A - bh}{h}$

4. $y = \dfrac{-3x + 4}{5}$ 5. 52 mph

6.a. 20%   b. $264.50

7. $20.625 million or $20,625,000

8.a. 18.5%   b. 15.7%

Objective 1

1. $B = \dfrac{3V}{h}$

3. $y = -\dfrac{5x}{2}$ or $y = -\dfrac{5}{2}x$

Objective 2

5. 6 ft   7. 8%

Objective 3

9. 3.5%

11. $19.880 million or $19,880,000

Objective 4

13. 6.5%   15. 25%

2.3 Applications of Linear Equations

Key Terms

1. sum, increased by, more than

2. product, double, of

3. quotient, per, ratio

4. difference, less than, decreased by

Now Try

1.a. $x(x-6) = 18$

b. $\dfrac{4+x}{9} = x+5$

c. $2x + 50 = x - 6$

2.a. expression; $5x - 13$

b. equation; $\dfrac{27}{16}$

3. length: 25 m; width: 6 m

4. 4860 walleyes, 6690 bass

5. $480

6. $4500 at 15%; $9000 at 12%

7. 18 lb   8. 15 ml

Objective 1, Objective 2

1. $6x = 7 + 5x$ 3. $\dfrac{x}{x-3} = 17$

Objective 3

5. equation; $\dfrac{17}{5}$

Objective 4

7. *Step 1*: Read the problem. What must be found? The number of votes received by each candidate.

What is given? Bush received five more votes than Gore. They received a combined total of 537 votes.

Step 2: Assign a variable. Let g = votes received by Gore. Then $g + 5$ = votes received by Bush.

Step 3: Write an equation.

$$\underset{\text{Gore}}{\text{Votes for}} + \underset{\text{Bush}}{\text{Votes for}} = \text{total votes.}$$

$$g \quad + \quad (g+5) \ = \ 537$$

Step 4: Solve the equation.

$$g + (g+5) = 537$$
$$2g + 5 = 537$$
$$2g + 5 - 5 = 537 - 5$$
$$2g = 532$$
$$\frac{2g}{2} = \frac{532}{2}$$
$$g = 266$$

Step 5: State the answer. Al Gore received 266 total electoral votes. George Bush received $g + 5 = 271$ total electoral votes.

Step 6: Check the answer by substituting these totals into the words of the original problem.

Objective 5

9. $42

Objective 6

11. $1600 at 8%; $3500 at 12%

Objective 7

13. 250 L

15. 6 L

2.4 Further Applications of Linear Equations

Key Terms

1. uniform motion

2. denomination

Now Try

1. 32 quarters, 18 nickels

2. $4\frac{1}{2}$ hr   3. 30 mi

4. 55°, 25°, 100°

Objective 1

1. 9 $5-bills; 5 $10-bills

3. 35 pennies, 29 nickels, 30 dimes

5. 50 44¢-stamps, 50 61¢-stamps, and 10 $4.95-stamps

Objective 2

7. 60 mph

9. plane A: 400 mph
   plane B: 360 mph

Objective 3

11. 20°, 115°, 45°

13. 60°, 30°, 90°

15. 40°, 58°, 82°

2.5 Linear Inequalities in One Variable

Key Terms

1. linear inequality in one variable

2. equivalent inequalities

3. three-part inequality

4. inequality

Now Try

1. $(9, \infty)$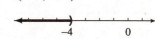

2. $(-\infty, -5)$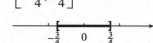

3.a. $(3, \infty)$

   b. $(-\infty, -10]$

4. $[19, \infty)$

5. $(1, \infty)$

6. $(5, 9)$

7. $[-8, 1]$

8. 55 students   9. 74

Objective 1

1. $[-2, \infty)$ 3. $(-\infty, -3)$

Objective 2

5. $(-\infty, -4)$

Objective 3

7. $(-8, 6)$

9. $\left[-\frac{3}{4}, \frac{3}{4}\right]$

Objective 4

11. All numbers less than or equal to 1.

13. The largest possible value is 10 feet.

15. The lowest score he can earn is 84 points

2.6 Set Operations and Compound Inequalities

Key Terms

1. union   2. intersection

3. compound inequality

Now Try

1. $\varnothing$

2. $[-2, 8]$

3. $[-5, 0)$

4. $\varnothing$

5. $\{-6, -5, -4, -3, -2, -1\}$

6. $(-\infty, -2) \cup [2, \infty)$

7. $[7, \infty)$

8. $(-\infty, \infty)$

9.a. {New Jersey, Pennsylvania, Ohio}

b. {New York, Pennsylvania, New Jersey, Maryland, Virginia, Ohio, Indiana, Kentucky}

Objective 1

1. $\{1, 3, 5\}$

3. $\varnothing$

Objective 2

5. $(-\infty, -5)$

7. $\varnothing$

Objective 3

9. $\{0, 1, 2, 3, 4, 5, 6, 7, 8, 9, 10\}$

Objective 4

11. $(-\infty, -4] \cup (4, \infty)$

13. $(-\infty, -1) \cup (5, \infty)$

15.a. {Indiana, Kentucky, Virginia}

b. {Delaware, Indiana, Kentucky, Maryland, New Jersey, New York, Pennsylvania, Ohio, Virginia, West Virginia}

2.7 Absolute Value Equations and Inequalities

Key Terms

1. absolute value equation

2. absolute value inequality

Now Try

1. $\{-4, 10\}$

2. $(-\infty, 4) \cup (8, \infty)$

3. $[-7, -3]$

4. $\left\{-\frac{5}{4}, -\frac{1}{4}\right\}$

5.a. $\left(-\infty, -\frac{3}{2}\right] \cup [2, \infty)$

b. $\left[-\frac{3}{2}, 2\right]$

6. $\left\{-\frac{3}{2}, 2\right\}$

7.a. $\varnothing$　　　　b. $\{-3\}$

8.a. $(-\infty, \infty)$　　b. $\varnothing$

c. $\{3\}$

Objective 1

1.

Objective 2

3. $\left\{\frac{7}{2}, -\frac{13}{2}\right\}$

Objective 3

5. $(-13, 3)$

7. $\left[-\frac{1}{4}, \frac{3}{4}\right]$

Objective 4

9.   ∅

Objective 5

11. $\left\{6, -\frac{2}{5}\right\}$ 13. $\{-1\}$

Objective 6

15. $(-\infty, \infty)$

Chapter 3
GRAPHS, LINEAR EQUATIONS, AND FUNCTIONS

3.1   The Rectangular Coordinate System

Key Terms

1.   rectangular coordinate system

2.   quadrant   3.   origin

4. x-intercept   5.   coordinates

6. x-axis   7.   ordered pair

8. y-intercept

9.   linear equation in two variables; standard form

10. y-axis

11. graph of an equation

12. plot

13. first-degree equation

14. components   15.   standard form

Now Try

1.

x	y
0	$-\frac{1}{2}$
1	0
3	1
-1	-1

2. $(5, 0), (0, -3)$

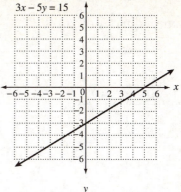

3.
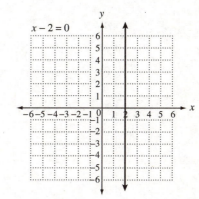

4.
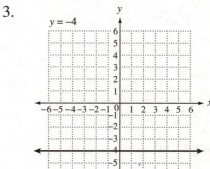

5.
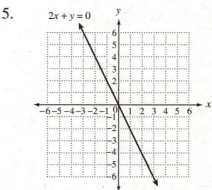

6. $(-4, 0)$

Objective 1

1.a. 220; second quarter of 2009

b. about 190; second quarter of 2006

c. fourth quarter of 2009

Objective 3

3.

x	y
0	−2
5	0
−5	−4
$-\frac{5}{2}$	−3

5.

x	y
0	5
−2	0
−3	$\frac{25}{2}$
−4	−5

Objectives 4 and 5

7. $\left(-\frac{8}{3},\ 0\right),\ \left(0,\ -\frac{8}{7}\right)$

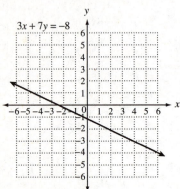

9. (−4, 0), (0, −4)

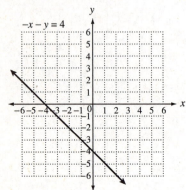

Objective 6

11.

Objective 7

13. $\left(\frac{3}{2}, \frac{5}{2}\right)$ 　　15. $\left(\frac{11}{4}, -1\right)$

3.2　The Slope of a Line

Key Terms

1. run　　　2. rise

3. slope

Now Try

1. $-\frac{1}{2}$ 　　　2. $\frac{4}{7}$

3.a. 0　　　b. undefined

4. $\frac{9}{4}$

5.

6. no　　　7. yes

8. neither　　9. $2.06

Objective 1

1. −8　　　3. 0

Objective 2

5. $\frac{2}{7}$

Objective 3

7.
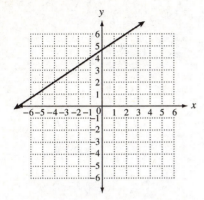

9.
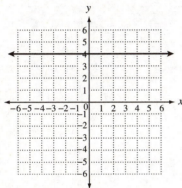

Objective 4

11. Parallel

Objective 5

13. 950

15.a. −350 students/yr

   b. The enrollment was decreasing.

3.3 Linear Equations in Two Variables

Key Terms

1. point-slope form

2. slope-intercept form

Now Try

1. $y = \frac{4}{5}x - 2$

2.
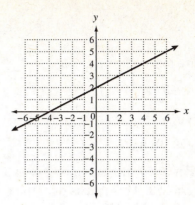

3. $y = -\frac{3}{2}x - 1$

4.a. $y = 2$ b. $x = -4$

5. $3x + 2y = 23$

6.a. $y = \frac{4}{3}x + 6$ b. $y = 2x + 8$

7. $y = \frac{208}{3}x + \frac{532}{3}$

   2007 expenditure: \$593.33

Objective 1

1. $2x - y = 5$

Objective 2

3. $x + y = -4$

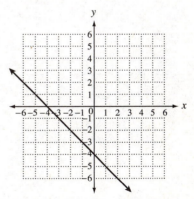

Objective 3

5. $4x + y = -3$

Objective 4

7. $x = 3$

Objective 5

9.a. $y = x + 5$ b. $x - y = -5$

11.a. $y = -\frac{2}{3}x + 1$ b. $2x + 3y = 3$

Objective 6

13. $x - 5y = 7$

Objective 7

15.a. $y = 19x + 25$

   b. For $x = 7$, the equation gives $y = 158$. This is less than the actual value of 160.

3.4 Linear Inequalities in Two Variables

Key Terms

1. linear inequality in two variables

2. test point

3. intersection of two or more inequalities

4. union of two inequalities

5. boundary line

Now Try

1.

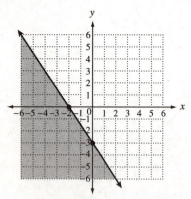

2.

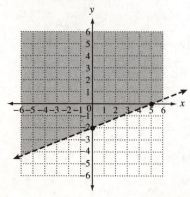

3.

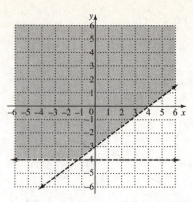

4.

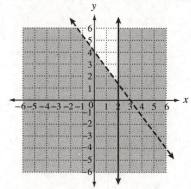

Objective 1

1.

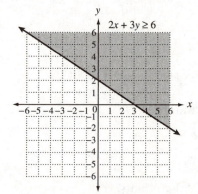

3.

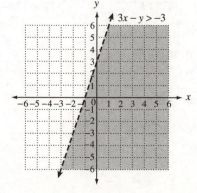

5.

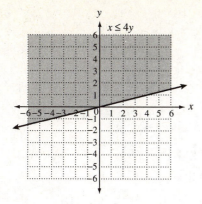

$x \le 4y$

Objective 2

7.

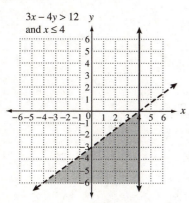

$3x - 4y > 12$
and $x \le 4$

9.

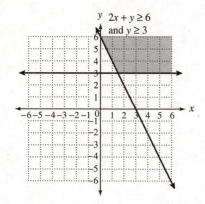

$2x + y \ge 6$
and $y \ge 3$

Objective 3

11.

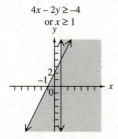

$4x - 2y \ge -4$
or $x \ge 1$

13.

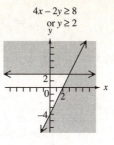

$4x - 2y \ge 8$
or $y \ge 2$

15.

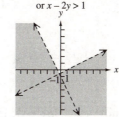

$2x + y < -1$
or $x - 2y > 1$

3.5   Introduction to Relations and Functions

Key Terms

1.   dependent variable

2.   range   3.   function

4.   independent variable

5.   domain   6.   relation

Now Try

1.a. not a function b.   function

2.a. function; domain: $\{1, 2, 3, 4\}$
      range: $\{4\}$

  b. function; domain: $\{1, 2, 3, 4\}$
      range: $\{2, 4, 8\}$

  c. not a function; domain: $\{2, 3, 4\}$
      range: $\{-2, -1, 1\}$

3.   domain: $[-4, \infty)$; range: $(-\infty, \infty)$

4.   not a function

5.a. function; $(-\infty, \infty)$

  b. function; $(-\infty, 0]$

  c. not a function; $[1, \infty)$

d. function; $\left(-\infty, -\frac{1}{3}\right) \cup \left(-\frac{1}{3}, \infty\right)$

e. not a function; $(-\infty, \infty)$

Objective 1

1. grades

Objective 2

3. not a function

Objective 3

5. function; domain: $\{-1, -2, 0\}$;
   range: $\{1, 2, 0\}$

7. function; domain: {A, B, C, D, E}
   range: {V, W, X, Z}

9. function; domain: {1, 2, 3, 4}
   range: {2}

Objective 4

11. not a function

13. function; $(-\infty, \infty)$

15. not a function; $(-\infty, \infty)$

3.6 Function Notation and Linear Functions

Key Terms

1. linear function

2. function notation

3. constant function

Now Try

1. -19

2.a. 1   b. $2s^2 + s - 5$

3. $5a + 17$

4.a. 2   b. 0

5.a. 3   b. -3

6. $f(x) = -\frac{1}{4}x - 1;\ f(3) = -\frac{7}{4};$
   $f(w) = -\frac{1}{4}w - 1$

7.

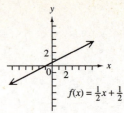

$f(x) = \frac{1}{2}x + \frac{1}{2}$

domain: $(-\infty, \infty)$

range: $(-\infty, \infty)$

Objective 1

1.a. 6   b. 18

3.a. -1 b. 3

5.a. -13 b. -7

  c. $-3x - 7$

7.a. 1   b. 3

  c. $|-2x + 3|$

9.a. $\frac{4}{5}$ b. 4

  c. $\frac{4}{1 + x^2}$

Objective 2

11. $f(x) = 2x + 2$

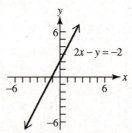

$2x - y = -2$

domain: $(-\infty, \infty)$

range: $(-\infty, \infty)$

13. $f(x) = -\dfrac{3}{2}x - 3$

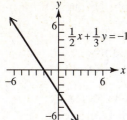

domain: $(-\infty, \infty)$

range: $(-\infty, \infty)$

15.

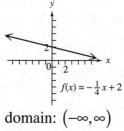

domain: $(-\infty, \infty)$

range: $(-\infty, \infty)$

Chapter 4
SYSTEMS OF LINEAR EQUATIONS

4.1   Systems of Linear Equations
      in Two Variables

Key Terms

1.   system of linear equations

2.   consistent system

3.   elimination method

4.   consistent system

5.   solution set of a linear system

6.   substitution method

7.   inconsistent system

8.   dependent equations

9.   independent equations

Now Try

1.a. yes   b.   no

2. $\{(-1, -2)\}$ 3. $\{(-1, -7)\}$

4. $\left\{\left(3, -\dfrac{2}{3}\right)\right\}$ 5. $\{(-12, 0)\}$

6. $\{(2, -4)\}$ 7. $\left(-\dfrac{13}{8}, \dfrac{11}{8}\right)$

8. $\{(x, y) \mid 6x - 5y = -1\}$

9. $\varnothing$

10.a. Both are $y = \dfrac{1}{2}x - \dfrac{5}{2}$.
       infinitely many solutions

   b. $y = 2x + 1$, $y = 2x + 7$
       parallel lines; no solution

Objective 1

1.a. not a solution   b.   solution

Objective 2

3.

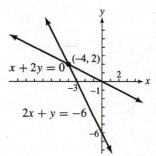

Objective 3

5. $\{(3, 1)\}$ 7. $\{(1, 3)\}$

Objective 4

9. $\{(-1, 2)\}$ 11. $\{(-5, 2)\}$

Objective 5

13. $\{(x, y) \mid 5x - 2y = 4\}$

15.a. $y = \dfrac{4}{3}x - 2$; $y = \dfrac{4}{3}x - \dfrac{5}{3}$
       no solution

   b. Both are $y = -\dfrac{4}{3}x + 4$.
       infinitely many solutions

4.2 Systems of Linear Equations in Three Variables

Key Terms

1. line in common
2. false statement
3. ordered triple
4. coincide
5. common point
6. no points common

Now Try

1. $\left\{(-3,\ 1,\ 2)\right\}$ 2. $\left\{(2,\ -5,\ 3)\right\}$
3. $\varnothing$
4. $\left\{(x,\ y,\ z)\mid x-5y+2z=0\right\}$
5. $\varnothing$

Objective 2

1. $\left\{(-1,\ 2,\ 1)\right\}$ 3. $\left\{(3,\ 1,\ -2)\right\}$
5. $\left\{(3,\ -6,\ 1)\right\}$

Objective 3

7. $\left\{(3,\ 5,\ 0)\right\}$ 9. $\left\{(4,\ 2,\ -1)\right\}$

Objective 4

11. $\varnothing$; inconsistent system
13. $\left\{(x,\ y,\ z)\mid -x+5y-2z=3\right\}$
    dependent equations
15. $\varnothing$; inconsistent system

4.3 Applications of Systems of Linear Equations

Key Terms

1. elimination method
2. substitution

Now Try

1. length: 28.125 in; width: 9.375 in.
2. 12 $5-bills; 20 $10-bills
3. 80 oz of 20% solution
   40 oz of 50% solution
4. Rick, 48 mph; Hilary, 78 mph
5. 20 lb of $4 candy
   30 lb of $6 candy
   50 lb of $10 candy
6. 1 capsule of Generic I
   3 capsules of Generic II
   1 capsule of Generic III

Objective 1

1. length: 17 ft; width: 12 ft
3. square: 16 cm; triangle: 12 cm

Objective 2

5. $10,000 at 4%; $5000 at 6%

Objective 3

7. $\dfrac{10}{9}$ L
9. 28 L of 75% solution
   42 L of 55% solution

Objective 4

11. 5.5 hr

Objective 5

13. $40,000 at 5%; $10,000 at 6%;
    $30,000 at 7%
15. arrangement I: 2
    arrangement II: 6
    arrangement III: 4

4.4 Solving Systems of Linear Equations by Matrix Methods

Key Terms

1. row
2. augmented matrix
3. matrix   4. row operations
5. elements of a matrix
6. square matrix
7. row echelon form

8. column

Now Try

1. $\{(4, -2)\}$ 2. $\{(1, -1, -1)\}$

3.a. $\varnothing$

b. $\{(x, y) \mid 2x + y = 10\}$

Objective 1

1.a. 2×3 b. 3×4

c. 2×2 d. 3×1

Objective 2

3.a. $\left[\begin{array}{cc|c} 1 & -1 & -4 \\ 3 & 1 & 1 \end{array}\right]$ b. $\left[\begin{array}{cc|c} 1 & 0 & -6 \\ 0 & 1 & 2 \end{array}\right]$

c. $\left[\begin{array}{ccc|c} 2 & 1 & -1 & 3 \\ 1 & -2 & 1 & 5 \\ 5 & -1 & 2 & 7 \end{array}\right]$

d. $\left[\begin{array}{ccc|c} 4 & -7 & 1 & 1 \\ 2 & 3 & -5 & -2 \\ 6 & -1 & 8 & 5 \end{array}\right]$

Objective 3

5. $\{(-2, 0)\}$ 7. $\left\{\left(\dfrac{1}{4}, 0\right)\right\}$

Objective 4

9. $\{(1, -2, 0)\}$ 11. $\{(2, 1, 2)\}$

Objective 5

13. $\{(x, y) \mid 2x - 3y = 61\}$

15. $\{(x, y) \mid 3x - y + 4z = 6\}$

Chapter 5
EXPONENTS, POLYNOMIALS, AND POLYNOMIAL FUNCTIONS

5.1 Integer Exponents and Scientific Notation

Key Terms

1. scientific notation

2. base; exponent

3. quotient rule for exponents

4. power rule for exponents

5. exponential (power)

6. product rule for exponents

Now Try

1.a. The product rule does not apply because the bases are not the same.

b. a^6 c. $20x^6 z^5$

2.a. 1 b. 1

c. -1 d. $1, x \neq 0$

e. 0

3.a. $\dfrac{1}{81}$ b. $\dfrac{1}{16a^2}$

c. $-\dfrac{3}{s^3}$ d. $\dfrac{1}{6}$

4.a. 16 b. $\dfrac{3}{8}$

5.a. k^3 b. $3^5 = 243$

c. This quotient rule does not apply because the bases are different.

6.a. $27x^{12}$ b. $\dfrac{64x^{18}}{z^{24}}$

7.a. $\dfrac{512}{27}$ b. $\dfrac{z^{16}}{16x^{12}}$

8.a. $\frac{3}{5}$ b. $\frac{8w^7}{9y^5}$

c. $\frac{25s^{16}}{9r^8t^8}$

9.a. 2.3651×10^{10} b. -4.7×10^{-4}

10.a. $72{,}000{,}000$ b. -0.000045

11. $10{,}000$

12. 9.5×10^{20} meters

Objective 1

1. 2^8 or 256

Objective 2

3. $-10p^2r^2$

Objective 3

5. r^7

Objective 4

7. $125m^{15}$

9. $\frac{1}{3^8 m^6}$ or $\frac{1}{6561m^6}$

Objective 5

11. $\frac{x^2 y}{72}$

Objective 6

13. 3.82×10^{-5}

15. 3.0×10^{-6} or 0.000003

5.2   Adding and Subtracting Polynomials

Key Terms

1. degree of a term

2. descending powers

3. term

4. trinomial   5. polynomial

6. numerical coefficient

7. monomial

8. degree of a polynomial

9. binomial

10. algebraic expression

11. negative of a polynomial

12. polynomial in x

13. leading term; leading coefficient

14. like terms

Now Try

1. $8x^4 - 2x^3 + x^2 - x - 1$
   leading term: $8x^4$
   leading coefficient: 8

2. none of these; degree 5

3.a. $-4c^3 + 12c^2 - 6c$

b. $-x^4 + 2x^3 - 7x^2 - 4$

4.a. $5m^3 - 2m^2 - 4m + 4$

b. $4x^2 + 6x - 18$

5.a. $-4m^2n - 4mn - 6m - 2n$

b. $4m^2 - 2$

Objective 1

1. $-3x^5 - x^3 + 2x^2 - 4$
   leading term: $-3x^5$
   leading coefficient: -3

3. none of these; 3

5. monomial; 0

Objective 2

7. $-18z^2 + 9z - 3$

9. $10x^2 + 3x + 4$

11. $5p + 19$

13. $-y^4 - 2y^3 + 7y^2 - 3y$

15. $-a^4 - 9a^2 + 14$

5.3 Polynomial Functions, Graphs, and Composition

Key Terms

1. polynomial function
2. composition of functions
3. cubing function
4. identity function
5. squaring function

Now Try

1. 1 2. $10,643.54

3.a. $3x^2 - 6x + 21$ b. $9x^2 - 8x + 3$

4.a. −23 b. 66

5. −15

6.a. $(f \circ g)(x) = -3x^2 + 12$

b. $(g \circ f)(x) = (-3x - 3)^2 - 5$
$$= 9x^2 + 18x + 4$$

7.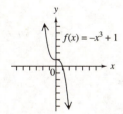

domain: $(-\infty, \infty)$;
range: $(-\infty, \infty)$

Objective 1

1.a. 22 b. 4

c. −6 d. 0

Objective 2

3.a. 52 ft b. 16 ft

5. 5.6%

Objective 3

7.a. $11x^2 - 5x$ b. $-5x^2 + 5x - 8$

9.a. 38 b. 15

Objective 4

11.a. $\frac{1}{5}$ b. 16

c. $\frac{3}{x^2} - \frac{4}{x} + 1$

Objective 5

13.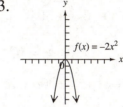

domain: $(-\infty, \infty)$;
range: $(-\infty, 0]$

15.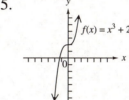

domain: $(-\infty, \infty)$;
range: $(-\infty, \infty)$

5.4 Multiplying Polynomials

Key Terms

1. FOIL
2. square a binomial
3. difference of the squares
4. intersection

Now Try

1. $108r^5 s^7$

2.a. $-35z^6 + 28z^5 - 14z^3$

b. $8a^4 - 10a^3 + 3a^2$

3. $6x^4 + 19x^3 - 9x^2 - 2x$

4.a. $2x^2 - 15x + 27$

b. $-6m^2 - mn + 12n^2$

5.a. $25m^2 - 16n^2$

b. $64x^4 - 9x^2$

6.a. $a^2 + 4ab + 4b^2$

b. $4m^2 - 12mp + 9p^2$

7.a. $36x^2 + 60x + 25 + 9y^2$
$\qquad\qquad + 36xy + 30y$

b. $125m^3 - 300m^2n + 240mn^2 - 64n^3$

c. $81a^4 - 216a^3 + 216a^2 - 96a + 16$

8. $(fg)(x) = 6x^3 - x^2 - 31x - 24$
$(fg)(3) = 36$

Objective 1

1. $72x^3y^4$

Objective 2

3. $-16z^6 - 32z^4$

5. $24s^3 + 11s^2 - 39s - 5$

Objective 3

7. $6z^2 - 7pz + 2p^2$

Objective 4

9. $16t^6 - 25$

Objective 5

11. $36r^2 - 48rs + 16s^2$

Objective 6

13. $(fg)(x) = 2x^3 - 25x^2 + 33x$
$(fg)(-1) = -60$

15. $(fg)(x) = 8x^3 - 47x^2 - 31x + 36$
$(fg)(-1) = 12$

5.5   Dividing Polynomials
Key Terms

1. dividend   2. quotient

3. divisor

Now Try

1.a. $4w^5 - 2w^3 + 3w$

b. $7b^2 - 3 - \dfrac{1}{9b^2}$

c. $-y^2 + 6x^2y - 6x^3$

2. $2r - 5$

3. $3x^3 - 3x^2 - 3x + 6$

4. $3x^2 - 6x + 2 + \dfrac{13x - 7}{2x^2 + 3}$

5. $4x^2 + \dfrac{2}{3}x - \dfrac{4}{3} - \dfrac{5}{6x + 3}$

6. $\left(\dfrac{f}{g}\right)(x) = 4x + 3,\ x \neq \dfrac{1}{2}$

$\left(\dfrac{f}{g}\right)(-2) = -5$

Objective 1

1. $9 + \dfrac{3}{q} + \dfrac{9q^2}{5}$ 3. $-n^2 + 3n - \dfrac{4}{n}$

Objective 2

5. $2q - 1 + \dfrac{4}{2q - 1}$

7. $4x - 3 + \dfrac{4x + 5}{3x^2 - 2x + 5}$

9. $y^2 - 1 + \dfrac{4}{y^2 - 1}$

Objective 3

11. $9x^2 + 6x + 4,\ x \neq \dfrac{2}{3}$

13. $2x^2 + 8 + \dfrac{-2x + 9}{x^2 - 2},\ x \neq \sqrt{2}$

15. undefined

Chapter 6   FACTORING

6.1   Greatest Common Factors; Factoring by Grouping

Key Terms

1. factoring by grouping

2. greatest common factor (GCF)

3. factoring

Now Try

1.a. $9(x-4)$ b. $3(4r+5)$

c. There is no common factor other than 1.

2.a. $3y(-5y+6)$ or $-3y(5y-6)$

b. $12a\left(8a^2-4a+5\right)$

c. $x^4z^2\left(1+x^2\right)$

d. $7x^2y\left(2xy+1-3x^3y^2\right)$

3.a. $(r-4s)\left(x^2+z^2\right)$

b. $(3x+1)(b+4)$

c. $(3r+4)^2\left(3rx^2+4x^2-y^2\right)$

4. $2t^2\left(-6t^3-3t+4\right)$ or
$-2t^2\left(6t^3+3t-4\right)$

5. $(x+6)(4y-1)$

6. $(3c-d)(4b-5x)$

7. $(x-4)\left(1+2y^2\right)$

8. $(3z+c)(3m-4n)$

Objective 1

1. $7x^2y\left(2xy+1-3x^3y^2\right)$

3. $5a^3\left(2a^2-3a+5\right)$

5. $(3x-1)(2x-1)$

7. $3x^3y\left(3y-4x^2+5x^3y^2\right)$

Objective 2

9. $(1-x)(1-y)$

11. $(5r-2q)(s-t)$

13. $\left(x^2+y^2\right)(3x+4y)$

15. $(x+y)(x-2)$

6.2   Factoring Trinomials

Key Terms

1. prime polynomial

2. product

3. common factor

4. FOIL   5.   sum

Now Try

1.a. $(x+2)(x+12)$

b. $(x+7)(x-4)$

2. prime

3. $(p-3m)(p-4m)$

4. $2t(s-10)(s+2)$

5. $(3m+4)(m-3)$

6.a. $(3y+1)(y+4)$

b. $(x-3)(5x-3)$

7. $(10z-3y)(4z+3y)$

8. $-(3r-4)(5r+6)$

9. $3a^2b(2a-1)(5a+2)$

10. $(19-4z)(7-2z)$

11. $\left(2y^2 - 5\right)\left(2y^2 + 9\right)$

Objective 1

1. $(x+7)(x-4)$

3. $(x+3y)(x+5y)$

Objectives 2 and 3

5. $(3z-4)(z+2)$

7. $(5q-4)(3q+1)$

9. $(4x-5)(2x+3)$

11. $(3y+2z)(2y-3z)$

Objective 4

13. $(3p-3q+7)(p-q+1)$

15. $\left(8m^2 + 1\right)\left(3m^2 - 5\right)$

6.3   Special Factoring

Key Terms

1. difference of squares

2. sum of cubes

3. perfect square trinomial

4. difference of cubes

Now Try

1.a. $(y+4)(y-4)$

b. $(5a+6)(5a-6)$

c. $9x^2(3x+10)(3x-10)$

d. $(a+7b)(a-5b)$

e. $\left(10x^2y - 9z\right)\left(10x^2y + 9z\right)$

2.a. $(4x-5)^2$ b. $(4a+9b)^2$

c. $(m-n-6)^2$

d. $(x-2+3y)(x-2-3y)$

3.a. $(x-7)\left(x^2 + 7x + 49\right)$

b. $3(5-3y)\left(25 + 15y + 9y^2\right)$

c. $2x(5x-4y)\left(25x^2 + 20xy + 16y^2\right)$

4.a. $(3x+5)\left(9x^2 - 15x + 25\right)$

b. $8(x+2b)\left(x^2 - 2bx + 4b^2\right)$

c. $\left(uv^2 + 6\right)\left(u^2v^4 - 6uv^2 + 36\right)$

d. $(2a-1)\left(a^2 - a + 1\right)$

Objective 1

1. $(6z+11)(6z-11)$

3. $\left(s^2 + 4\right)(s+2)(s-2)$

Objective 2

5. $(4t+7)^2$ 7. $(y+z+7)^2$

Objective 3

9. $(2a-5b)\left(4a^2 + 10ab + 25b^2\right)$

11. $(2n)\left(3m^2 + n^2\right)$

Objective 4

13. $(5p+q)\left(25p^2 - 5pq + q^2\right)$

15. $2(t+1)\left(t^2 + 2t + 4\right)$

6.4   A General Approach to Factoring

Key Terms

1. difference of squares

2. factor out a common factor

3. perfect square trinomial

Now Try

1.a. $5rt(r-2+t)$

b. $2(x+1)^2$

2.a. $2(4x-y)(16x^2+4xy+y^2)$

b. $-4ab$ c.   prime

3.a. $(a-6)(2a-5)$

b. $(2x+3y)^2$

c. $(b-1)(b+2)(4b+1)$

4.a. $(s+3)(t+2)$

b. $(r-s)(r^2+s^2)$

c. $(2-3k)(4-3k^3)$

Objective 1

1. $6a^2b^3(2a+b^2)$

3. $-11(2q+1)$

Objective 2

5. $a^4(a+1)(a-1)$

7. $8s$

Objective 3

9. $(3s+t)(s+2t)$

11. $(3p-1)(2p-3)$

Objective 4

13. $8a(x+y)$

15. $(x-9)(y+z)$

6.5 Solving Equations by Factoring

Key Terms

1. zero-factor property

2. standard form

3. quadratic equation

Now Try

1. $\{0, -5\}$

2.a. $\{-6, -3\}$ b. $\left\{-\frac{2}{7}, \frac{3}{2}\right\}$

3. $\{0, -3\}$ 4. $\left\{-\frac{7}{3}, \frac{7}{3}\right\}$

5. $\left\{-3, \frac{1}{2}\right\}$ 6. $\{-7, 0, 2\}$

7. base: 12 cm; height: 7 cm

8. $1\frac{3}{8}$ sec or $1\frac{1}{2}$ sec

9. $b = \dfrac{3a^2-2a}{3a-1}$

Objective 1

1. $\{0, 2\}$ 3. $\left\{-\frac{5}{4}, \frac{1}{3}\right\}$

5. $\left\{-\frac{3}{2}, \frac{5}{4}\right\}$ 7. $\left\{-\frac{3}{5}, -2\right\}$

Objective 2

9. $-4, -3$ or $3, 4$

11. 40 items or 110 items

Objective 3

13. $\dfrac{-bx-25}{ax+20}$ 15. $y = \dfrac{8x-5}{2x+2}$

Chapter 7
Rational Expressions and Functions

7.1 Rational Expressions and Functions; Multiplying and Dividing

Key Terms

1. reciprocal   2. opposites

3. rational function

4. fundamental property of rational numbers

5. rational expression

Now Try

1.a. $\{x \mid x \neq 0\}$

b. $\{x \mid x \text{ is a real number}\}$

c. $\{x \mid x \neq -2, 2\}$

2.a. $\dfrac{3y+2}{2y+1}$

b. already in lowest terms

c. $\dfrac{r^2 + rs + s^2}{r+s}$

3.a. -1 b. $-x-5$

4.a. $\dfrac{18z^4}{r}$ b. $\dfrac{x+4}{2(x-4)}$

5.a. $\dfrac{2(m-5)}{m+3}$ b. $-\dfrac{6(a+1)}{a}$

Objectives 1 and 2

1. $\dfrac{7}{4}$ 3. $1, 2$

Objective 3

5. $-\dfrac{11r^2}{6}$ 7. $\dfrac{r^2 + rs + s^2}{r+s}$

Objective 4

9. $\dfrac{x+4}{2(x-4)}$

11. $-\dfrac{x+4}{x-4}$ or $\dfrac{x+4}{4-x}$

Objectives 5 and 6

13. $\dfrac{2k-3}{k-1}$ 15. $\dfrac{a-3}{2a-3}$

7.2 Adding and Subtracting Rational Expressions

Key Terms

1. least common denominator (LCD)

2. equivalent expressions

Now Try

1.a. $\dfrac{1}{b-2}$ b. $\dfrac{2}{m-5}$

c. 5

2.a. $24x^3y^5$

b. $a^2 - b^2$ or $b^2 - a^2$

c. $(x-4)(x-3)(x+3)$

3.a. $\dfrac{46}{45z}$ b. $\dfrac{30}{s(s-6)}$

4.a. $\dfrac{1}{x+5}$ b. $\dfrac{16}{3(x+4)}$

5. $\dfrac{-x+4}{x-8}$ or $\dfrac{x-4}{8-x}$

6. $\dfrac{1}{x}$

7. $\dfrac{2z^2 + 5z + 4}{(z+2)(z+3)(z+4)}$

8. $\dfrac{3y^2 + 10y - 12}{(y+4)(y+2)(y-2)}$

Objective 1

1. $\dfrac{4n-7}{m+3}$ 3. $\dfrac{1}{k+5}$

Objective 2

5. $r(r+4)(r+1)$

Objective 3

7. $2r + 2$

9. $\dfrac{2(3-p)}{(p+3)^2}$

11. $\dfrac{3x^2 - 18x + 7}{(2x+1)(2x-1)(x+2)}$

13. $\dfrac{-4m - 9}{(m+3)(m-3)}$ or

$-\dfrac{4m+9}{(m+3)(m-3)}$

15. $\dfrac{2\left(a^2 + 3ab + 4b^2\right)}{(b+a)^2 (3b+a)}$

7.3 Complex Fractions

Key Terms

1. complex fraction

2. least common denominator

Now Try

1.a. 2

b. $\dfrac{2(2a+3)}{3a+5}$

2.a. $\dfrac{15(a-2)}{3a^2 + 5}$

b. $\dfrac{8x - 18}{30x^2 + 3x}$ or $\dfrac{2(4x-9)}{3x(10x+1)}$

3.a. $\dfrac{1-t}{1+t}$

b. $\dfrac{4(x-4)}{3}$

4. $\dfrac{(x+y)(x+1)}{x}$

Objective 1

1. $\dfrac{2}{k}$

3. $\dfrac{24}{w-3}$

Objective 2

5. $\dfrac{(x-2)^2}{x(x+2)}$

Objective 3

7. $\dfrac{7(5k-m)}{4}$

9. $\dfrac{3(8-x)}{2(15+x)}$ or $-\dfrac{3(x-8)}{2(x+15)}$

11. $\dfrac{2(s-9)}{5}$

Objective 4

13. $\dfrac{2z^3 + xy^2 z^3}{x}$

15. $\dfrac{2y^3}{x^2 y^3 + 3x^2}$

7.4 Equations with Rational Expressions and Graphs

Key Terms

1. vertical asymptote

2. domain of the variable in a rational equation

3. discontinuous

4. horizontal asymptote

5. reciprocal function

Now Try

1.a. $\{x \mid x \neq -8, 7\}$

b. $\{x \mid x \neq -2, 0, 2\}$

2. $\{-1\}$

3. $\varnothing$

4. $\{-7\}$

5. $\left\{-\dfrac{3}{2}, \dfrac{3}{5}\right\}$

6.

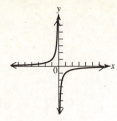

Vertical asymptote: $x = 0$
Horizontal asymptote: $y = 0$

Objective 1

1.a. $-6, -1$ b. $\{x \mid x \neq -6, -1\}$

3.a. $-1, -2, -3$

b. $\{x \mid x \neq -3, -2, -1\}$

Objective 2

5. $\{4\}$ 7. $\left\{\dfrac{1}{3}\right\}$

9. $\{-2, 1\}$

Objective 3

11.

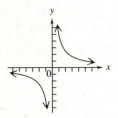

Vertical asymptote: $x = 0$
Horizontal asymptote: $y = 0$

13.

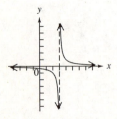

Vertical asymptote: $x = 3$
Horizontal asymptote: $y = 0$

15.

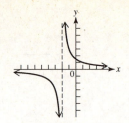

Vertical asymptote: $x = -2$
Horizontal asymptote: $y = 0$

7.5   Applications of Rational Expressions

Key Terms

1. ratio   2. rate of work

3. proportion

Now Try

1. $\dfrac{11}{3}$ 2. $R = \dfrac{R_1 R_2}{R_1 + R_2}$

3. $R_r = \dfrac{R_1 (R_2 - A)}{A}$ or

$R_r = \dfrac{R_1 R_2}{A} - R_1$

4. 39.91 million African-Americans

5. 4 gallons   6. 2 mph

7. 3 mph   8. $\dfrac{10}{7}$ or $1\dfrac{3}{7}$ hr

Objective 1

1. $m = \dfrac{75}{8}$ 3. $h = 12$

Objective 2

5. $a_n = \dfrac{2S_n}{n} - a_1$ or $a_n = \dfrac{2S_n - a_1 n}{n}$

Objective 3

7. 1750 crimes   9. $28.80

Objective 4

11. 2 mph

Objective 5

13. $\frac{12}{7}$ or $1\frac{5}{7}$ hr   15.   60 hours

7.6 Variation

Key Terms

1.   constant of variation

2.   varies inversely

3.   varies directly

4.   proportional

Now Try

1. $25   2.   125 psi

3.   153.86 sq cm

4.   90 revolutions per minute

5. $\frac{9}{8}$ 6.   1280 psi

7.   about 63.5 cm^3

Objective 2

1.   36   3.   100 newtons

Objective 3

5. $\frac{400}{27}$ foot-candles

7.   90 revolutions/minute

Objective 4

9.   243 ergs   11.   750°

Objective 5

13.   1.105 L

15.   6,000,000 dynes

Chapter 8
Roots, Radicals, and Root Functions

8.1 Radical Expressions and Graphs

Key Terms

1.   radical

2.   radicand

3.   cube root function

4.   square root function

5.   index (order)

6.   radical expression

7.   principal root

Now Try

1.a. 5   b.   10

 c. $\frac{8}{3}$ d.   0.2

2.a. 22   b.   −18

 c. −4

 d. not a real number

3.a.

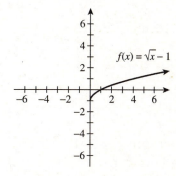

 Domain: $[0, \infty)$
 Range: $[-1, \infty)$

 b.

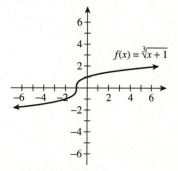

 Domain: $(-\infty, \infty)$
 Range: $(-\infty, \infty)$

4.a. 73   b.   37

 c. $|a|$ d. $|a|$

5.a. −5   b.   4

c. −2   d. $-x^4$

e. $\left|x^5\right|$ f. w^{10}

6.a. −11.747   b. 3.733

   c. 5.189

7.   23.2 miles per hour

Objectives 1 and 2

1. $-\dfrac{25}{22}$

3.   Not a real number

Objective 3

5.   Domain: $[-6, \infty)$; range: $[0, \infty)$

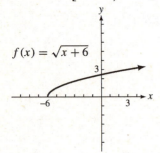

$f(x) = \sqrt{x+6}$

7.   Domain: $(-\infty, \infty)$; range: $(-\infty, \infty)$

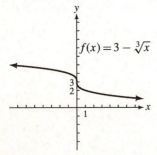

$f(x) = 3 - \sqrt[3]{x}$

Objective 4

9. 9   11. a^3

Objective 5

13. 8.883   15. 2.221

8.2   Rational Exponents

Key Terms

1.   radical form of $a^{m/n}$

2.   power rule for exponents

3.   product rule for exponents

4.   quotient rule for exponents

Now Try

1.a. 6   b. −6

   c. 2

   d. not a real number

   e. −2   f. $\dfrac{1}{2}$

2.a. 9   b. 8

   c. −216

   d. not a real number

   e. 9

3.a. $\dfrac{1}{4}$ b. $-\dfrac{1}{625}$

   c. $\dfrac{16}{9}$

4.a. $\sqrt[4]{216}$ b. $\left(\sqrt[3]{18}\right)^2$

   c. $5\left(\sqrt[4]{x}\right)^5$

   d. $\left(\sqrt[3]{2x}\right)^4 - 3\left(\sqrt[5]{x}\right)^2$

   e. $\dfrac{1}{\left(\sqrt{x}\right)^3}$ f. $\sqrt[4]{x^2 - y^2}$

   g. $10^{1/2}$ h. $x^{1/4}$

5.a 5^3 b. $a^{2/15}$

   c. $\dfrac{x^8}{y}$ d. $a^{5/6}$

   e. $r^{7/6} + r^{19/6}$

6.a. $x^{7/6}$ b. $\dfrac{1}{y^{3/4}}$

   c. $x^{1/4}$

Objective 1

1.   −2   3.   −12

Objective 2

5. $\frac{1}{36}$ 7. 27

Objective 3

9. $\dfrac{1}{\left(\sqrt[3]{2x^4-3y^2}\right)^4}$

11. $a^{1/4}$

Objective 4

13. $d^{3/28}+4d^{13/7}$

15. $x^{5/16}$

8.3 Simplifying Radical Expressions

Key Terms

1. Pythagorean theorem
2. product rule for radicals
3. quotient rule for radicals
4. distance formula
5. simplified radical

Now Try

1.a. $\sqrt{14}$ b. $\sqrt{21x}$

2.a. $\sqrt[3]{21}$ b. $\sqrt[3]{35xy}$

c. $\sqrt[5]{8w^4}$

d. cannot be simplified using the product rule for radicals

3.a. $\frac{6}{7}$ b. $\frac{\sqrt{13}}{9}$

c. $-\frac{7}{5}$ d. $\frac{\sqrt[5]{7x}}{2}$

e. $-\frac{a^2}{5}$

4.a. $2\sqrt{21}$ b. $9\sqrt{2}$

c. cannot be simplified

d. $-4\sqrt[3]{4}$ e. $2\sqrt[5]{16}$

5.a. $10y^6$ b. $4m^2r^4\sqrt{3mr}$

c. $-2n^2t\sqrt[3]{4nt^2}$ d. $-3y^2\sqrt[4]{5x^3y}$

6.a. $\sqrt[4]{9^3}$ or $\sqrt[4]{729}$ b. $\sqrt[5]{z^4}$

7. $\sqrt[6]{63}$

8.a. $12\sqrt{2}$ b. $3\sqrt{5}$

9. $\sqrt{34}$

Objective 1

1. $\sqrt{42xt}$

3. Cannot be simplified

Objective 2

5. $\frac{a^2}{5}$

Objective 3

7. $3x^3\sqrt[3]{2x^2}$

9. $2x^3y^2\sqrt[4]{y^2}$ or $2x^3y^2\sqrt{y}$

Objective 4

11. $2\sqrt[10]{972}$

Objective 5

13. $2\sqrt{2}$

Objective 6

15. $4\sqrt{3}$

8.4 Adding and Subtracting Radical Expressions

Key Terms

1. unlike radicals
2. like radicals

Now Try

1.a. $-\sqrt{6}$ b. $13\sqrt{2z}$

c. cannot be simplified

2.a. $-3\sqrt[3]{2}$ b. $5z\sqrt[4]{2y^2z}$

c. $6\sqrt{5x}+2\sqrt[3]{5x}$

3.a. $\dfrac{2\sqrt{5}}{3}$ b. $\dfrac{17}{w^2}$

Objective 1

1. $27\sqrt{3}$

3. cannot be simplified

5. $-13\sqrt[4]{3}$

7. $(10+12rz)\sqrt[3]{3r^2z}$

9. $5\sqrt[3]{2r}$ 11. $-\dfrac{\sqrt{3}}{35}$

13. $-\dfrac{k\sqrt{k}}{6}$ 15. $\dfrac{2t\sqrt{3y}-10\sqrt[3]{y}}{t^2}$

8.5 Multiplying and Dividing Radical Expressions

Key Terms

1. conjugate

2. rationalizing the denominator

Now Try

1.a. $-55-25\sqrt{6}$ b. -38

c. $9-12\sqrt{x}+4x$

d. $16-\sqrt[3]{4}$

2.a. $\dfrac{2\sqrt{15}}{3}$ b. $-\dfrac{3\sqrt{7}}{7}$

c. $\dfrac{3\sqrt{5}}{5}$

3.a $-\dfrac{3\sqrt{10}}{8}$ b. $\dfrac{9x\sqrt{2xt}}{t^3}$

4.a. $\dfrac{\sqrt[3]{10}}{5}$ b. $\dfrac{2t\sqrt[4]{4t^2x}}{x^2}$

5.a. $-4\sqrt{3}+8$ b. $-\dfrac{5\sqrt{5}+5\sqrt{3}}{2}$

c. $5+2\sqrt{6}$ d. $\dfrac{3\sqrt{x}+3\sqrt{3y}}{x-3y}$

6.a. $\dfrac{3+2\sqrt{15}}{4}$ or $\dfrac{3}{4}+\dfrac{\sqrt{15}}{2}$

b. $\dfrac{1-\sqrt{2x}}{2}$

Objective 1

1. $6+3\sqrt{7}+2\sqrt{2}+\sqrt{14}$

3. $9-\sqrt[3]{25}$

Objective 2

5. $\dfrac{\sqrt{38}}{8}$ 7. $\dfrac{5y^2\sqrt{6y}}{4x}$

Objective 3

9. $3\sqrt{7}-9$

11. $-\dfrac{3\sqrt{3}+21-\sqrt{21}-7\sqrt{7}}{46}$

Objective 4

13. $\dfrac{1+2\sqrt{3}}{5}$

15. $\dfrac{1+x^2\sqrt{2x}}{2x}$ or $\dfrac{1}{2x}+\dfrac{x\sqrt{2x}}{2}$

8.6 Solving Equations With Radicals

Key Terms

1. power rule

2. radical equation

3. extraneous solution

Now Try

1. $\{10\}$ 2. $\varnothing$

3. $\{5\}$ 4. $\{3\}$

5. $\{4\}$ 6. $\{2\}$

7. $h=\dfrac{3V}{\pi r^2}$

Objective 1

1. $\{11\}$ 3. $\varnothing$

Objective 2

5. $\{3, 4\}$ 7. $\{1\}$

9. $\{3\}$

Objective 3

11. $\{0\}$

Objective 4

13. $L = Z^2 C$ 15. $r = \dfrac{a}{4\pi^2 N^2}$

8.7   Complex Numbers

Key Terms

1. complex number

2. complex conjugate

3. imaginary part

4. real part

5. standard form (of a complex number)

6. pure imaginary number

Now Try

1.a. $4i$ b. $-12i$

   c. $i\sqrt{11}$ d. $8i\sqrt{2}$

2.a. -33 b. $-\sqrt{21}$

   c. -12 d. $i\sqrt{21}$

3.a. 5 b. $4i$

4.a. $2 - 9i$ b. $-10 - 4i$

5.a. $11 - i$ b. $5 + 3i$

   c. $4i$

6.a. $-14 + 8i$ b. $-14 + 10i$

   c. $-8 + 6i$

7.a. $\dfrac{18}{29} + \dfrac{13i}{29}$ b. $-\dfrac{1}{4} - \dfrac{i}{2}$

8.a. 1 b. i

   c. -1 d. $-i$

Objective 1

1. $12i\sqrt{2}$ 3. -6

Objective 2

5. nonreal complex

Objective 3

7. $-4 + i$

Objective 4

9. $11 + 13i$

Objective 5

11. $\dfrac{1}{5} + \dfrac{3i}{5}$ 13. $\dfrac{35}{29} - \dfrac{14i}{29}$

Objective 6

15. $-i$

Chapter 9
Quadratic Equations, Inequalities, and Functions

9.1   The Square Root Property and Completing the Square

Key Terms

1. standard form

2. quadratic equation

3. square root property

4. zero-factor property

Now Try

1. $\left\{-\dfrac{7}{3}, 4\right\}$

2.a. $\left\{-2\sqrt{21}, 2\sqrt{21}\right\}$

   b. $\left\{-\dfrac{\sqrt{3}}{2}, \dfrac{\sqrt{3}}{2}\right\}$

3. About 2.7 seconds

4. $\{-3, 5\}$

    557

5. $\left\{\dfrac{-5-4\sqrt{2}}{2}, \dfrac{-5+4\sqrt{2}}{2}\right\}$ or

$\left\{-\dfrac{5}{2}-2\sqrt{2}, -\dfrac{5}{2}+2\sqrt{2}\right\}$

6. $\left\{3-2\sqrt{2}, 3+2\sqrt{2}\right\}$

7. $\left\{\dfrac{11-\sqrt{89}}{2}, \dfrac{11+\sqrt{89}}{2}\right\}$

8. $\left\{-\dfrac{4}{3}, \dfrac{2}{3}\right\}$

9.a. $\pm 4i\sqrt{2}$ b. $\left\{-2\pm 7i\right\}$

c. $\left\{\dfrac{1}{2}\pm 2i\right\}$

Objective 1

1. $\left\{-\dfrac{5}{2}, -\dfrac{2}{3}\right\}$

Objective 2

3. $\left\{1, -1\right\}$

5. 3.9 seconds

Objective 3

7. $\left\{\dfrac{-3-\sqrt{6}}{2}, \dfrac{-3+\sqrt{6}}{2}\right\}$

Objective 4

9. $\left\{-4-2\sqrt{3}, -4+2\sqrt{3}\right\}$

11. $\left\{\dfrac{2-\sqrt{10}}{3}, \dfrac{2+\sqrt{10}}{3}\right\}$

Objective 5

13. $\left\{\dfrac{1-i\sqrt{5}}{7}, \dfrac{1+i\sqrt{5}}{7}\right\}$

15. $\left\{\dfrac{1-i\sqrt{5}}{6}, \dfrac{1+i\sqrt{5}}{6}\right\}$

9.2   The Quadratic Formula
Key Terms

1. quadratic formula   2. discriminant

Now Try

1. $\left\{\dfrac{4}{3}, \dfrac{3}{2}\right\}$ 2. $\left\{\dfrac{1-\sqrt{7}}{2}, \dfrac{1+\sqrt{7}}{2}\right\}$

3. $\left\{\dfrac{2\pm i\sqrt{2}}{2}\right\}$ or $\left\{1\pm\dfrac{i\sqrt{2}}{2}\right\}$

4.a. 81; two rational solutions; factoring

b. −48; two nonreal complex solutions; quadratic formula

c. 0; one rational solution; factoring

d. 80; two irrational solutions; quadratic formula

5. −28, 28

Objective 2

1. $\left\{\dfrac{3}{5}, 2\right\}$

3. $\left\{1-\sqrt{2}, 1+\sqrt{2}\right\}$

5. $\left\{\dfrac{-3-2\sqrt{11}}{5}, \dfrac{-3+2\sqrt{11}}{5}\right\}$

7. $\left\{\dfrac{-1-2i}{2}, \dfrac{-1+2i}{2}\right\}$

Objective 3

9. B; factoring

11. C; quadratic formula

13. D; quadratic formula

15. −56, 56

9.3   Equations in Quadratic Form
Key Terms

1. quadratic in form

2. standard form

Now Try

1. $\left\{\dfrac{2}{3}\right\}$ 2. 2 mph

3. Tom: 12 hours; Huck: 24 hours

4.a. $\{2\}$ b. $\left\{\frac{1}{4}, \frac{1}{3}\right\}$

5.a. $\{\pm 1, \pm 2\}$ b. $\{-2, 10\}$

c. $\{-1, 27\}$

6. $\left\{\pm\sqrt{2 - 2\sqrt{3}}, \pm\sqrt{2 + 2\sqrt{3}}\right\}$

Objective 1

1. $\{-3, 2\}$ 3. $\left\{-3, -\frac{1}{5}\right\}$

Objective 2

5. Bike: 12 mph; hike: 2 mph

7. 9 hr

Objective 3

9. $\{3,5\}$ 11. $\left\{\frac{1}{4}\right\}$

Objective 4

13. $\left\{-\frac{17}{3}, -4\right\}$

15. $\left\{2, -2, i\sqrt{5}, -i\sqrt{5}\right\}$

9.4   Formulas and Further Applications

Key Terms

1. quadratic function

2. Pythagorean theorem

Now Try

1.a. $z = \dfrac{p\sqrt{6}}{y}$ b. $t = \pm\dfrac{\sqrt{mxF}}{F}$

2. $q = \dfrac{-k \pm k\sqrt{5}}{2p}$

3. south: 72 mi; east: 54 mi

4. The mounted border will be 2.5 in.

5. The ball will hit the ground at about 4.1 seconds.

6.a. $4200

b. 40 items or 60 items

Objective 1

1. $c = \dfrac{(a-1)^2}{b}$

3. $x = \dfrac{\pm\sqrt{4m - 2m^2}}{2}$

Objective 2

5. 36 ft

7. 8 ft

Objective 3

9. 3 cm   11.   3 ft

Objective 4

13. about 1.2 sec

15. 2 units or 8 units

9.5   Graphs of Quadratic Functions

Key Terms

1. axis   2.   vertex

3. quadratic function

4. parabola

Now Try

1. vertex: $(0, -1)$; axis: $x = 0$
domain: $(-\infty, \infty)$; range: $[-1, \infty)$

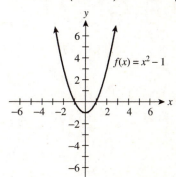

2. vertex: $(-3, 0)$; axis: $x = -3$
domain: $(-\infty, \infty)$; range: $[0, \infty)$

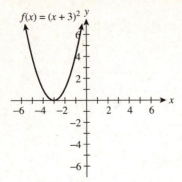

3. vertex: $(-2, -1)$; axis: $x = -2$
domain: $(-\infty, \infty)$; range: $[-1, \infty)$

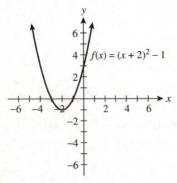

4. vertex: $(0, 0)$; axis: $x = 0$
domain: $(-\infty, \infty)$; range: $(-\infty, 0]$

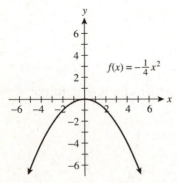

5. vertex: $(1, 1)$; axis: $x = 1$
domain: $(-\infty, \infty)$; range: $[1, \infty)$

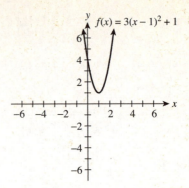

6.a.

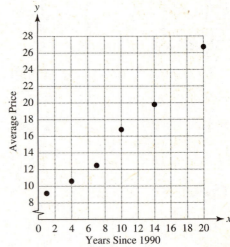

b. positive

c. $y = 0.01x^2 + 0.77x + 8.36$

d. $33.86

Objectives 1 and 2

1. vertex: $(0, 3)$; axis: $x = 0$
domain: $(-\infty, \infty)$; range: $[3, \infty)$

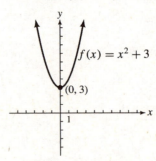

3. vertex: $(-3, -1)$; axis: $x = -3$
    domain: $(-\infty, \infty)$; range: $[-1, \infty)$

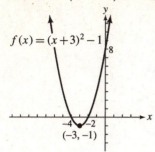

$f(x) = (x+3)^2 - 1$

$(-3, -1)$

5. vertex: $(1, 2)$; axis: $x = 1$
    domain: $(-\infty, \infty)$; range: $[2, \infty)$

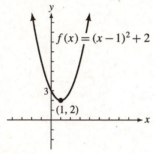

$f(x) = (x-1)^2 + 2$

$(1, 2)$

Objective 3

7. Upward; wider
    vertex: $(1, 0)$

9. Upward; narrower
    vertex: $(1, 7)$

Objective 4

11. quadratic; negative

13. quadratic; positive

15.a.

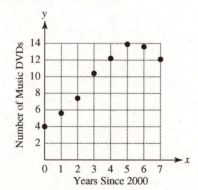

b. quadratic; negative

c. $y = -0.411x^2 + 4.037x + 4$

d. 3270

9.6   More About Parabolas and Their Applications

Key Terms

1. discriminant

2. vertex

Now Try

1. $(3, -5)$ 2. $\left(\dfrac{3}{2}, \dfrac{1}{2}\right)$

3. $(1, 1)$

4.

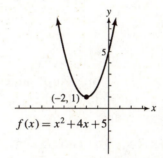

$(-2, 1)$

$f(x) = x^2 + 4x + 5$

vertex: $(-2, 1)$; axis: $x = -2$
domain: $(-\infty, \infty)$; range: $[1, \infty)$

5.a. 2   b.   0

  c. 1

6. maximum area: 125,000 sq yd
    length: 500 yd; width: 250 yd

7. maximum height: 286 feet after 1.5 seconds

8. vertex: $(1, 2)$; axis: $y = 2$
    domain: $[1, \infty)$ range: $(-\infty, \infty)$

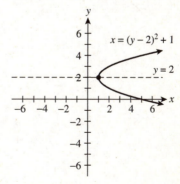

$x = (y-2)^2 + 1$

$y = 2$

9. vertex: $(1, -1)$; axis: $y = -1$
    domain: $[1, \infty)$ range: $(-\infty, \infty)$

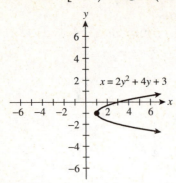

$x = 2y^2 + 4y + 3$

Objective 1
Objective 3

1. vertex: $(1, 3)$, opens up, same shape, discriminant: -12, no x-intercepts

3. vertex: $(-1, 3)$, opens down, narrower, discriminant: 24, two x-intercepts

5. vertex: $(1, 1)$; opens left; narrower

Objective 2

7. vertex: $(-4, 4)$; axis: $x = -4$
    domain: $(-\infty, \infty)$; range: $[4, \infty)$

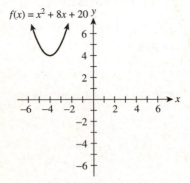

$f(x) = x^2 + 8x + 20$

9. vertex: $(-2, -2)$; axis: $x = -2$
    domain: $(-\infty, \infty)$; range: $[-2, \infty)$

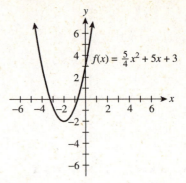

$f(x) = \frac{5}{4}x^2 + 5x + 3$

Objective 4

11. 50 pots; $200

Objective 5

13. vertex: $(0, 3)$; axis: $y = 3$
     domain: $(-\infty, 0]$; range: $(-\infty, \infty)$

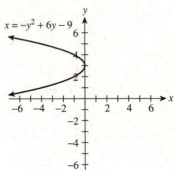

$x = -y^2 + 6y - 9$

15. vertex: $(1, 1)$; axis: $y = 1$
     domain: $[1, \infty)$; range: $(-\infty, \infty)$

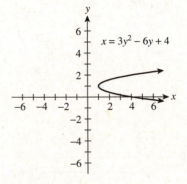

$x = 3y^2 - 6y + 4$

9.7 Quadratic and Rational Inequalities

Key Terms

1. quadratic inequality
2. rational inequality

Now Try

1.a. $(-\infty, 2) \cup (6, \infty)$

 b. $(2, 6)$

2. $(-1, 2)$

3.a. $\varnothing$ b. $(-\infty, \infty)$

4. $\left(-\infty, -\frac{3}{2}\right] \cup \left[-\frac{1}{3}, \frac{1}{2}\right]$

5. $\left(-\frac{3}{2}, -1\right)$

6. $(-\infty, 3) \cup [8, \infty)$

Objective 1

1. $(-\infty - 4) \cup (-3, \infty)$

3. $\left(-3, \frac{1}{2}\right)$

5. $\varnothing$

Objective 2

7. $(-\infty, -5] \cup [-3, 1]$

9. $\left[\frac{3}{4}, \frac{10}{3}\right] \cup \left[\frac{7}{2}, \infty\right)$

Objective 3

11. $\left(\frac{2}{3}, \frac{8}{3}\right]$

13. $(-\infty, -2) \cup \left(-\frac{1}{3}, \infty\right)$

15. $(-2, \infty)$

Chapter 10
Inverse, Exponential, and Logarithmic Functions

10.1 Inverse Functions
Key Terms

1. one-to-one function
2. inverse of a function

Now Try

1.a. Not one-to-one

 b. One-to-one
 $G^{-1} = \{(2, 3), (-2, -3), (3, 2), (-3, -2)\}$

 c. Not one-to-one

2.a. One-to-one

 b. Not one-to-one

3.a. $f^{-1}(x) = x^2 + 1; \quad x \geq 0$

 b. $f^{-1}(x) = \sqrt[3]{\dfrac{x+3}{2}}$

 c. Not one-to-one

4.

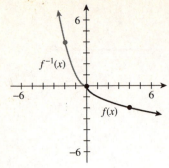

Now Try

1.

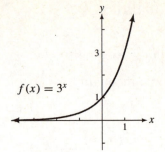

$f(x) = 3^x$

Objective 1

1. $\{(1, -3), (2, -2), (3, -1), (4, 0)\}$

3. $\{(0, 0), (1, 1), (-1, -1), (2, 2), (-2, -2)\}$

Objective 2

5. One-to-one

7. Not a function; not one-to-one

Objective 3

9. $f^{-1}(x) = \dfrac{x+5}{2}$

11. $f^{-1}(x) = \sqrt[3]{x+1}$

Objective 4

13.

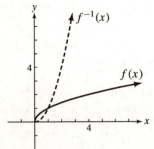

15. Not one-to-one

10.2 Exponential Functions
Key Terms

1. exponential function

2. exponential equation

3. asymptote

2.

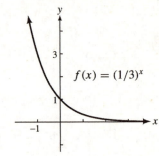

$f(x) = (1/3)^x$

3.

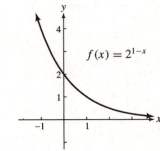

$f(x) = 2^{1-x}$

4. $\left\{ \dfrac{3}{2} \right\}$

5.a. $\{-2\}$ b. $\{-5\}$

  c. $\{-4\}$

6. 256,000

7.a. about 32,656   b. in 2010

Objective 1

1.a Exponential function

  b. Not an exponential function

  c. Not an exponential function

  d. Exponential function

Objective 2

3.

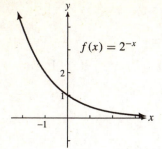

$f(x) = 2^{-x}$

5.

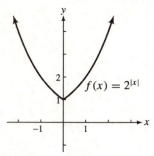

$f(x) = 2^{|x|}$

Objective 3

7. $\left\{\dfrac{1}{2}\right\}$

9. $\left\{\dfrac{3}{2}\right\}$

Objective 4

11. 896 geese

13. 7750 bacteria

15. 200 bacteria

10.3 Logarithmic Functions
Key Terms

1. logarithmic function with base a

2. logarithm

3. logarithmic equation

Now Try

1.a. $\log_8 64 = 2$ b. $\log_{81} 27 = \dfrac{3}{4}$

c. $16^{-1/2} = \dfrac{1}{4}$ d. $\left(\dfrac{1}{2}\right)^{-2} = 4$

2.a. $\left\{\dfrac{1}{64}\right\}$ b. $\{40\}$

c. $2\sqrt{3}$ d. $\left\{\dfrac{1}{6}\right\}$

3.a. 1 b. 1

c. 0 d. 0

4.

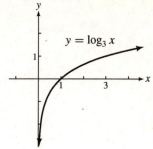

$y = \log_3 x$

5.

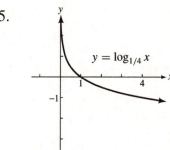

$y = \log_{1/4} x$

6. 5 fish

Objective 1
Objective 2

1.a. 3 b. 2

c. −1 d. $\dfrac{1}{2}$

e. −2 f. $\dfrac{1}{4}$

3.a. $3^3 = 27$ b. $16^{1/4} = 2$

c. $4^{-2} = \dfrac{1}{16}$ d. $10^{-3} = 0.001$

Objective 3

5. $\left\{\dfrac{1}{2}\right\}$ 7. $\left\{\dfrac{1}{8}\right\}$

Objective 4

9.

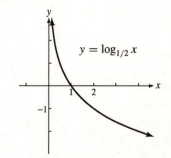

$y = \log_{1/2} x$

11.

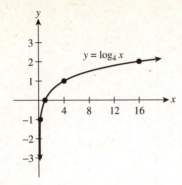

$y = \log_4 x$

Objective 5

13.　$600　　　15.　110 decibels

10.4　Properties of Logarithms

Key Terms

1.　special properties

2.　quotient rule for logarithms

3.　product rule for logarithms

4.　power rule for logarithms

Now Try

1.a.　$\log_6 5 + \log_6 3$

　　b.　$\log_5 21$　　c.　$1 + \log_4 x$

　　d.　$3\log_4 x$

2.a.　$\log_5 4 - \log_5 9$

　　b.　$\log_6\left(\frac{x}{3}\right)$　　c.　$2 - \log_4 11$

3.a.　$3\log_6 4$　　b.　$6\log_b x$

　　c.　$\frac{1}{2}\log_b 13$　　d.　$\frac{3}{4}\log_3 x$

4.a.　7　　b.　3

　　c.　5

5.a.　$2 + 3\log_6 x$

　　b.　$\frac{1}{2}\left(\log_b x - \log_b 3\right)$

　　c.　$\log_b \frac{xy^4}{z}$　　d.　$\log_2 \frac{x^2(x-1)}{\sqrt[3]{x^2+1}}$

　　e.　cannot be rewritten

6.a.　3.8671

　　b.　-0.3247

　　c.　3.5424

7.a.　False　　　　b.　False

Objective 1

1.　$\log_3 6 + \log_3 5$

3.　$\log_{10} 63$

Objective 2

5.　$\log_2 \frac{r}{s}$

Objective 3

7.　$7\log_m 2$　　　9.　1

Objective 4

11.　$\log_5 7 + 3\cdot\log_5 m - \log_5 8 - \log_5 y$

13.　$\log_{10} \frac{4k}{3j}$

15.　$\log_2 1$ or 0

10.5　Common and Natural Logarithms

Key Terms

1.　universal constant

2.　common logarithm

3.　natural logarithm

Now Try

1.a.　2.9928　　b.　4.8994

　　c.　-0.4776

2.　pH $= 7.2076$; rich fen

3.　2.5×10^{-4}

4.　85 dB

5.a.　4.5850　　b.　6.6758

　　c.　-1.0996

6.　11.9 years　　7.　3.2266

8. 72.38 trillion ft^3

Objective 1

1. 1.7576   3. 5.4472

Objective 2

5.a. 6.3×10^{-6} b. 5.0×10^{-2}

Objective 3

7. −2.1203   9. 1.7918

Objective 4

11. about 12 years

Objective 5

13. 1.1887   15. −2.3219

10.6 Exponential and Logarithmic Equations; Further Applications

Key Terms

1. continuous compounding

2. compound interest

Now Try

1. 2.465   2. 6.770

3. $-1 + \sqrt[3]{36}$ 4. $\left\{\dfrac{1}{2}\right\}$

5. $\left\{\dfrac{1}{4}, 1\right\}$

6. $12,201.90; interest: $2201.90

7. 13.89 years

8.a. 5309.18   b. 34.7 years

9.a. 11.3 mg   b. 30.0 years

Objective 1

1. {3.936}   3. {−1.710}

Objective 2

5. $\left\{\dfrac{1}{4}\right\}$ 7. {5}

Objective 3

9. $55,201   11. 18.6 years

Objective 4

13.a. 13.0 mg   b. 8.02 days

15.a. 5214 bacteria   b. 49.5 hours

Chapter 11
Nonlinear Functions, Conic Sections, and Nonlinear Systems

11.1 Additional Graphs of Functions

Key Terms

1. step function

2. greatest integer function

3. asymptotes

4. absolute value function

5. square root function

6. squaring function

7. reciprocal function

Now Try

1.

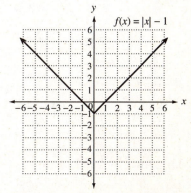

Domain: $(-\infty, \infty)$; range: $[0, \infty)$

2.

Domain: $(-\infty, \infty)$; range: $[-1, \infty)$

3.

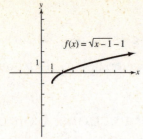

Domain: $[1, \infty)$; range: $[-1, \infty)$

4.a. 8   b.   −3

   c. 5   d.   −7

5.

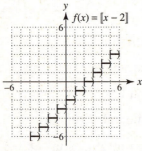

The domain is $(-\infty, \infty)$. The range is $\{\ldots, -2, -1, 0, 1, 2, \ldots\}$ or the set of integers.

6.

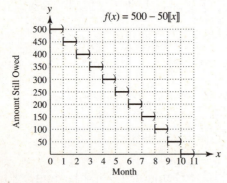

Objective 1

1.

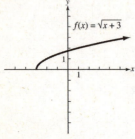

Domain: $[-3, \infty)$
Range: $[0, \infty)$

3.

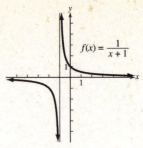

Domain:
$(-\infty, -1) \cup (-1, \infty)$
Range: $(-\infty, \infty)$

5.

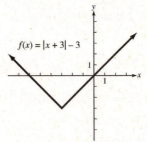

Domain: $(-\infty, \infty)$
Range: $[-3, \infty)$

7.

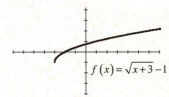

Domain: $[-3, \infty)$
Range: $[-1, \infty)$

9.

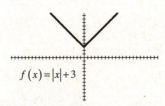

Domain: $(-\infty, \infty)$
Range: $[3, \infty)$

Objective 2

11.

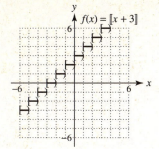

13.

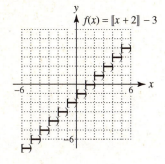

15.

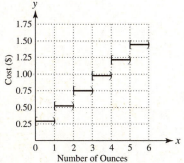

11.2 The Circle and the Ellipse
Key Terms

1. circle   2. ellipse

3. conic sections

4. center of the circle

5. radius

6. center-radius form

7. foci

8. center (of an ellipse)

Now Try

1. $x^2 + y^2 = 25$

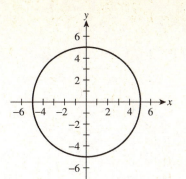

2. $(x+5)^2 + (y-4)^2 = 16$

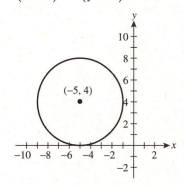

3. $x^2 + (y-3)^2 = 2$

4. Center: (4, 1); radius: $\sqrt{2}$

5.

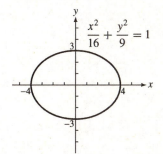

6.

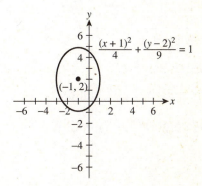

Objective 1

1. $(x-2)^2 + (y+3)^2 = 25$

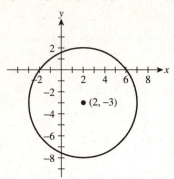

3. $(x+2)^2 + (y+3)^2 = 3$

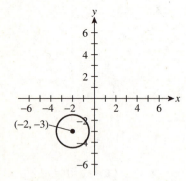

5. $x^2 + (y-5)^2 = 9$

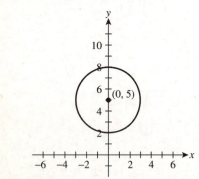

Objective 2

7. center: (1, 3); radius: 5

9. center: (3, −2); radius: $\frac{3}{2}$

Objective 3

11.

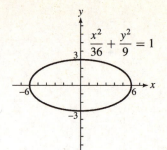

13.

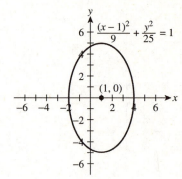

15.

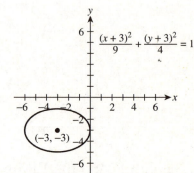

11.3 The Hyperbola and Functions Defined by Radicals

Key Terms

1. hyperbola   2. asymptotes

3. fundamental rectangle

4. transverse axis

5. generalized square root function

Now Try

1.

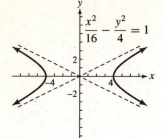

$$\frac{x^2}{16} - \frac{y^2}{4} = 1$$

2.

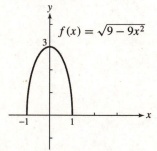

$$\frac{y^2}{4} - \frac{x^2}{9} = 1$$

3.a. hyperbola   b. ellipse

  c. parabola

4.

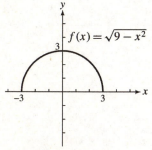

$f(x) = \sqrt{9 - x^2}$

Domain: [−3, 3]
Range: [0, 3]

5.

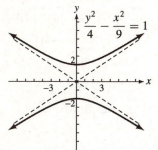

$f(x) = \sqrt{9 - 9x^2}$

Domain: [−1, 1]
Range: [0, 3]

Objectives 1 and 2

1.

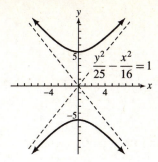

$$\frac{y^2}{25} - \frac{x^2}{16} = 1$$

3.

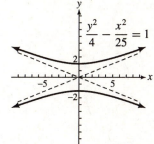

$$\frac{y^2}{4} - \frac{x^2}{25} = 1$$

5.

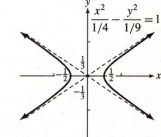

$$\frac{x^2}{1/4} - \frac{y^2}{1/9} = 1$$

Objective 3

7. Ellipse   9. Hyperbola

Objective 4

11.

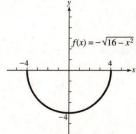

$f(x) = -\sqrt{16 - x^2}$

Domain: [−4, 4]
Range: [−4, 0]

13.

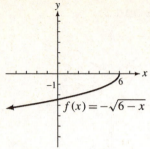

Domain: $(-\infty, 6]$

Range: $(-\infty, 0]$

15.

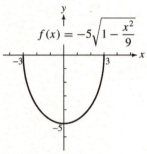

Domain: $[-3, 3]$

Range: $[-5, 0]$

11.4 Nonlinear Systems of Equations

Key Terms

1. nonlinear equation

2. nonlinear system of equations

Now Try

1. $\{(-4, 1), (1, -4)\}$

2. $\{(1, 1)\}$

3. $\left\{(0, -4), \left(2\sqrt{3}, 2\right), \left(-2\sqrt{3}, 2\right)\right\}$

4. $\{(-4, 1), (-1, 4), (1, -4), (4, -1)\}$

Objective 1

1. $\left\{(4, -1), \left(\frac{16}{5}, -\frac{13}{5}\right)\right\}$

3. $\left\{\left(\frac{1}{4}, \frac{3}{2}\right), (-1, 1)\right\}$

5. $\{(3, 8), (-4, -6)\}$

Objective 2

7. $\{(3, 1), (3, -1), (-3, 1), (-3, -1)\}$

9. $\{(1, 1), (1, -1), (-1, 1), (-1, -1)\}$

Objective 3

11. $\{(2, -3), (-2, 3), (3, -2), (-3, 2)\}$

13. $\left\{(3, -2), (-3, 2), \left(\frac{2\sqrt{6}}{3}, -\frac{3\sqrt{6}}{2}\right), \left(-\frac{2\sqrt{6}}{3}, \frac{3\sqrt{6}}{2}\right)\right\}$

15. $\{(2, 2), (-2, -2), (2i, -2i), (-2i, 2i)\}$

11.5 Second-Degree Inequalities and Systems of Inequalities

Key Terms

1. second-degree inequality

2. system of inequalities

Now Try

1.

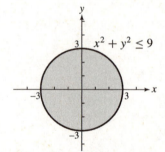

2.

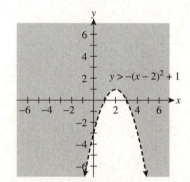

3.

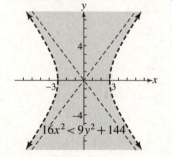

4.

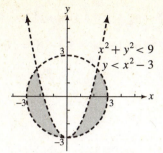

$x^2 + y^2 < 9$
$y < x^2 - 3$

5.

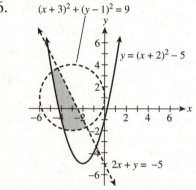

$x \leq 3$

$4x - y > 2$

$y > -x - 2$

6.

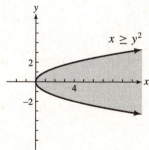

$(x + 3)^2 + (y - 1)^2 = 9$

$y = (x + 2)^2 - 5$

$2x + y = -5$

Objective 1

1.

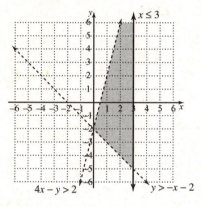

$x \geq y^2$

3.

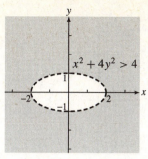

$x^2 + 4y^2 > 4$

5.

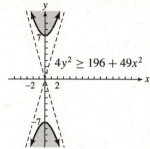

$4y^2 \geq 196 + 49x^2$

7.

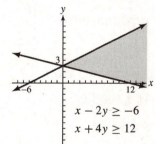

$y < x^2 - 4x + 5$

Objective 2

9.

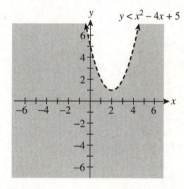

$x - 2y \geq -6$
$x + 4y \geq 12$

11.

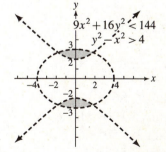

$9x^2 + 16y^2 < 144$
$y^2 - x^2 > 4$

13.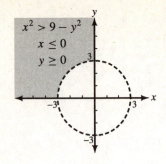

15.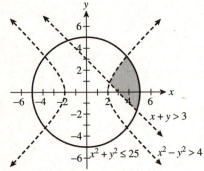

Chapter 12
Sequences and Series

12.1 Sequences and Series
Key Terms

1. infinite sequence

2. series

3. summation notation

4. index of summation

5. general term

6. arithmetic mean (average)

7. terms of a sequence

8. finite sequence

Now Try

1. $a_6 = \frac{8}{30} = \frac{4}{15}$

2. $a_n = (-1)^{n-1} 2n$

3. Month 1: $620
   Month 2: $607.60
   Month 3: $595.45
   Month 4: 583.54
   Remaining balance: $3593.41

4.a. $1 + 5 + 9 + 13 + 17 + 21 + 25 = 91$

b. $-10 + (-6) + 0 + 8 + 18 = 10$

c. $2 + 4 + 8 + 16 = 30$

5.a $\displaystyle\sum_{i=1}^{6} 4i$ b. $\displaystyle\sum_{i=1}^{5} \frac{i}{i+1}$

6. 3.38 in.

Objective 1

1. $-1, 1, -1, 1, -1$

3. $\frac{11}{21}$

Objective 2

5. $a_n = \left(\sqrt{3}\right)^n$

Objective 3

7. 16 g

Objective 4

9. $5 + 7 + 9 + 11 = 32$

Objective 5

Other answers may be possible.

11. $\displaystyle\sum_{i=1}^{5} i^2$ 13. $\displaystyle\sum_{i=1}^{5} i \cdot x^{i-1}$

Objective 6

15. $\frac{5}{16}$

12.2 Arithmetic Sequences
Key Terms

1. arithmetic sequence (arithmetic progression)

2. common difference

Now Try

1. 2

2. $9, 4, -1, -6, -11$

3. $a_n = 31 + 4n;\ a_{20} = 111$

4. $41,800

5.a. 93　　　b. −44

6. 16　　　7. 39

8. 806　　　9. 365

Objective 1

1. 4

3. 6, 3, 0, −3, −6

Objective 2

5. $a_n = \frac{1}{3}n + \frac{1}{6}$

Objective 3

7. $10,900　　9. $11.75

Objective 4

11. $\frac{41}{2}$

Objective 5

13. −145　　　15. 140

12.3 Geometric Sequences
Key Terms

1. common ratio　2. annuity

3. geometric sequence (geometric progression)

4. future value of an annuity

5. ordinary annuity

6. term of the annuity

7. payment period

Now Try

1. $\frac{1}{4}$　　　2. $a_n = \left(\sqrt{3}\right)^n$

3.a. −729　　　b. $\frac{1}{384}$

4. 0.8, −4, 20, −100, 500

5. $\frac{728}{3}$　　　6. 682

7.a. $328,988.05　b. $271,983.18

8. $\frac{25}{6}$

9. $|r| > 1,$ so the sum does not exist.

Objective 1

1. $\sqrt{2}$

Objective 2

Other answers may be possible.

3. $a_n = 6 \cdot 2^{n-1}$

5. $a_n = -4\left(\frac{1}{5}\right)^n$

Objective 3

7. −162

Objective 4

9. 2186　　　11. −258

Objective 5

13. $3,464.34

Objective 6

15. $|r| > 1,$ so the sum does not exist.

12.4 The Binominal Theorem
Key Terms

1. binomial theorem (general binomial expansion)

2. Pascal's triangle

Now Try

1. 3,628,800

2.a. 1   b. 36   c. 9   d. 126

3. 45

4. $81r^4 - 216r^3s + 216r^2s^2 - 96rs^3 + 16s^4$

5. $-\frac{x^5}{32} + \frac{5}{8}x^4 - 5x^3 + 20x^2 - 40x + 32$

6. $-5103m^6r^5$

Objective 1

1. 5040   3. 1   5. 455   7. $\dfrac{z^3}{8} - \dfrac{z^2x}{4} + \dfrac{zx^2}{6} - \dfrac{x^3}{27}$

9. $32y^5 + 240y^4z + 720y^3z^2 + 1080y^2z^3 + 810yz^4 + 243z^5$

Objective 2

11. $90m^3$ 13. $135r^2$ 15. $-324k^2p^{14}$